MODERN EXPERIMENTAL BIOCHEMISTRY

THIRD EDITION

Of related interest – from the Benjamin/Cummings Series in the Life Sciences and Chemistry

Biochemistry and Cell Biology

W. M. Becker, L. J. Kleinsmith, and J. Hardin
The World of the Cell, Fourth Edition (2000)

D. J. Holme and H. Peck
Analytical Biochemistry, Third Edition (1998)

C. K. Mathews, K. E. van Holde, and K. G. Ahern
Biochemistry, Third Edition (2000)

Ecology and Evolution

M. G. Barbour, J. H. Burk, W. D. Pitts, F. S. Gilliam, and M. W. Schwartz
Terrestrial Plant Ecology, Third Edition (1999)

C. J. Krebs
Ecological Methodology, Second Edition (1999)

C. J. Krebs
Ecology: The Experimental Analysis of Distribution and Abundance, Fourth Edition (1994)

J. W. Nybakken
Marine Biology: An Ecological Approach, Fourth Edition (1997)

E. R. Pianka
Evolutionary Ecology, Sixth Edition (2000)

Molecular Biology and Genetics

M. V. Bloom, G. A. Freyer, and D. A. Micklos
Laboratory DNA Science (1996)

R. J. Brooker
Genetics: Analysis and Principles (1999)

J. P. Chinnici and D. J. Matthes
Genetics: Practice Problems and Solutions (1999)

R. J. B. King
Cancer Biology (1996)

R. P. Nickerson
Genetics: A Guide to Basic Concepts and Problem Solving (1990)

P. J. Russell
Fundamental of Genetics, Second Edition (2000)

P. J. Russell
Genetics, Fifth Edition (1998)

J. D. Watson, N. H. Hopkins, J. W. Roberts, J. A. Steitz, and A. M. Weiner
Molecular Biology of the Gene, Fourth Edition (1987)

MODERN EXPERIMENTAL BIOCHEMISTRY

THIRD EDITION

Rodney Boyer

Hope College

Benjamin Cummings

An imprint of Addison Wesley Longman

San Francisco • Boston • New York
Capetown • Hong Kong • London • Madrid • Mexico City
Montreal • Munich • Paris • Singapore • Sydney • Tokyo • Toronto

Disclaimer

The experiments in this book have been exhaustively tested for safety and all attempts have been made to use the least hazardous chemicals and procedures possible. However, the author and publisher cannot be held liable for any injury or damage which may occur during the performance of the experiments. It is assumed that before an experiment is initiated, a Material Safety Data Sheet (MSDS) for each chemical used will have been studied by the instructor and students to ensure its safe handling and disposal.

Sponsoring Editor	*Ben Roberts*
Publishing Associate	*Claudia Herman*
Production Editor	*Jean Lake*
Cover Design	*Emi Koike*
Cover Illustration	*Quade Paul, FiVth.com*
Text Design	*Robert Gray/Synapse Creative*
Copy Editor	*Mary Prescott*
Composition and Illustrations	*Pre-Press Company, Inc.*
Printing and Binding	*Maple-Vail Press*
Cover Printer	*Lehigh Press*

Library of Congress Cataloging-in-Publication Data

Boyer, Rodney
    Modern experimental biochemistry / Rodney Boyer. – 3rd ed.
        p. cm.
    Includes bibliographical references and index.
    ISBN 0-8053-3111-5
    1. Biochemistry–Methodology.   2. Biochemistry–Laboratory manuals.   3. Molecular biology–Methodology.   4. Molecular biology–Laboratory manuals.   I. Title.   II. Series.
QP519.7.B68 2000
572'.028–dc21

00-044528

2 3 4 5 6 7 8 9 10–MV–04 03 02 01 00

Benjamin/Cummings, an imprint of Addison Wesley Longman
1301 Sansome Street, San Francisco, CA 94111

To: H. B.

The undergraduate biochemistry teaching laboratory has become an essential feature in the training of students majoring in biochemistry, chemistry, molecular biology, and related biological sciences. Students training for careers in chemistry and the molecular life sciences must acquire extensive experience working with biomolecules in the laboratory, and a formal laboratory course is usually the first step to gain that experience. This step prepares students for participation in future research and development projects.

The purpose of the third edition of this book is to provide junior-senior science students with a modern, balanced, and thorough practice in experimental biochemistry and molecular biology. With such a broad array of potential topics and techniques available, it is difficult to select those that students should experience and master. It is a certainty that one instructor's list of essential topics and techniques to include in a biochemistry teaching lab would vary from another instructor's list. However, there are techniques and concepts that most of us would agree form a "core" in biochemistry laboratory. In selecting experiments for this book I have used my judgment developed from over 25 years of teaching and research and also advice from the Committee on Professional Training (CPT) of the American Chemical Society (ACS). In the Fall 1998 issue of the *CPT Newsletter*, the core topics of biochemistry suggested for classroom and lab were:

Biological Structures and Interactions that Stabilize Biological Molecules. Fundamental building blocks (amino acids, carbohydrates, lipids, nucleotides), organic and inorganic prosthetic groups, biopolymers (nucleic acids, peptides/ proteins, polysaccharides), membranes.

Biological Reactions. Biosynthesis and catabolism of biological molecules (amino acids, carbohydrates, lipids, nucleic acids, peptides/proteins), metabolic cycles, biological catalysis and kinetics, mechanisms, organic and inorganic cofactors.

Biological Equilibria and Energetics. pH/buffers, binding/recognition, proton and electron transport, oxidation/reduction, macromolecular conformations. Some of these topics may be covered in laboratory courses. The experiments that are used for this purpose should emphasize techniques including error and statistical analysis of experimental data, spectroscopic methods, electrophoretic techniques, chromatographic separations, isolation and identification of macromolecules.

All of the topics and techniques are included in this edition of the text.

As with the first two editions, the book is organized into two parts: I. Theory and Experimental Techniques and II. Experiments. Part I introduces students to theoretical and background material for the experiments. This part may also serve as a supplement for instructors who use their own experiments. In Part II there are 15 experiments that represent all areas of biochemistry, including working with proteins and nucleic acid isolation and characterization. The number of experiments has been reduced from earlier editions at the request of instructors and students who believed the book had more experiments than needed for a typical one-semester course. There are, however, still sufficient experiments for a two-semester course sequence. The reduction in the number of experiments has also been achieved by combining some experiments.

Many other changes have been made in the new edition. Approximately one-third of the experiments are entirely new, covering topics such as the use of the Internet in literature/structure searching, Western blotting, ligand-protein interactions, and analysis of amino acids by HPLC or CE. The remaining experiments have been thoroughly revised and updated in written directions and experimental methods. Study problems for student practice are now included at the end of each chapter in Part I as well as in each experiment. Each of the 7 chapters and 15 experiments has 10 study problems, many of which have answers in the Appendix. In addition, the list of literature references at the end of chapters and experiments contains World Wide Web sites for student and instructor use.

Both Parts I and II have been completely rewritten and reflect the many advances in biochemistry–molecular biology theory and techniques. Especially noteworthy have been the technical advances in chromatography (perfusion, FPLC, bioaffinity), electrophoresis (pulsed gel, capillary, nucleic acid sequencing), spectrophotometry (nmr, ms, and diode array detectors), and molecular biology (microsequencing of proteins and nucleic acids, blotting, restriction enzymes).

In the development of an experiment, primary consideration was given to the use of modern procedures and techniques. Therefore, students will learn procedures that are now used in actual laboratory settings, whether academic or industrial. Each experiment presents a challenging laboratory situation or problem to be solved by the student. In each experiment, a technique is introduced that allows students to obtain and evaluate properties of a biomolecule, a biochemical reaction, or a biological process.

The outline for each experiment is as follows.

Introduction and Theory.   In this section it is assumed that a student has studied the general subject in the classroom. Only a summary or review of significant aspects of background is included. The general discussion includes theoretical and practical information that is generally not available in biochemistry textbooks. The general thrust of the experiment is explained, and a flowchart of the experiment is often presented if appropriate. This outline of the experiment allows the student to recognize the importance of each part of the experiment to the achievement of the overall objective.

Materials and Supplies.   A complete list of all materials, supplies, and equipment required for the experiment is provided. An Instructor's Manual describing the preparation of all reagents and solutions and advice on how to set up a biochemistry laboratory is available from the publisher.

Experimental Procedure.   This section contains a step-by-step detailed procedure for the experiment. It is divided into logical parts for ease of completion and to facilitate the interruption of an experiment, if necessary. To enhance student awareness, an icon (⚑) is used to alert students and instructors to possible safety concerns.

Analysis of Results.   In this section the student is instructed in the proper collection and handling of data from the experiment. Each table or graph that is to be constructed is explained and sample calculations are outlined. Typical data for the experiment may be disclosed but only to aid the student in interpretation of results.

Study Problems.   Ten study problems are provided for student practice at the end of each experiment. Some questions will deal with various details of the experiment and numerical problems are emphasized. This symbol (▶) indicates questions that are answered in Appendix IX.

Further Reading.   Each experiment ends with a complete list of references that provide either a more detailed theoretical background or an expanded explanation of procedures and techniques. In addition, each chapter and experiment now has Web sites related to topics and techniques.

ACKNOWLEDGMENTS

Writing and publishing this textbook required the assistance of many creative, talented, and dedicated individuals. Ben Roberts, Senior Chemistry Editor at Benjamin/Cummings, guided the project through all the necessary steps of writing, reviewing, and publishing. Claudia Herman, Publishing Associate, skillfully assisted in these efforts. I am especially indebted to my Production Editor, Jean Lake, for her constant diligence, patience, and encouragement during the publication stage. I also thank Mrs. Norma Plasman (Hope College), who with her typical efficiency and skill, typed many drafts of the manuscript.

Developing a textbook in a rapidly evolving discipline such as biochemistry requires the input of knowledgeable scientists and dedicated teachers. These qualities were present in reviewers of the manuscript including:

First Edition

Hugh Akers, Lamar University

John Cronin, Arizona State University

Patricia Dwyer-Hallquist, Appleton Paper Inc.

Robert Lindquist, San Francisco State University

Kenneth Marx, Dartmouth College

Richard Paselk, Humboldt State University

William Scovell, Bowling Green State University

Ev Trip, University of British Columbia

Dennis Vance, University of British Columbia

Ronald Watanabe, San Jose State University

Second Edition

Larry Byers, Tulane University

Stephen Dahms, San Diego State University

Anthony Gawienowski, University of Massachusetts, Amherst

Rebecca Jurtshuk, University of Houston

Robert Lindquist, San Francisco State University

Denise Magnuson, Texas A&M University

Henry Mariani, University of Massachusetts, Boston

William Meyer, University of Vermont

Ilka Nemere, University of California, Riverside

Robert Sanders, University of Kansas

Anthony Toste, Southwest Missouri State University

Steven Wietstock, Alma College

Third Edition

Linda Brunauer, Santa Clara University

Jeffrey A. Cohlberg, California State University, Long Beach

Diane Dottavio, University of Houston

Donald Fox, University of Houston

Colleen Johnson, New Mexico State University

Robert D. Kuchta, University of Colorado at Boulder

Robert Lindquist, San Francisco State University

Ray Lutgring, University of Evansville

Chris Makaroff, Miami University

Susan Martinis, University of Houston

Douglas D. McAbee, California State University, Long Beach

Linda Roberts, California State University, Sacramento

Tim Sherwood, Arkansas Tech University

Alexander Volkov, Oakwood College

Kimberly Waldron, Regis University

William R. Widger, University of Houston

I also thank my wife, Christel, who patiently tolerated the lifestyle changes associated with writing a book. In addition, she designed experimental procedures and searched for ideal laboratory conditions for several experiments. I am happy to report that Mäusi, our blue-point Himalayan, is still napping under the desk lamp.

I encourage all users of this book to send comments that will assist in the preparation of future editions.

Rodney Boyer
boyer@hope.edu

TABLE OF CONTENTS

P A R T T W O E X P E R I M E N T S

Appendices

THEORY AND EXPERIMENTAL TECHNIQUES

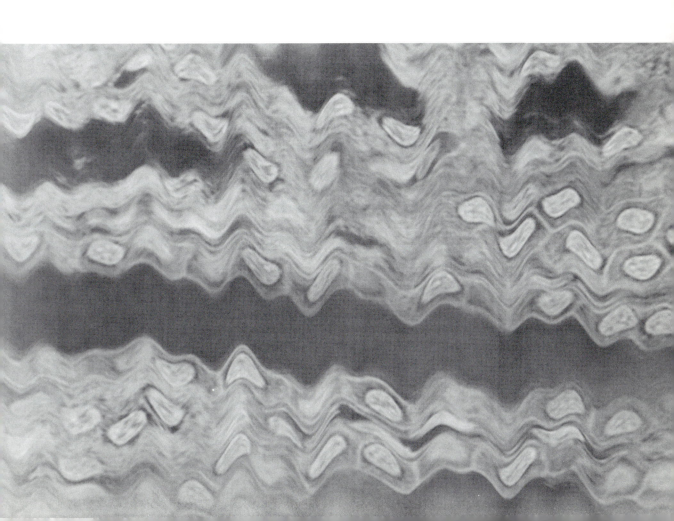

1

INTRODUCTION TO THE BIOCHEMISTRY LABORATORY

Welcome to your biochemistry laboratory course! This is not the first chemistry laboratory course for most of you, but I believe you will find it to be among the most exciting and dynamic of those in which you have enrolled. Most of the experimental techniques and skills that you have acquired over the years will be of great value in this laboratory. However, you will be introduced to several new procedures and instruments. Your success in the biochemistry laboratory will depend on your mastery of these specialized techniques, use of equipment, and understanding of chemical-biochemical principles.

As you proceed through the schedule of experiments for this term, you will, no doubt, compare your work with previous laboratory experiences. In biochemistry laboratory you will seldom run reactions and isolate several hundred milligrams or a few grams of solid and liquid products as you did in organic laboratory. Rather, you will work with milligram or even microgram quantities, and in most cases the biomolecules will be extracted from plant, animal, or bacterial sources and dissolved in solution so you never really "see" the materials under study. But, you will observe the dynamic chemical and biological changes brought about by biomolecules. The techniques and procedures introduced in the laboratory will be your "eyes" and will monitor the occurrence of biochemical events.

This chapter is an introduction to procedures that are of utmost importance for the safe and successful completion of a biochemical project. It is recommended that you become familiar with the following sections before you begin laboratory work.

A. SAFETY IN THE LABORATORY

The concern for laboratory safety can never be overemphasized. Most students have progressed through at least two years of college laboratory work without even a minor accident. This record is, indeed, something to be proud of; however, it should not lead to overconfidence. You must always be aware that chemicals used in the laboratory are potentially toxic, irritating, or flammable. Such chemicals are a hazard, however, only when they are mishandled or improperly disposed of. It is my experience that accidents happen least often to students who come to each laboratory session mentally prepared and with a complete understanding of the experimental procedures to be followed. Since dangerous situations can develop unexpectedly, though, you must be familiar with general safety practices, facilities, and emergency action. Students must have a special concern for the safety of classmates. Carelessness on the part of one student can often cause injury to other students.

The experiments in this book are designed with an emphasis on safety. However, no amount of planning or pretesting of experiments substitutes for awareness and common sense on the part of the student. All chemicals used in the experiments outlined here must be handled with care and respect. The use of chemicals in all U.S. workplaces, including academic research and teaching laboratories, is now regulated by the Federal Hazard Communication Standard, a document written by the Occupational Safety and Health Administration (OSHA).[1] Specifically, the OSHA standard requires all workplaces where chemicals are used to do the following: (1) develop a written hazard communication program, (2) maintain files of Material Safety Data Sheets (MSDS) on all chemicals used in that workplace, (3) label all chemicals with information regarding hazardous properties and procedures for handling, and (4) train employees in the proper use of these chemicals. Several states have passed "right to know" legislation which amends and expands the federal OSHA standard. If you have an interest in or concern about any chemical used in the laboratory, the MSDS for that chemical may be obtained from your instructor or laboratory manager. The actual form of an MSDS for a chemical may vary, but certain specific information must be present. Figure 1.1 is a partial copy of the MSDS for glacial acetic acid, a reagent often used in biochemical research. Different systems for labeling chemical reagent bottles are commercially available. One of the most widely used is the Hazardous Materials Identification System (HMIS). A copy of the actual label for acetic acid is shown in Figure 1.2A. The health, flammability, reactivity, and personal protection codes are defined in Figure 1.2B.

It is easy to overlook some of the potential hazards of working in a biochemistry laboratory. Students often have the impression that they are working less with chemicals and more with natural biomolecules; therefore,

---

[1] Federal Register, *Vol. 48, Nov. 25, 1983, p. 53280; Federal Register, Vol. 50, Nov. 27, 1985, p. 48758.*

Figure 1.1

Partial MSDS for glacial acetic acid. *Courtesy of Sigma Chemical Co.*

Section 2—Composition/Information on Ingredient

Substance Name	CAS #	SARA 313
ACETIC ACID	64-19-7	No

Formula $C2H4O2$

Synonyms Acetic acid (ACGIH:OSHA), Acetic acid, glacial, Acide acetique (French), Acido acetico (Italian), Azijnzuur (Dutch), Essigsaeure (German), Ethanoic acid, Ethylic acid, Glacial acetic acid, Kyselina octova (Czech), Methanecarboxylic acid, Octowy kwas (Polish), Vinegar acid

---

Section 4—First Aid Measures

Oral Exposure

If swallowed, wash out mouth with water provided person is conscious. Call a physician immediately.

Inhalation Exposure

If inhaled, remove to fresh air. If not breathing give artificial respiration. If breathing is difficult, give oxygen.

Dermal Exposure

In case of skin contact, flush with copious amounts of water for at least 15 minutes. Remove contaminated clothing and shoes. Call a physician.

Eye Exposure

In case of contact with eyes, flush with copious amounts of water for at least 15 minutes. Assure adequate flushing by separating the eyelids with fingers. Call a physician.

Section 7—Handling and Storage

Handling

  User Exposure

  Do not breathe vapor. Do not get in eyes, on skin, on clothing. Avoid prolonged or repeated exposure.

Storage

  Suitable

  Keep tightly closed. Store in a cool dry place.

Section 9—Physical/Chemical Properties

Appearance	Color	Form
	Colorless	Clear liquid

Molecular Weight: 60.05 AMU

Property	Value	At Temperature or Pressure
pH	N/A	
BP/BP Range	117–118°C	760 mmHg
MP/MP Range	4°C	
Freezing Point	N/A	
Vapor Pressure	11.4 mmHg	20°C
Vapor Density	2.07 g/f	
Saturated Vapor Conc.	N/A	
SG/Density	1.06 g/cm3	

Section 11—Toxicological Information

Route of Exposure

  Skin Contact

  Causes burns.

  Skin Absorption

  Harmful if absorbed through skin.

  Eye Contact

  Causes burns.

  Inhalation

  May be harmful if inhaled.

  Ingestion

  May be harmful if swallowed.

Target Organ(s) or System(s)

Teeth. Kidneys.

Signs and Symptoms of Exposure

Material is extremely destructive to tissue of the mucous membranes and upper respiratory tract, eyes, and skin. Inhalation may result in spasm, inflammation and edema of the larynx and bronchi, chemical pneumonitis, and pulmonary edema. Symptoms of exposure may include burning sensation, coughing, wheezing, laryngitis, shortness of breath, headache, nausea, and vomiting. Ingestion or inhalation of concentrated acetic acid causes damage to tissues of the respiratory and digestive tracts. Symptoms include: hematemesis, bloody diarrhea, edema and/or perforation of the esophagus and pylorus, hematuria, anuria, uremia, albuminuria, hemolysis, convulsions, bronchitis, pulmonary edema, pneumonia, cardiovascular collapse, shock, and death. Direct contact or exposure to high concentrations of vapor with skin or eyes can cause: erythema, blisters, tissue destruction with slow healing, skin blackening, hyperkeratosis, fissures, corneal erosion, opacification, iritis, conjunctivitis, and possible blindness. To the best of our knowledge, the chemical, physical, and toxicological properties have not been thoroughly investigated.

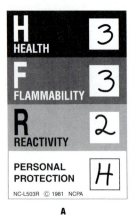

A

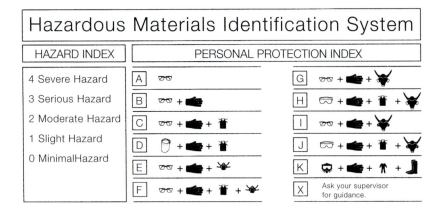

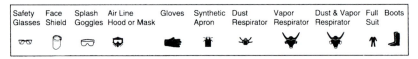

B

Figure 1.2

A Hazardous Materials Identification System (HMIS) for glacial acetic acid.
B Definition of the hazard index and personal protection index. *HMIS is copyrighted by the National Paint and Coatings Association and marketed exclusively through Labelmaster, Chicago, IL.*

there is less need for caution. However, this is not true, since many reagents used are flammable and toxic. In addition, materials such as fragile glass (disposable pipets), sharp objects (needles), and potentially infectious biological materials (blood, bacteria) must be used and disposed of with caution. The extensive use of electrical equipment including hot plates, stirring motors, and high-voltage power supplies presents special hazards.

Proper disposal of all waste chemicals, sharp objects, and infectious agents is essential not only to maintain safe laboratory working conditions but also to protect the general public and your local environment. Some of the liquid chemical reagents and reaction mixtures from each experiment are relatively safe and may be disposed of in the laboratory drainage system without causing environmental damage. However, special procedures must be followed for the use and disposal of most reagents and materials. Often this means that the instructor will provide detailed information on the proper use procedures. In some cases proper disposal will require the collection of waste materials from each laboratory worker and the institution will be responsible for removal. For each experiment in this book, instructions will be given for the proper handling and disposal of all excess reagents, reaction mixtures, and other materials. At the beginning of each experiment, a precautionary note highlighted by the logo ⏩ will appear. Here the dangerous properties of each chemical or procedure will be described along with recommendations for safe handling, first aid, and proper disposal.

So that all students are aware of potential laboratory hazards, it is imperative that a set of rules be developed for each laboratory situation. The following can serve as guidelines:

1. Some form of eye protection is required at all times. Safety glasses with wide side shields are recommended, but normal eyeglasses with safety lenses may be permitted.

    In many workplaces where hazardous chemicals are used or handled, the wearing of contact lenses is prohibited or discouraged. A careful study of the literature by knowledgeable consultants has refuted these risks. Recent studies and experience have suggested that, in fact, contact lenses do not increase risks but can actually minimize or prevent injury in many situations. Because of the ever-increasing use of contact lenses and the benefits they provide, the American Chemical Society Committee on Chemical Safety, having studied and reviewed the issue, is of the consensus that contact lenses can be worn in most work environments provided the same approved eye protection is worn as required of other workers in the area. Clearly, the type of eye protection needed depends upon the circumstances. It should be stressed that contact lenses, by themselves, do not provide adequate protection in any environment in which the chance of an accidental splash of a chemical can reasonably be anticipated.[1]

2. Never work alone in the laboratory.

3. Be familiar with the properties of all chemicals used in the laboratory. This includes their flammability, reactivity, toxicity, and proper disposal. This information may be obtained from your instructor or from an MSDS. Always wear disposable gloves when using potentially dangerous chemicals or infectious agents.

4. Eating, drinking, and smoking in the laboratory are strictly prohibited.

5. Unauthorized experiments are not allowed.

6. Mouth suction should never be used to fill pipets or to start siphons.

7. Become familiar with the location and use of standard safety features in your laboratory. All chemistry laboratories should be equipped with fire extinguishers, eyewashes, safety showers, fume hoods, chemical spill kits, first-aid supplies, and containers for chemical disposal. Any questions regarding the use of these features should be addressed to your instructor or teaching assistant.

Rules of laboratory safety and chemical handling are not designed to impede productivity, nor should they instill a fear of chemicals or laboratory procedures. Rather, their purpose is to create a healthy awareness of potential laboratory hazards, to improve the efficiency of each student

[1] *American Chemical Society Committee on Chemical Safety, 1998.*

worker, and to protect the general public and the environment from waste contamination. The list of references at the end of this chapter includes books and manuals describing proper and detailed safety procedures.

B. THE LABORATORY NOTEBOOK AND EXPERIMENT REPORTS

The biochemistry laboratory experience is not finished when you complete the experimental procedure and leave the laboratory. All scientists have the obligation to prepare written reports of the results of experimental work. Since this record may be studied by many individuals, it must be completed in a clear, concise, orderly, and accurate manner. This means that procedural details, observations, and results must be recorded in a laboratory notebook *while* the experiment is being performed. The notebook should be hardbound with quadrille-ruled (gridded) pages and used only for the biochemistry laboratory. This provides a durable, permanent record and the potential for construction of graphs, charts, etc. It is recommended that the first one or two pages of the notebook be used for a constantly updated table of contents. Although your instructor may have his or her own rules for preparation of the notebook, the most readable notebooks are those in which only the right-hand pages are used for record keeping. The left-hand pages may be used for your own notes, reminders, and calculations. The outline in Figure 1.3 may be used as a guide for an experimental write-up.

Figure 1.3

Outline of experimental write-up.

- -

Prelab
- I. Introduction
  - (a) Objective or purpose
  - (b) Theory
- II. Experimental
  - (a) Table of materials and reagents
  - (b) List of equipment
  - (c) Flowchart
  - (d) Record of procedure
- III. Data and Calculations
  - (a) Record of all raw data including printouts
  - (b) Method of calculation with statistical analysis
  - (c) Present final data in tables, graphs, or figures when appropriate
- IV. Results and Discussion
  - (a) Conclusions
  - (b) Compare results with known values
  - (c) Discuss the significance of the data
  - (d) Was the original objective achieved?
  - (e) Literature references

The basic outline follows that required by most biochemical research journals. Note that Parts I–IIc are labeled Prelab and should be completed before you enter the laboratory. The final report may be handwritten in your notebook or printed on separate paper using a word processor.

Details of Experimental Write-Up

I. Introduction

This section begins with a three- or four-sentence statement of the objective or purpose of the experiment. For preparing this statement, ask yourself, "What are the goals of this experiment?" This statement is followed by a brief discussion of the theory behind the experiment. If a new technique or instrumental method is introduced, give a brief description of the method. Include chemical or biochemical reactions when appropriate.

II. Experimental

Begin this section with a list of all reagents and materials used in the experiment. The sources of all chemicals and the concentrations of solutions should be listed. Instrumentation is listed with reference to company name and model number. A flowchart to describe the stepwise procedure for the experiment should be included after the list of equipment. For the early experiments, a flowchart is provided (for example, see Figure E4.4 in Experiment 4). Flowcharts for later experiments should be designed by the student.

The write-up to this point is to be completed as a Prelab assignment. The experimental procedure followed is then recorded in your notebook as you proceed through the experiment. The detail should be sufficient so that a fellow student can use your notebook as a guide. You should include observations, such as color changes or gas evolution, made during the experiment. If you obtain a recorder printout of numbers, a spectrum from a spectrophotometer, or a photograph, these records must be saved and handed in with your report.

III. Data and Calculations

All raw data from the experiment are to be recorded directly in your notebook, not on separate sheets of paper or paper towels. Calculations involving the data must be included for at least one series of measurements. Proper statistical analysis must be included in this section.

For many experiments, the clearest presentation of data is in a tabular or graphical form. The Analysis of Results section following each experimental procedure in this book describes the preparation of graphs and tables. A graph may be prepared directly on the gridded pages of your notebook, or one prepared by computer software may be provided.

IV. Results and Discussion

This is the most important section of your write-up, because it answers the questions, "Did you achieve your proposed goals and objectives?" and "What is the significance of the data?" Any conclusion that you make must be supported by experimental results. It is often possible to compare your data with known values and results from the literature. If this is feasible, calculate percentage error and explain any differences. If problems were encountered in the experiment, these should be outlined with possible remedies for future experiments.

All library references (books, journal articles, and Web sites) that were used to write up the experiment should be listed at the end. The standard format to follow for a reference listing is shown at the end of this chapter in the reference section.

Everyone has his or her own writing style, some better than others. It is imperative that you continually try to improve your writing skills. When your instructor reviews your write-up, he or she should include helpful writing tips in the grading. Read the works listed in the references at the end of this chapter for further instructions in scientific writing.

C.   CLEANING LABORATORY GLASSWARE

The results of your experiments will depend, to a great extent, on the cleanliness of your equipment. There are at least two reasons for this. (1) Many of the chemicals and biochemicals will be used in milligram or microgram amounts. Any contamination, whether on the inner walls of a beaker, in a pipet, or in a glass cuvette, could be a significant percentage of the total experimental sample. (2) Many biochemicals and biochemical processes are sensitive to one or more of the following common contaminants: metal ions, detergents, and organic residues. In fact, the objective of many experiments is to investigate the effect of a metal ion, organic molecule, or other chemical agent on a biochemical process. Contaminated glassware will virtually ensure failure.

Glassware

Many contaminants, including organics and metal ions, adhere to the inner walls of glass containers. Washing glassware, including pipets, with dilute detergent (0.5% in water) followed by five to ten water rinses is probably sufficient for most purposes. The final rinse should be with distilled or deionized water. Metal ion contamination can be greatly reduced from glassware by rinsing with concentrated nitric acid followed by extensive rinsing with purified water.

Dry equipment is required for most processes carried out in biochemistry laboratory. When you needed dry glassware in organic laboratory, you probably rinsed the piece of equipment with acetone, which rapidly evaporated, leaving a dry surface. Unfortunately, that surface is coated with an organic residue consisting of nonvolatile contaminants in the acetone. Since this residue could interfere with your experiment, it is best to refrain from acetone washing. Glassware and plasticware should be rinsed well with purified water and dried in an oven designated for glassware, not one used for drying chemicals.

Quartz and Glass Cuvettes

Never clean cuvettes or any optically polished glassware with ethanolic KOH or other strong base, as this will cause etching. All cuvettes should be cleaned carefully with 0.5% detergent solution, in a sonicator bath, or in a cuvette washer.

D.   PREPARATION AND STORAGE OF SOLUTIONS

Water Quality

The most common solvent for solutions used in the biochemical laboratory is water. Ordinary tap water contains a variety of impurities including particulate matter (sand, silt, etc.); dissolved organics, inorganics, and gases; and microorganisms (bacteria, viruses, protozoa, algae). In addition, the natural degradation of microorganisms leads to the presence of by-products called pyrogens. Tap water should never be used for the preparation of any reagent solutions. For most laboratory procedures, it is recommended that some type of purified water be used.

There are five basic water purification technologies—distillation, ion exchange, carbon adsorption, reverse osmosis, and membrane filtration. Most academic laboratories are equipped with "in-house" purified water, which typically is produced by a combination of the above purifying technologies. For most procedures carried out in a biochemistry teaching laboratory, water purified by deionization, reverse osmosis, or distillation usually is acceptable. For special procedures such as buffer standardization, liquid chromatography, and tissue culture, ultrapure water should be used.

The water quality necessary will depend on the solutions to be prepared and on the biochemical procedures to be investigated. Water that is purified only by ion exchange will be low in metal ion concentration, but may contain certain organics that are washed from the ion-exchange resin. These contaminants will increase the ultraviolet absorbance properties of water. If sensitive ultraviolet absorbance measurements are to be made, distilled water is better than deionized.

Solution Preparation

The concentrations for solutions used in biochemistry may be expressed in many different units. In your biochemistry laboratory, the most common units will be:

- *Molarity (M):* concentration based on the number of moles of solute per liter of solution.

In biochemistry, it is more common to use concentration ranges that are millimolar (mM, 1×10^{-3} M), micromolar (μM, 1×10^{-6} M), or nanomolar (nM, 1×10^{-9} M).

*   *Percent by weight (% wt/wt):* concentration based on the number of grams of solute per 100 g of solution.

*   *Percent by volume (% wt/vol):* concentration based on the number of grams of solute per 100 mL of solution.

*   *Weight per volume (wt/vol):* concentration based on the number of grams, milligrams, or micrograms of solute per unit volume; for example, mg/mL, g/liter, mg/100 mL, etc.

Example 1   Many solutions you use will be based on molarity. For practice, assume you require 1 liter of solution that is 0.1 M (100 mM) glucose:

MW glucose = 180.2 amu

1 mole of glucose = 180.2 g

0.1 mole of glucose = 18.02 g

To prepare a 0.1 M glucose solution, weigh 18.02 g of glucose and transfer to a 1-liter volumetric flask. Add about 700–800 mL of purified water and swirl to dissolve. Then add water so that the bottom of the meniscus is at the etched line on the flask. Stopper and mix well. The flask must be labeled with solution contents (0.1 M glucose), date prepared, and name of preparer.

In general, solid solutes should be weighed on weighing paper or plastic weighing boats, using an analytical or top-loading balance. Liquids are more conveniently dispensed by volumetric techniques; however, this assumes that the density is known. If a small amount of a liquid is to be weighed, it should be added to a tared flask by means of a disposable Pasteur pipet with a latex bulb. The hazardous properties of all materials should be known before use and the proper safety precautions obeyed.

The storage conditions of reagents and solutions are especially critical. Although some will remain stable indefinitely at room temperature, it is good practice to store all solutions in a closed container. Often it is necessary to store some solutions in a refrigerator at 4°C. This inhibits bacterial growth and slows decomposition of the reagents. Some solutions may require storage below 0°C. If these are aqueous solutions or others that will freeze, be sure there is room for expansion inside the container. Stored solutions must always have a label containing the name and concentration of the solution, the date prepared, and the name of the preparer.

All stored containers, whether at room temperature, 4°C, or below freezing, must be properly sealed. This reduces contamination by bacteria and vapors in the laboratory air (CO_2, NH_3, HCl, etc.). Volumetric flasks, of course, have glass stoppers, but test tubes, Erlenmeyer flasks, bottles, and other containers should be sealed with screw caps, corks, or hydrocarbon foil (Parafilm). Remember that hydrocarbon foil, a wax, is dissolved by solutions containing nonpolar organic solvents.

Bottles of pure chemicals and reagents should also be properly stored. Many manufacturers now include the best storage conditions for a reagent on the label. The common conditions are: store at room temperature, store at 0–4°C, store below 0°C, or store in a desiccator at 0–4°C or below 0°C. Many biochemical reagents form hydrates by taking up moisture from the air. If the water content of a reagent increases, the molecular weight and purity of the reagent change. For example, when NAD^+ is purchased, the label usually reads "Anhydrous molecular weight = 663.5; when assayed, contained 3 H_2O per mole." The actual molecular weight that should be used for solution preparation is 663.5 + (18)(3) = 717.5. However, if this reagent is stored in a moist refrigerator or freezer outside a desiccator, the moisture content may change to an unknown value.

E.   QUANTITATIVE TRANSFER OF LIQUIDS

Practical biochemistry is highly reliant on analytical methods. Many analytical techniques must be mastered, but few are as important as the quantitative transfer of solutions. Some type of pipet will almost always be used in liquid transfer. Since students may not be familiar with the many types of pipets and the proper techniques in pipetting, this instruction is included here.

Filling a Pipet

Figure 1.4 illustrates the various types of pipets and fillers. The use of any pipet requires some means of drawing reagent into the pipet. *Liquids should never be drawn into a pipet by mouth suction on the end of the pipet.* Small latex bulbs are available for use with disposable pipets (Figure 1.4A). For volumetric and graduated pipets, two types of bulbs are available. One type (Figure 1.4B) features a special conical fitting that accommodates common sizes of pipets. To use these, first place the pipet tip below the surface of the liquid. Squeeze the bulb with your left hand (if you are a right-handed pipettor) and then hold it tightly to the end of the pipet. Slowly release the pressure on the bulb to allow liquid to rise to 2 or 3 cm above the top graduated mark. Then, remove the bulb and quickly grasp the pipet with your index finger over the top end of the pipet. The level of solution in the pipet will fall slightly, but not below the top graduated mark. If it does fall too low, use the bulb to refill.

Mechanical pipet fillers (sometimes called safety pipet fillers, propipets, or pi-fillers) are more convenient than latex bulbs (Figure 1.4C,D). Equipped with a system of hand-operated valves, these fillers can be used for the complete transfer of a liquid. The use of a safety pipet filler is outlined in Figure 1.5. *Never allow any solvent or solution to enter the pipet bulb.* To avoid this, two things must be kept in mind: (1) always maintain careful control while using valve S to fill the pipet, and (2) never use valve S unless the pipet tip is

Figure 1.4

Examples of pipets and pipet fillers.
A Latex bulb courtesy of VWR Scientific Division of Univar.
B Pipet filler courtesy of Curtin Matheson Scientific, Inc.
C Mechanical pipet filler courtesy of VWR Scientific, Division of Univar.
D Pipettor pump courtesy of Cole-Parmer Instrument Co.
E Pasteur pipet courtesy of VWR Scientific, Division of Univar.
F Volumetric pipet courtesy of VWR Scientific, Division of Univar.
G Mohr pipet courtesy of VWR Scientific, Division of Univar.
H Serological pipet courtesy of VWR Scientific, Division of Univar.

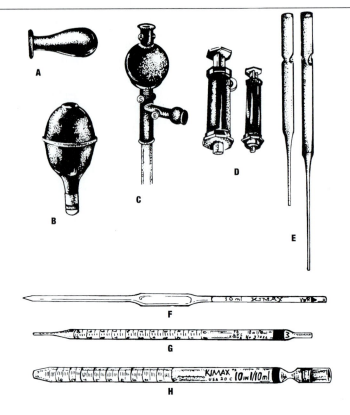

below the surface of the liquid. If the tip moves above the surface of the liquid, air will be sucked into the pipet and solution will be flushed into the bulb. Other pipet fillers are used in a similar fashion.

Disposable Pasteur Pipets

Often it is necessary to perform a semiquantitative transfer of a small volume (1 to 10 mL) of liquid from one vessel to another. Since pouring is not efficient, a **Pasteur pipet** with a small latex bulb may be used (Figure 1.4A,E). Pasteur pipets are available in two lengths (15 cm and 23 cm) and hold about 2 mL of solution. These are especially convenient for the transfer of nongraduated amounts to and from test tubes. Typical recovery while using a Pasteur pipet is 90 to 95%. If dilution is not a problem, rinsing the original vessel with a solvent will increase the transfer yield. Used disposable pipets should be discarded in special containers for broken glass.

Calibrated Pipets

Although most quantitative transfers are now done with automatic pipetting devices, which are described later in the chapter, instructions will be

Figure 1.5

How to use a Spectroline safety pipet filler. *Courtesy of Spectronics Corporation, Westbury, NY 11590.*

----

1. Using thumb and forefinger, press on valve A and squeeze bulb with other fingers to produce a vacuum for aspiration. Release valve A leaving bulb compressed.

2. Insert pipet into liquid. Press on valve S. Suction draws liquid to desired level.

3. Press on valve E to expel liquid.

4. To deliver the last drop, maintain pressure on valve E, cover E inlet with middle finger, and squeeze the small bulb.

given for the use of all types of pipets. If a quantitative transfer of a specific and accurate volume of liquid is required, some form of calibrated pipet must be used. **Volumetric pipets** (Figure 1.4F) are used for the delivery of liquids required in whole milliliter amounts (1, 2, 3, 4, 5, 10, 15, 20, 25, 50, and 100 mL). To use these pipets, draw liquid with a latex bulb or mechanical pipet filler to a level 2 to 3 cm above the "fill line." Touch the tip of the pipet to the inside of the glass wall of the container from which it was filled.

Release liquid from the pipet until the bottom of the meniscus is directly on the fill line. Transfer the pipet to the inside of the second container and release the liquid. Hold the pipet vertically, allow the solution to drain until the flow stops, and then wait an additional 5 to 10 seconds. Touch the tip of the pipet to the inside of the container to release the last drop from the outside of the tip. Remove the pipet from the container. Some liquid may still remain in the tip. Most volumetric pipets are calibrated as "TD" (to deliver), which means the intended volume is transferred *without final blow-out,* i.e., the pipet delivers the correct volume.

Fractional volumes of liquid are transferred with **graduated pipets,** which are available in two types—Mohr and serological. **Mohr pipets** (Figure 1.4G) are available in long- or short-tip styles. Long-tip pipets are especially attractive for transfer to and from vessels with small openings. Virtually all Mohr pipets are TD and are available in many sizes (0.1 to 10.0 mL). The marked subdivisions are usually 0.01 or 0.1 mL, and the markings end a few centimeters from the tip. Selection of the proper size of pipet is especially important. For instance, do not try to transfer 0.2 mL with a 5 or 10 mL pipet. Use the smallest pipet that is practical.

The use of a Mohr pipet is similar to that of a volumetric pipet. Draw the liquid into the pipet with a pipet filler to a level about 2 cm above the "0" mark. Lower the liquid level to the 0 mark. Remove the last drop from the tip by touching it to the inside of the glass container. Transfer the pipet to the receiving container and release the desired amount of solution. The solution should not be allowed to move below the last graduated mark on the pipet. Touch off the last drop.

Serological pipets (Figure 1.4H) are similar to Mohr pipets, except that they are graduated downward to the very tip and are designed for blow-out. Their use is identical to that of a Mohr pipet except that the last bit of solution remaining in the tip must be forced out into the receiving container with a rubber bulb. This final blow-out should be done after 15 to 20 seconds of draining.

Automatic Pipetting Systems

For most quantitative transfers, including many identical small-volume transfers, a mechanical microliter pipettor (Eppendorf type) is ideal. This allows accurate, precise, and rapid dispensing of fixed volumes from 1 to 5000 μL (5 mL). The pipet's push-button system can be operated with one hand, and it is fitted with detachable polypropylene tips (Figure 1.6). Other useful information about pipetting is available at the Web site www.gilson.com/ pipe.htm. The advantage of polypropylene tips is that the reagent film remaining in the pipet after delivery is much less than for glass tips. Mechanical pipettors are available in up to 25 different sizes. Newer models offer continuous volume adjustment, so a single model can be used for delivery of specific volumes within a certain range.

To use the pipettor, choose the proper size and place a polypropylene pipet tip firmly onto the cone as shown in Figure 1.6. Tips for pipets are

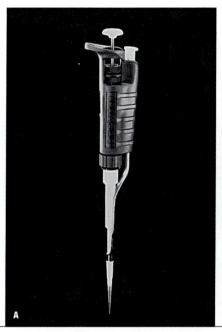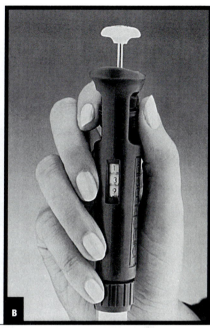

Figure 1.6   A   B

A How to use an adjustable pipetting device.   **B** Set the digital micrometer to the desired volume using the adjustment knob. Attach a new disposable tip to the shaft of the pipet. Press on firmly with a slight twisting motion.   **C** Depress the plunger to the first positive stop, immerse the disposable tip into the sample liquid to a depth of 2 to 4 mm, and allow the pushbutton to return slowly to the up position and wait I to 2 seconds.   **D** To dispense sample, place the tip end against the side wall of the receiving vessel and depress the plunger slowly to the first stop. Wait 2 to 3 seconds, and then depress the plunger to the second stop to achieve final blow-out. Withdraw the device from the vessel carefully with the tip sliding along the inside wall of the vessel. Allow the plunger to return to the up position. Discard the tip by depressing the tip ejector button. *Photos courtesy of Rainin Instrument Company, Inc., Woburn, MA. Pipetman is a registered trademark of Gilson Medical Electronics. Exclusive license to Rainin Instrument Company, Inc.*

available in several sizes, for 1–20 µL, 20–250 µL, 200–1000 µL, and 1000–5000 µL capacity. Details of the operation of an adjustable pipet are given in Figure 1.6.

For rapid and accurate transfer of volumes greater than 5 mL, automatic repetitive dispensers are commercially available. These are particularly useful for the transfer of corrosive materials. The dispensers, which are available in several sizes, are simple to use. The volume of liquid to be dispensed is mechanically set; the syringe plunger is lifted for filling and pressed downward for dispensing. Hold the receiving container under the spout while depressing the plunger. Touch off the last drop on the inside wall of the receiving container.

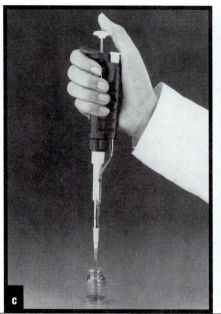 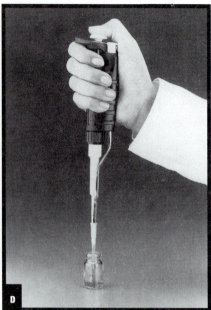

Figure 1.6

continued

Cleaning and Drying Pipets

Special procedures are required for cleaning glass pipets. Immediately after use, every pipet should be placed, tip up, in a vertical cylinder containing a dilute detergent solution (less than 0.5%). The pipet must be completely covered with solution. This ensures that any reagent remaining in the pipet is forced out through the tip. If reagents are allowed to dry inside a pipet, the tips can easily become clogged and are very difficult to open. After several pipets have accumulated in the detergent solution, the pipets should be transferred to a pipet rinser. Pipet rinsers continually cycle fresh water through the pipets. Immediately after detergent wash, tap water may be used to rinse the pipets, but distilled water should be used for the final rinse. Pipets may then be dried in an oven.

F.   STATISTICAL ANALYSIS OF EXPERIMENTAL DATA

The purpose of each laboratory exercise in this book is to observe and measure characteristics of a biomolecule or a biological system. The characteristic is often quantitative, a single number or a group of numbers. These measured characteristics may be the molecular weight of a protein, the pH of a buffer solution, the absorbance of a colored solution, the rate of an enzyme-catalyzed reaction, or the radioactivity associated with a molecule. If you measure a quantitative characteristic many times under identical conditions, a slightly different result will most likely be obtained each time. For example, if a radioactive sample is counted twice under identical experimental and instrumental conditions, the second measurement immediately following the first, the probability is very low that the numbers of counts will be identical. If the absorbance of a solution is determined several times at a specific wavelength, the value of each measurement will surely vary from the others. Which measurements, if any, are correct? Before this question can be answered, you must understand the source and treatment of numerical variations in experimental measurements.

Analysis of Experimental Data

An **error** in an experimental measurement is defined as a deviation of an observed value from the true value. There are two types of errors, **determinate** and **indeterminate.** Determinate errors are those that can be controlled by the experimenter and are associated with malfunctioning equipment, improperly designed experiments, and variations in experimental conditions. These are sometimes called human errors because they can be corrected or at least partially alleviated by careful design and performance of the experiment. Indeterminate errors are those that are random and cannot be controlled by the experimenter. Specific examples of indeterminate errors are variations in radioactive counting and small differences in the successive measurements of glucose in a serum sample.

Two statistical terms involving error analysis that are often used and misused are **accuracy** and **precision.** Precision refers to the extent of agreement among repeated measurements of an experimental value. Accuracy is defined as the difference between the experimental value and the true value for the quantity. Since the true value is seldom known, accuracy is better defined as the difference between the experimental value and the accepted true value. Several experimental measurements may be precise (that is, in close agreement with each other) without being accurate.

If an infinite number of identical, quantitative measurements could be made on a biosystem, this series of numerical values would constitute a **statistical population.** The average of all of these numbers would be the **true value** of the measurement. It is obviously not possible to achieve this in practice. The alternative is to obtain a relatively small **sample of data,** which is a subset of the infinite population data. The significance and precision of these data are then determined by statistical analysis.

This section explores the mathematical basis for the statistical treatment of experimental data. Most measurements required for the completion of the experiments can be made in duplicate, triplicate, or even quadruplicate, but it would be impractical and probably a waste of time and materials to make numerous determinations of the same measurement. Rather, when you perform an experimental measurement in the laboratory, you will collect a small sample of data from the population of infinite values for that measurement. To illustrate, imagine that an infinite number of experimental measurements of the pH of a buffer solution are made, and the results are written on slips of paper and placed in a container. It is not feasible to calculate an average value of the pH from all of these numbers, but it is possible to draw five slips of paper, record these numbers, and calculate an average pH. By doing this, you have collected a sample of data. By proper statistical manipulation of this small sample, it is possible to determine whether it is representative of the total population and the amount of confidence you should have in these numbers.

The data analysis will be illustrated here primarily with the counting of radioactive materials, although it is not limited to such applications. Any replicate measurements made in the biochemistry laboratory can be analyzed by these methods.

Determination of the Mean, Sample Deviation, and Standard Deviation

Radioactive decay with emission of particles is a random process. It is impossible to predict with certainty when a radioactive event will occur. Therefore, a series of measurements made on a radioactive sample will result in a series of different count rates, but they will be centered around an **average** or **mean** value of counts per minute. Table 1.1 contains such a series of count rates obtained with a scintillation counter on a single radioactive sample. A similar table could be prepared for other biochemical measurements, including the rate of an enzyme-catalyzed reaction or the protein concentration of a solution as determined by the Bradford method. The arithmetic average or mean of the numbers is calculated by totaling all the experimental values observed for a sample (the counting rates, the velocity of the reaction, or protein concentration) and dividing the total by the number of times the measurement was made. The mean is defined by Equation 1.1.

$$\bar{x} = \frac{\sum\limits_{i}^{n} x_i}{n}$$

Equation 1.1

where

$\bar{x}$ = arithmetic average or mean

x_i = the value for an individual measurement

n = the total number of experimental determinations

The mean counting rate for the data in Table 1.1 is 1222. If the same radioactive sample were again counted for a series of ten observations, that series of counts would most likely be different from those listed in the table, and a different mean would be obtained. If we were able to make an infinite number of counts on the radioactive sample, then a **true mean** could be calculated. The true mean would be the actual amount of radioactivity in the sample. Although it would be desirable, it is not possible experimentally to measure the true mean. Therefore, it is necessary to use the average of the

Table 1.1

The Observed Counts and Sample Deviation from a Typical Radioactive Sample

Counts per Minute	Sample Deviation $x_i - \bar{x}$
1243	+21
1250	+28
1201	−21
1226	+4
1220	−2
1195	−27
1206	−16
1239	+17
1220	−2
1219	−3
Mean = 1222	

counts as an approximation of the true mean and to use statistical analysis to evaluate the precision of the measurements (that is, to assess the agreement among the repeated measurements).

Since it is not usually practical to observe and record a measurement many times as in Table 1.1, what is needed is a means to determine the reliability of an observed measurement. This may be stated in the form of a question. How close is the result to the true value? One approach to this analysis is to calculate the **sample deviation,** which is defined as the difference between the value for an observation and the mean value, $\bar{x}$ (Equation 1.2). The sample deviations are also listed for each count in Table 1.1.

$$\text{Sample deviation} = x_i - \bar{x}$$

Equation 1.2

A more useful statistical term for error analysis is **standard deviation,** a measure of the spread of the observed values. Standard deviation, s, for a sample of data consisting of n observations may be estimated by Equation 1.3.

$$s = \sqrt{\frac{\sum (x_i - \bar{x})^2}{n - 1}}$$

Equation 1.3

It is a useful indicator of the probable error of a measurement. Standard deviation is often transformed to **standard deviation of the mean** or **standard error.** This is defined by Equation 1.4, where n is the number of measurements.

$$s_m = \frac{s}{\sqrt{n}}$$

Equation 1.4

Figure 1.7

The normal distribution curve.

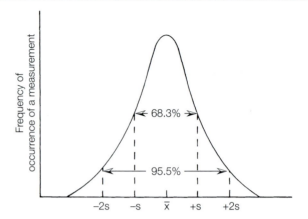

It should be clear from this equation that as the number of experimental observations becomes larger, s_m becomes smaller, or the precision of a measurement is improved.

Standard deviation may also be illustrated in graphical form (Figure 1.7). The shape of the curve in Figure 1.7 is closely approximated by the **Gaussian distribution** or **normal distribution curve.** This mathematical treatment is based on the fact that a plot of relative frequency of a given event yields a dispersion of values centered about the mean, $\bar{x}$. The value of $\bar{x}$ is measured at the maximum height of the curve. The normal distribution curve shown in Figure 1.7 defines the spread or dispersion of the data. The probability that an observation will fall under the curve is unity or 100%. By using an equation derived by Gauss, it can be calculated that for a single set of sample data, 68.3% of the observed values will occur within the interval $\bar{x} \pm s$, 95.5% of the observed values within $\bar{x} \pm 2s$, and 99.7% of the observed values within $\bar{x} \pm 3s$. Stated in other terms, there is a 68.3% chance that a single observation will be in the interval $\bar{x} \pm s$.

For many experiments, a single measurement is made so a mean value, $\bar{x}$, is not known. In these cases, error is expressed in terms of s but is defined as the **percentage proportional error,** $\%E_x$, in Equation 1.5.

$$\%E_x = \frac{100k}{\sqrt{n}}$$

Equation 1.5

The parameter k is a proportional constant between E_x and the standard deviation. The percent proportional error may be defined within several probability ranges. **Standard error** refers to a confidence level of 68.3%; that is, there is a 68.3% chance that a single measurement will not exceed the $\%E_x$. For standard error, $k = 0.6745$. **Ninety-five hundredths error** means there is a 95% chance that a single measurement will not exceed the $\%E_x$. The constant k then becomes 1.45.

The previous discussion of standard deviation and related statistical analysis placed emphasis on estimating the reliability or precision of experimentally observed values. However, standard deviation does not give specific information about how close an experimental mean is to the true mean. Statistical analysis may be used to estimate, within a given probability, a range within which the true value might fall. The range or confidence interval is defined by the experimental mean and the standard deviation. This simple statistical operation provides the means to determine quantitatively how close the experimentally determined mean is to the true mean. **Confidence limits** (L_1 and L_2) are created for the sample mean as shown in Equations 1.6 and 1.7.

>> $$L_1 = \bar{x} + (t)(s_m)$$ **Equation 1.6**

>> $$L_2 = \bar{x} - (t)(s_m)$$ **Equation 1.7**

where

t = a statistical parameter that defines a distribution between a sample mean and a true mean

The parameter t is calculated by integrating the distribution between percent confidence limits. Values of t are tabulated for various confidence limits. Such a table is in Appendix VIII. Each column in the table refers to a desired confidence level (0.05 for 95%, 0.02 for 98%, and 0.01 for 99% confidence). The table also includes the term **degrees of freedom,** which is represented by $n - 1$, the number of experimental observations minus 1. The values of $\bar{x}$ and s_m are calculated as previously described in Equations 1.1 and 1.4.

Statistical Analysis in Practice

The equations for statistical analysis that have been introduced in this chapter are of little value if you have no understanding of their practical use, meaning, and limitations. A set of experimental data will first be presented, and then several statistical parameters will be calculated using the equations. This example will serve as a summary of the statistical formulas and will also illustrate their application.

Example 2   Ten identical protein samples were analyzed by the Bradford method for protein analysis. The following values for protein concentration were obtained.

Observation Number	Protein Concentration (mg/mL), x
1	1.02
2	0.98
3	0.99
4	1.01
5	1.03
6	0.97
7	1.00
8	0.98
9	1.03
10	1.01

Sample mean

$$\bar{x} = \frac{\sum x}{n} = \frac{10.02}{10} = 1.00 \text{ mg/mL}$$

Sample deviation

Sample deviation $= x_i - \bar{x}$

Observation	$x_i - \bar{x}$
1	+0.02
2	−0.02
3	−0.01
4	+0.01
5	+0.03
6	−0.03
7	0.00
8	−0.02
9	+0.03
10	+0.01

Calculation of the sample deviation for each measurement gives an indication of the precision of the determinations.

Standard deviation

$$s = \sqrt{\frac{\sum (x_i - \bar{x})^2}{n - 1}}$$

$$s = 0.02$$

The mean can now be expressed as $\bar{x} \pm s$ (for this specific example, 1.00 ± 0.02 mg/mL). The probability of a single measurement falling within these limits is 68.3%. For 95.5% confidence (2s), the limits would be 1.00 ± 0.04 mg/mL.

Standard error of the mean

$$S_m = \frac{s}{\sqrt{n}}$$

$$S_m = \frac{0.02}{\sqrt{10}}$$

$$S_m = 0.006$$

This statistical parameter can be used to gauge the precision of the experimental data.

Confidence limits

The desired confidence limits will be set at the 95% confidence level. Therefore we will choose a value for t from Appendix VIII in the column labeled $t_{0.05}$ and $n - 1 = 9$.

$$L_1 = \bar{x} + (t_{0.05})(s_m)$$

$$L_2 = \bar{x} - (t_{0.05})(s_m)$$

$$s_m = 0.006$$

$$t_{0.05} = 2.262$$

$$\bar{x} = 1.00$$

$$L_1 = 1.00 + (2.262)(0.006)$$

$$L_1 = 1.01$$

$$L_2 = 0.99$$

We can be 95% confident that the true mean falls between 0.99 and 1.01 mg/mL.

Study Problems

1. Define each of the following terms.
   - (a) OSHA
   - (b) MSDS
   - (e) Purified water
   - (f) Error

(c) Flowchart   (g) Standard deviation

(d) Pasteur pipet   (h) Molarity

► 2. What personal protection items must be worn when handling glacial acetic acid?

3. Draw a schematic picture of your biochemistry lab and mark locations of the following safety features: eyewashes, first-aid kit, shower, fire extinguisher, chemical spill kits, and direction to nearest exit.

► 4. Describe how you would prepare a 1-liter aqueous solution of each of the following reagents:

(a) 1 M glycine

(b) 0.5 M glucose

(c) 10 mM ethanol

(d) 100 nM hemoglobin

► 5. Describe how you would prepare just 10 mL of each of the solutions in Problem 4.

► 6. If you mix 1 mL of the 1 M glycine solution in Problem 4 with 9 mL of water, what is the final concentration of this diluted solution in mM?

► 7. Convert each of the concentrations below to mM and μM.

(a) 10 mg of glucose per 100 mL

(b) 100 mL of a solution 2% in alanine

► 8. You have just prepared a solution by weighing 20 g of sucrose, transferring it to a 1-liter volumetric flask, and adding water to the line. Calculate the concentration of the sucrose solution in terms of mM, mg/mL, and % (wt/vol).

► 9. The concentrations of cholesterol, glucose, and urea in blood from a fasting individual are listed below in units of mg/100 mL (sometimes called mg%). These are standard concentration units used in the clinical chemistry lab. Convert the concentrations to mM.

cholesterol–200 mg%

glucose–75 mg%

urea–20 mg%

► 10. The following optical rotation readings were taken by a polarimeter on a solution of an unknown carbohydrate.

(a) Calculate the sample mean.

(b) Calculate the standard deviation.

(c) Calculate the 95% confidence levels for the measurement.

α_{obs} (degrees)			
+3.24	+3.20	+3.17	+3.25
+3.15	+3.21	+3.23	
+3.30	+3.19	+3.20	

Further Reading

Data Analysis

P. Meir and R. Zund, *Statistical Methods in Analytical Chemistry* (1993), John Wiley & Sons (New York).

J. Miller and J. Miller, *Statistics for Analytical Chemistry,* 3rd ed. (1992), Prentice-Hall (Englewood Cliffs, NJ).

J. Robyt and B. White, *Biochemical Techniques: Theory and Practice* (1987), Waveland Press (Prospect Heights, IL).

Reagents and Solutions

M. Brush, *The Scientist,* June 8, 1998, pp. 18–23. A discussion of water purification.

S. Kegley and J. Andrews, *The Chemistry of Water* (1997), University Science Books (Sausalito, CA). A discussion of water purity and analysis.

J. Risley, *J. Chem. Educ.* **68,** 1054–1055 (1991). "Preparing Solutions in the Biochemistry Lab."

Safety

M. Armour, *Hazardous Laboratory Chemicals Disposal Guide,* 2nd ed. (1996), CRC Press (Boca Raton, FL).

K. Barker, Editor, *At the Bench: A Laboratory Navigator* (1998), Cold Spring Harbor Laboratory Press (Cold Spring Harbor, NY).

C. Gorman, Editor, *Working Safely with Chemicals in the Laboratory,* 2nd ed. (1994), Genium Publishing Corp. (Schenectady, NY).

G. Lowry and R. Lowry, *Lowry's Handbook of Right-to-Know and Emergency Planning* (1990), Lewis Publishers (Chelsea, MI).

Prudent Practices in the Laboratory: Handling and Disposal of Chemicals (1995), National Research Council, National Academy Press (Washington, DC).

Safety in Academic Chemical Laboratories, 6th ed. (1995), American Chemical Society (Washington, DC).

J. Young, Editor, *Improving Safety in the Chemical Laboratory—A Practical Guide,* 2nd ed. (1991), Wiley-Interscience (New York).

Writing Laboratory Reports

K. Barker, Editor, *At the Bench: A Laboratory Navigator* (1998), Cold Spring Harbor Laboratory Press (Cold Spring Harbor, NY). Covers several important topics including lab orientation, keeping a notebook, and lab procedures.

H. Beall and J. Trimber, *A Short Guide to Writing about Chemistry* (1996), Benjamin-Cummings (Menlo Park, CA).

R. Day, *How to Write and Publish a Scientific Paper,* 2nd ed. (1988), Oryx Press (Phoenix).

R. Dodd, Editor, *The ACS Style Guide: A Manual for Authors and Editors,* 2nd ed. (1997), American Chemical Society (Washington, DC).

H. Ebel, C. Bliefert, and W. Russey, *The Art of Scientific Writing: From Student Reports to Professional Publications in Chemistry and Related Fields* (1997), John Wiley & Sons (New York).

C. Lobban and M. Schefter, *Successful Lab Reports: A Manual for Science Students* (1992), Cambridge University Press (Cambridge).

J. Walker, *Biochem. Educ.* **19,** 31–32 (1991). "A Student's Guide to Practical Write-ups."

On the Web

http://www.graphpad.com/prism/Prism.htm
  Software for statistics and curve fitting.

http://www.statistics.com/
  Click on Free Web-based software for data analysis.

http://www.gilson.com/pipe.htm
  Information on automatic pipets, procedures for use, and helpful hints.

http://ehs.clemson.edu/bsm-spil.html
  Biological Safety Manual.

http://www.hendrix.edu/chemistry/chemsafe.htm
  Information on chemical hygiene and safety with links to MSDS searches.

http://research.nwfsc.noaa.gov/msds.html
  Links to MSDS searches.

http://www.osha.gov/
  Review of functions and regulatory procedures by OSHA.

http://www.dixie.edu/mort/manual/mechanics/Notebook.html.
  How and why to keep a notebook. Procedures for use and helpful hints.

http://practicingsafescience.org
  Advice from the Howard Hughes Medical Institute.

GENERAL LABORATORY PROCEDURES

All biochemical laboratory activities, whether in education, research, or industry, are replete with techniques that must be carried out almost on a daily basis. This chapter outlines the theoretical and practical aspects of some of these general and routine procedures, including use of buffers, pH and other electrodes, dialysis, membrane filtration, lyophilization, centrifugal concentration, and quantitative methods for protein and nucleic acid measurement.

A. PH, BUFFERS, ELECTRODES, AND BIOSENSORS

Most biological processes in the cell take place in a water-based environment. Water is an **amphoteric** substance; that is, it may serve as a proton donor (acid) or a proton acceptor (base). Equation 2.1 shows the ionic equilibrium of water.

$$\text{>>} \qquad H_2O \rightleftharpoons H^+ + OH^- \qquad\qquad \textit{Equation 2.1}$$

In pure water, $[H^+] = [OH^-] = 10^{-7}$; in other words, the pH or $-\log [H^+]$ is 7. Acidic and basic molecules, when dissolved in water in a biological cell or test tube, react with either H^+ or OH^- to shift the equilibrium of Equation 2.1 and result in a pH change of the solution.

Biochemical processes occurring in cells and tissues depend on strict regulation of the hydrogen ion concentration. Biological pH is maintained at a constant value by natural buffers. When biological processes are studied *in vitro,* artificial media must be prepared that mimic the cell's natural environment. Because of the dependence of biochemical reactions on pH, the accurate determination of hydrogen ion concentration has always been of major interest. Today, we consider the measurement and control of pH to be a simple and rather mundane activity. However, an inaccurate pH measurement or a poor choice of buffer can lead to failure in the biochemistry laboratory. You should become familiar with several aspects of pH measurement, electrodes, and buffers.

Measurement of pH

A pH measurement is usually taken by immersing a glass combination electrode into a solution and reading the pH directly from a meter. At one time, pH measurements required two electrodes, a pH-dependent glass electrode sensitive to H^+ ions and a pH-independent calomel reference electrode. The potential difference that develops between the two electrodes is measured as a voltage as defined by Equation 2.2.

$$V = E_{constant} + \frac{2.303RT}{F}\, \text{pH}$$

Equation 2.2

where

$$V = \text{voltage of the completed circuit}$$
$$E_{constant} = \text{potential of reference electrode}$$
$$R = \text{the gas constant}$$
$$T = \text{the absolute temperature}$$
$$F = \text{the Faraday constant}$$

A pH meter is standardized with buffer solutions of known pH before a measurement of an unknown solution is taken. It should be noted from Equation 2.2 that the voltage depends on temperature. Hence, pH meters must have some means for temperature correction. Older instruments usually have a knob labeled "temperature control," which is adjusted by the user to the temperature of the measured solution. Newer pH meters automatically display a temperature-corrected pH value.

Most pH measurements today are obtained using a single **combination electrode** (Figure 2.1). Both the reference and the pH-dependent electrode are contained in a single glass or plastic tube. Although these are more expensive than dual electrodes, they are much more convenient to use, especially for smaller volumes of solution. Using a pH meter with a combination electrode is relatively easy, but certain guidelines must be followed. A pH meter not in use is left in a "standby" position. Before use, check the level of saturated KCl in the electrode. If it is low, check with your instructor for the filling procedure. Turn the temperature control, if available, to the temperature of the standard calibration buffers and the test solutions. Be sure the function dial is set to pH. Lift the electrode out of the storage solution, rinse it with distilled water from a wash bottle, and *gently* clean and dry the elec-

Figure 2.1

The combination pH electrode. *Courtesy of Hanna Instruments.*

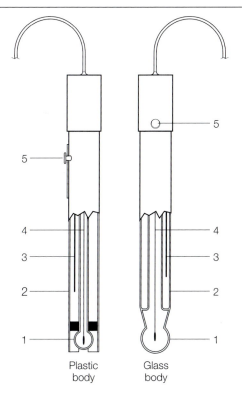

Plastic
body

Glass
body

Electrodes are housed in either plastic or an all-glass body configuration. They can be either single cells or as shown in the diagram, combined into one body for ease of use. Regardless of the configuration, there are several features common to all electrodes.

1. Sensing membrane glass: Performs actual measurement.

2. Reference junction: Acts as a liquid path electrical conductor.

3. Internal reference: Supplies a constant equilibrium voltage.

4. pH internal element: Supplies a voltage based on the pH value of the sample.

5. Reference fill hole: Used to replace the reference electrolyte solution.

trode with a tissue. Immerse the electrode in a standard buffer. Common standard buffers are pH 4, 7, and 10 with accuracy of ± 0.02 pH unit. The standard buffer should have a pH within two pH units of the expected pH of the test solution. The bulb of the electrode must be completely covered with solution. Turn the pH meter to "on" or "read" and adjust the meter with the "calibration dial" (sometimes called "intercept") until the proper pH of the standard buffer is indicated on the dial. Turn the pH meter to standby position. Remove the electrode and again rinse with distilled water and carefully blot dry with tissue. Immerse the electrode in a standard buffer of different pH and turn the pH meter to "read." The dial should read within ± 0.05 pH unit of the known value. If it does not, adjust to the proper pH and again check the first standard pH buffer. Clean the electrode and immerse it in the test solution. Record the pH of the test solution.

As with all delicate equipment, the pH meter and electrode must receive proper care and maintenance. All electrodes should be kept clean and stored in solutions suggested by manufacturers. Glass electrodes are fragile and expensive, so they must be handled with care. If pH measurements of protein solutions are often taken, a protein film may develop on the electrode; it can be removed by soaking in 5% pepsin in 0.1 M HCl for 2 hours and rinsing well with water.

Measurements of pH are always susceptible to experimental errors. Some common problems are:

1. The Sodium Error   Most glass combination electrodes are sensitive to Na^+ as well as H^+. The sodium error can become quite significant at high pH values, where 0.1 M Na^+ may decrease the measured pH by 0.4 to 0.5 unit. Several things may be done to reduce the sodium error. Some commercial suppliers of electrodes provide a standard curve for sodium error correction. Newer electrodes that are virtually Na^+ impermeable are now commercially available. If neither a standard curve nor a sodium-insensitive electrode is available, potassium salts may be substituted for sodium salts.

2. Concentration Effects   The pH of a solution varies with the concentration of buffer ions or other salts in the solution. This is because the pH of a solution depends on the *activity* of an ionic species, not on the concentration. **Activity,** you may recall, is a thermodynamic term used to define species in a nonideal solution. At infinite dilution, the activity of a species is equivalent to its concentration. At finite dilutions, however, the activity of a solute and its concentration are not equal.

It is common practice in biochemical laboratories to prepare concentrated "stock" solutions and buffers. These are then diluted to the proper concentration when needed. Because of the concentration effects described above, it is important to adjust the pH of these solutions *after* dilution.

3. Temperature Effects   The pH of a buffer solution is influenced by temperature. This effect is due to a temperature-dependent change of the dissociation constant (pK_a) of ions in solution. The pH of the commonly used buffer Tris is greatly affected by temperature changes, with a $\Delta pK_a/C°$ of -0.031. This means that a pH 7.0 Tris buffer made up at 4°C would have a pH of 5.95 at 37°C. The best way to avoid this problem is to prepare the buffer solution at the temperature at which it will be used and to standardize the electrode with buffers at the same temperature as the solution you wish to measure.

Biochemical Buffers

Buffer ions are used to maintain solutions at constant pH values. The selection of a buffer for use in the investigation of a biochemical process is of critical importance. Before the characteristics of a buffer system are discussed, we will review some concepts in acid-base chemistry.

Weak acids and bases do not completely dissociate in solution but exist as equilibrium mixtures (Equation 2.3).

>>

$$HA \underset{k_2}{\overset{k_1}{\rightleftharpoons}} H^+ + A^-$$

Equation 2.3

HA represents a weak acid and A^- represents its conjugate base; k_1 represents the rate constant for dissociation of the acid and k_2 the rate constant for association of the conjugate base and hydrogen ion. The equilibrium constant, K_a, for the weak acid HA is defined by Equation 2.4.

$$\text{>>}\qquad K_a = \frac{k_1}{k_2} = \frac{[H^+][A^-]}{[HA]}$$

Equation 2.4

which can be rearranged to define $[H^+]$ (Equation 2.5).

$$\text{>>}\qquad [H^+] = \frac{K_a[HA]}{[A^-]}$$

Equation 2.5

The $[H^+]$ is often reported as pH, which is $-\log [H^+]$. In a similar fashion, $-\log K_a$ is represented by pK_a. Equation 2.5 can be converted to the $-\log$ form by substituting pH and pK_a:

$$\text{>>}\qquad pH = pK_a + \log \frac{[A^-]}{[HA]}$$

Equation 2.6

Equation 2.6 is the familiar Henderson-Hasselbalch equation, which defines the relationship between pH and the ratio of acid and conjugate base concentrations. The Henderson-Hasselbalch equation is of great value in buffer chemistry because it can be used to calculate the pH of a solution if the molar ratio of buffer ions ($[A^-]/[HA]$) and the pK_a of HA are known. Also, the molar ratio of HA to A^- that is necessary to prepare a buffer solution at a specific pH can be calculated if the pK_a is known.

A solution containing both HA and A^- has the capacity to resist changes in pH; i.e., it acts as a **buffer.** If acid (H^+) were added to the buffer solution, it would be neutralized by A^- in solution:

$$\text{>>}\qquad H^+ + A^- \longrightarrow HA$$

Equation 2.7

Base (OH^-) added to the buffer solution would be neutralized by reaction with HA:

$$\text{>>}\qquad OH^- + HA \longrightarrow A^- + H_2O$$

Equation 2.8

The most effective buffering system contains equal concentrations of the acid, HA, and its conjugate base, A^-. According to the Henderson-Hasselbalch equation (2.6), when $[A^-]$ is equal to $[HA]$, pH equals pK_a. Therefore, the pK_a of a weak acid-base system represents the center of the buffering region. The effective range of a buffer system is generally two pH units, centered at the pK_a value (Equation 2.9).

$$\text{>>}\qquad \text{Effective pH range for a buffer} = pK_a \pm 1$$

Equation 2.9

Selection of a Biochemical Buffer

Virtually all biochemical investigations must be carried out in buffered aqueous solutions. The natural environment of biomolecules and cellular organelles is under strict pH control. When these components are extracted from cells, they are most stable if maintained in their normal pH range, usually 6 to 8. An artificial buffer system is found to be the best substitute for the natural cell milieu. It should also be recognized that many biochemical processes (especially some enzyme processes) produce or consume hydrogen ions. The buffer system neutralizes these solutions and maintains a constant chemical environment.

　　　Although most biochemical solutions require buffer systems effective in the pH range 6 to 8, there is occasionally a need for buffering over the pH range 2 to 12. Obviously, no single acid–conjugate base pair will be effective over this entire range, but several buffer systems are available that may be used in discrete pH ranges. Figure 2.2 compares the effective buffering ranges of common biological buffers. It should be noted that some buffers (phosphate, succinate, and citrate) have more than one pK_a value, so they may be used in different pH regions. Many buffer systems are effec-

---

Figure 2.2

Effective buffering ranges of several common buffers.

- -

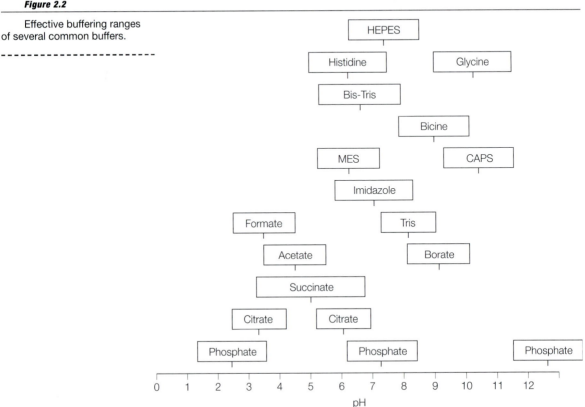

tive in the usual biological pH range (6 to 8); however, there may be major problems in their use. Several characteristics of a buffer must be considered before a final selection is made. The molecular weights and pK values of several common buffer compounds are listed in Appendix II. Following is a discussion of the advantages and disadvantages of the commonly used buffers.

Phosphate Buffers

The phosphates are among the most widely used buffers. These solutions have high buffering capacity and are very useful in the pH range 6.5 to 7.5. Because phosphate is a natural constituent of cells and biological fluids, its presence affords a more "natural" environment than many buffers. Sodium or potassium phosphate solutions of all concentrations are easy to prepare using the Henderson-Hasselbalch equation. The major disadvantages of phosphate solutions are (1) precipitation or binding of common biological cations (Ca^{2+} and Mg^{2+}), (2) inhibition of some biological processes, including some enzymes, and (3) limited useful pH range.

Zwitterionic Buffers (Good's Buffers)

In the mid-1960s, N. E. Good and his colleagues recognized the need for a set of buffers specifically designed for biochemical studies (Good and Izawa, 1972). He and others noted major disadvantages of the established buffer systems. Good outlined several characteristics essential in a biological buffer system:

1. pK_a between 6 and 8.
2. Highly soluble in aqueous systems.
3. Exclusion or minimal transport by biological membranes.
4. Minimal salt effects.
5. Minimal effects on dissociation due to ionic composition, concentration, and temperature.
6. Buffer–metal ion complexes nonexistent or soluble and well defined.
7. Chemically stable.
8. Insignificant light absorption in the ultraviolet and visible regions.
9. Readily available in purified form.

Good investigated a large number of synthetic zwitterionic buffers and found many of them to meet these criteria. Table 2.1 lists several of these buffers and their properties. Good's buffers are widely used, but their main disadvantage is high cost. Some Good's buffers, such as Tris, HEPES, and PIPES, have been shown to produce radicals under a variety of experimental conditions, so they should be avoided if biological redox processes or radical-based reactions are being studied. Radicals are not produced from MES or MOPS buffers.

Table 2.1

Structures and Properties of Several Synthetic Zwitterionic Buffers

Buffer Name	Abbreviation	Structure (All structures shown in salt form)	pK_a (20°C)	Useful pH Range	$\Delta pK_a/C°$	Concentration of a Saturated Solution (M, 0°C)
N-2-Acetamido-2-amino-ethanesulfonic acid	ACES	$H_2NCOCH_2\overset{+}{N}H_2CH_2CH_2SO_3^-$	6.9	6.4–7.4	−0.020	0.22
N-2-Acetamidoiminodiacetic acid	ADA	$H_2NCOCH_2\overset{+}{N}\overset{CH_2COO^-}{\underset{H}{|}}CH_2COO^-$	6.6	6.2–7.2	−0.011	–
N,N-Bis(2-hydroxyethyl)-2-aminoethanesulfonic acid	BES	$(HOCH_2CH_2)_2\overset{+}{N}HCH_2CH_2SO_3^-$	7.15	6.6–7.6	−0.016	3.2
N,N-Bis(2-hydroxyethyl)-glycine	Bicine	$(HOCH_2CH_2)_2\overset{+}{N}HCH_2COO^-$	8.35	7.5–9.0	−0.018	1.1
3-(Cyclohexylamino)-propanesulfonic acid	CAPS	 $-NHCH_2CH_2CH_2SO_3^-$	10.4	10.0–11.0	−0.009	0.85
Cyclohexylaminoethane-sulfonic acid	CHES	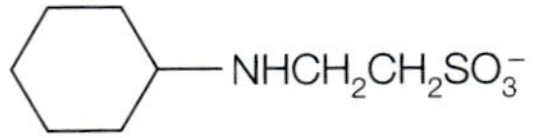 $-NHCH_2CH_2SO_3^-$	9.5	9.0–10.0	−0.009	0.85
Glycylglycine	Gly-Gly	$H_3\overset{+}{N}CH_2CONHCH_2COO^-$	8.4	7.5–9.5	−0.028	1.1
N-2-Hydroxyethyl-piperazine-N'-2-ethanesulfonic acid	HEPES	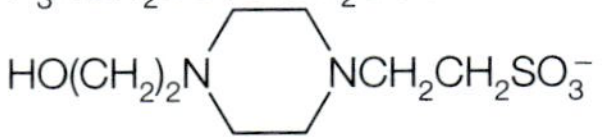 $HO(CH_2)_2N\!\!-\!\!NCH_2CH_2SO_3^-$	7.55	7.0–8.0	−0.014	2.25
2-(N-Morpholino)ethane-sulfonic acid	MES	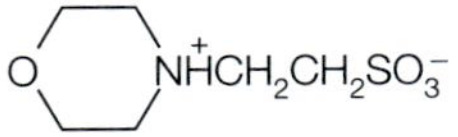 $\overset{+}{N}HCH_2CH_2SO_3^-$	6.15	5.8–6.5	−0.011	0.65
3-(N-Morpholino)-propanesulfonic acid	MOPS	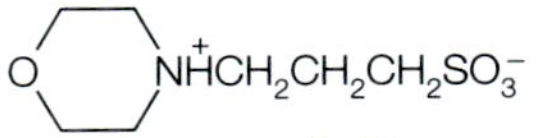 $\overset{+}{N}HCH_2CH_2CH_2SO_3^-$	7.20	6.5–7.9	–	–
Piperazine-N,N'-bis-2-ethanesulfonic acid	PIPES	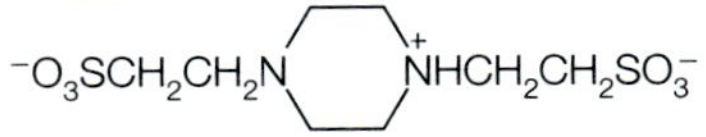 $^-O_3SCH_2CH_2\overset{+}{N}\!\!-\!\!\overset{+}{N}HCH_2CH_2SO_3^-$	6.8	6.4–7.2	−0.0085	–
N-Tris(hydroxymethyl)-methyl-2-aminoethane-sulfonic acid	TES	$(HOCH_2)_3C\overset{+}{N}HCH_2CH_2SO_3^-$	7.5	7.0–8.0	−0.020	2.6
N-Tris(hydroxymethyl)-methylglycine	Tricine	$(HOCH_2)_3C\overset{+}{N}H_2CH_2COO^-$	8.15	7.5–8.5	−0.021	0.8
Tris(hydroxymethyl)-aminomethane	Tris	$(HOCH_2)_3C\overset{+}{N}H_3$	8.3	7.5–9.0	−0.031	2.4

The use of the synthetic zwitterionic buffer Tris [tris(hydroxymethyl) aminomethane] is now probably greater than that of phosphate. It is useful in the pH range 7.5 to 8.5. Tris is available in a basic form as highly purified crystals, which makes buffer preparation especially convenient. To prepare solutions, the appropriate amount of Tris base is weighed and dissolved in water. For 1 L of a 0.1 M solution, 12.11 g (0.1 mole) of Tris base is weighed and dissolved in 950 to 975 mL of distilled water. The pH is adjusted by addition of acid (concentrated hydrochloric if Tris-HCl is desired), with stirring, until the appropriate pH is attained. Water is added to a final volume of 1 L and a final pH check is made. Although Tris is a primary amine, it causes minimal interference with biochemical processes and does not precipitate calcium ions. However, Tris has several disadvantages, including (1) pH dependence on concentration, since the pH decreases 0.1 pH unit for each 10-fold dilution; (2) interference with some pH electrodes; and (3) a large $\Delta pK_a/C°$ compared to most other buffers. Most of these drawbacks can be minimized by (1) adjusting the pH *after* dilution to the appropriate concentration, (2) purchasing electrodes that are compatible with Tris, and (3) preparing the buffer at the temperature at which it will be used.

Carboxylic Acid Buffers

The most widely used buffers in this category are acetate, formate, citrate, and succinate. This group is useful in the pH range 3 to 6, a region that offers few other buffer choices. All of these acids are natural metabolites, so they may interfere with the biological processes under investigation. Also, citrate and succinate may interfere by binding metal ions (Fe^{3+}, Ca^{2+}, Zn^{2+}, Mg^{2+}, etc.). Formate buffers are especially useful because they are volatile and can be removed by evaporation under reduced pressure.

Borate Buffers

Buffers of boric acid are useful in the pH range 8.5 to 10. Borate has the major disadvantage of complex formation with many metabolites, especially carbohydrates.

Amino Acid Buffers

The most commonly used amino acid buffers are glycine (pH 2 to 3 and 9.5 to 10.5), histidine (pH 5.5 to 6.5), glycine amide (pH 7.8 to 8.8), and glycylglycine (pH 8 to 9). These provide a more "natural" environment to cellular components and extracts; however, they may interfere with some biological processes, as do the carboxylic acid and phosphate buffers.

The Oxygen Electrode

Second in popularity only to the pH electrode is the oxygen electrode. This device is a polarographic electrode system that can be used to measure oxygen concentration in a liquid sample or to monitor oxygen uptake or evolution by a chemical or biological system.

The basic electrode was developed by L. C. Clark, Jr. in 1953. The popular Clark-type oxygen electrode with biological oxygen monitor is shown in Figure 2.3. The electrode system, which consists of a platinum cathode and silver anodes, is molded into an epoxy block and surrounded by a Lucite holder. A thin Teflon membrane is stretched over the electrode end of the probe and held in place with an O-ring. The membrane separates the electrode elements, which are bathed in an electrolyte solution (saturated KCl), from the test solution in the sample chamber. The membrane is permeable to oxygen and other gases, which then come into contact with the surface of the platinum cathode. If a suitable polarizing voltage (about 0.8 volt) is applied across the cell, oxygen is reduced at the cathode.

$$O_2 + 2e^- + 2H_2O \longrightarrow [H_2O_2] + 2OH^-$$

$$\underline{[H_2O_2] + 2e^- \longrightarrow 2OH^-}$$

$$\text{Total} \quad O_2 + 4e^- + 2H_2O \longrightarrow 4OH^-$$

The reaction occurring at the anode is

$$4Ag^\circ + 4Cl^- \longrightarrow 4AgCl + 4e^-$$

The overall electrochemical process is

$$4Ag^\circ + O_2 + 4Cl^- + 2H_2O \longrightarrow 4AgCl + 4OH^-$$

Figure 2.3

YSI Model 5300 Biological Oxygen Monitor. *Courtesy of YSI, Inc.*

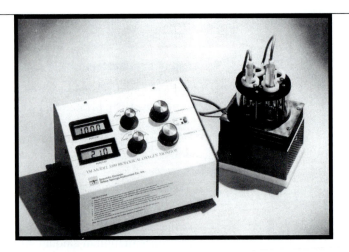

The current flowing through the electrode system is, therefore, directly proportional to the amount of oxygen passing through the membrane. According to the laws of diffusion, the rate of oxygen flowing through the membrane depends on the concentration of dissolved oxygen in the sample. Therefore, the magnitude of the current flow in the electrode is directly related to the concentration of dissolved oxygen. The function of the electrode can also be explained in terms of oxygen pressure. The oxygen concentration at the cathode (inside the membrane) is virtually zero because oxygen is rapidly reduced. Oxygen pressure in the sample chamber (outside the membrane) will force oxygen to diffuse through the membrane at a rate that is directly proportional to the pressure (or concentration) of oxygen in the test solution. The current generated by the electrode system is electronically amplified and transmitted to a meter, which usually reads in units of % oxygen saturation.

Many biochemical processes consume or evolve oxygen. The oxygen electrode is especially useful for monitoring such changes in the concentration of dissolved oxygen. If oxygen is consumed, for example, by a suspension of mitochondria or by an oxygenase enzyme system, the rate of oxygen diffusion through the probe membrane will decrease, leading to less current flow. A linear relationship exists between the electrode current and the amount of oxygen consumed by the biological system.

The use of an oxygen electrode is not free of experimental and procedural problems. Probably the most significant source of trouble is the Teflon membrane. A torn, damaged, or dirty membrane is easily recognized by recorder noise or electronic "spiking." Temperature control of the sample chamber and electrode is essential because the rate of oxygen diffusion through the membrane is temperature dependent. The sample chamber is equipped with a bath assembly for temperature regulation and magnetic stirrer in order to maintain a constant oxygen concentration throughout the solution. Air bubbles in the sample chamber lead to considerable error in measurements and must be removed.

Ion-Selective Electrodes and Biosensors

Although electrodes that respond to H^+ and O_2 are the most widely used, other ions and gases may be measured with specific electrodes and potentiometric methods. Table 2.2 lists some of the electrodes that are valuable in biochemical measurements. Each type of electrode is made from a specially prepared material that has increased permeability toward a single type of ion. Ion-specific electrodes are sensitive (some are useful in the parts-per-billion range), require simple, inexpensive equipment (may be connected to a pH meter set in the millivolt mode), and save time (after electrode calibration, an analysis takes about 1 minute). A major difficulty that has slowed the adoption of ion-selective electrodes is interference by other ions. In most cases, the serious interfering ions are known (see Table 2.2), and procedures are available for masking or eliminating some competing ions.

Table 2.2

Ion-Specific Electrodes

Ions	Concentration Range (M)	Interferences[1]
Ammonium	10^0 to 10^{-6}	Volatile amines
Bromide	10^0 to 5×10^{-6}	S^{2-}, I^-
Cadmium	10^0 to 10^{-7}	Ag^+, Hg^{2+}, Cu^{2+}, Pd^{2+}, Fe^{3+}
Carbon dioxide	10^{-2} to 10^{-4}	Sr^{2+}, Mg^{2+}, Ba^{2+}, Ni^{2+}, volatile weak acids
Chloride	10^0 to 8×10^{-6}	ClO_4^-, I^-, NO_3^-, SO_4^{2-}, Br^-, OH^-, OAc^-, HCO_3^-, F^-
Cupric	Saturated to 10^{-8}	S^{2-}, Hg^{2+}, Ag^+
Cyanide	10^{-2} to 10^{-6}	S^{2-}, I^-, Br^-, Cl^-
Divalent cations	10^0 to 6×10^{-6}	Na^+, Cu^{2+}, Zn^{2+}, Fe^{2+}, Ni^{2+}, Sr^{2+}, Bi^{2+}
Fluoride	Saturated to 10^{-6}	OH^-
Iodide	10^0 to 2×10^{-5}	S^{2-}
Lead	10^0 to 10^{-7}	Ag^+, Hg^{2+}, Cu^{2+}, Cd^{2+}, Fe^{3+}
Nitrate	10^0 to 6×10^{-6}	I^-, Br^-, NO_2^-
Nitrite	10^{-2} to 5×10^{-7}	CO_2
Potassium	10^0 to 10^{-5}	Cs^+, NH_4^+, H^+
Silver/sulfide	10^0 to $\times 10^{-7}$	Hg^{2+}
Sodium	Saturated to 10^{-6}	Cs^+, Li^+, K^+
Thiocyanate	10^0 to 5×10^{-6}	OH^-, Br^-, Cl^-

[1] The presence of these ions does not rule out the use of a specific-ion electrode, since many ions interfere only at relatively high concentrations.

Electrodes or electrode-like devices are now being developed for the specific measurement of physiologically important molecules such as urea, carbohydrates, enzymes, antibodies, and metabolic products. This type of device, now referred to as a **biosensor,** is an analytical tool or system consisting of an immobilized biological material (such as an enzyme, antibody, whole cell, organelle, or combinations thereof) in intimate contact with a suitable transducer device which will convert the biochemical signal into a quantifiable electric signal. The important components of a biosensor as shown in Figure 2.4 are (1) a reaction center consisting of a membrane or gel containing the biochemical system to be studied, (2) a transducer, (3) an amplifier, and (4) a computer system for data acquisition and processing. When biomolecules in the reaction center interact, a physicochemical change occurs. This change in the molecular system, which may be a modification of concentration, absorbance, mass, conductance, or redox state, is converted into an electrical signal by the transducer. The signal is then amplified and displayed on a computer screen. Each biosensor is specifically designed for a type of molecule or biological interaction; therefore, details of construction and function vary. Some specific examples of biosensors include:

1. Electrodes based on enzyme activity. These are selective and sensitive devices that may be used to measure substrate concentrations. A biosensor based on glucose oxidase is used to measure the concentration of glucose by detecting the production of H_2O_2.

$$\text{Glucose} + O_2 \rightleftharpoons \text{gluconic acid} + H_2O_2$$

Components of a biosensor system for determining the identity and concentration of biomolecule A. A is stoichiometrically converted to B by a specific enzyme in the reaction center. The concentration of B is measured by the sensor.

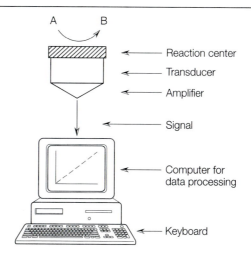

A   B

Reaction center

Transducer

Amplifier

Signal

Computer for data processing

Keyboard

2.   Optical biosensors that respond to the absorption of light from a laser. The optical response is transmitted to a computer by a fiber-optic system.

B.   MEASUREMENT OF PROTEIN SOLUTIONS

Biochemical research often requires the quantitative measurement of protein concentrations in solutions. Several techniques have been developed; however, most have limitations because either they are not sensitive enough or they are based on reactions with specific amino acids in the protein. Since the amino acid content varies from protein to protein, no single assay will be suitable for all proteins. In this section we discuss five assays: three older, classical methods that are occasionally used today and two newer methods that are widely used. In four of the methods, chemical reagents are added to protein solutions to develop a color whose intensity is measured in a spectrophotometer. A "standard protein" of known concentration is also treated with the same reagents and a calibration curve is constructed. The other assay relies on a direct spectrophotometric measurement. None of the methods is perfect because each is dependent on the amino acid content of the protein. However, each will provide a satisfactory result if the proper experimental conditions are used and/or a suitable standard protein is chosen. Other important factors in method selection include the sensitivity and accuracy desired, the presence of interfering substances, and the time available for the assay. The various methods are compared in Table 2.3.

The Biuret and Lowry Assays

When substances containing two or more peptide bonds react with the biuret reagent, alkaline copper sulfate, a purple complex is formed. The colored product is the result of coordination of peptide nitrogen atoms with

Table 2.3

Methods for Protein Measurement

Method	Sensitivity	Time	Principle	Interferences	Comments
Biuret	Low 1–20 mg	Moderate 20–30 min	Peptide bonds + alkaline $Cu^{2+} \longrightarrow$ purple complex	Zwitterionic buffers, some amino acids	Similar color with all proteins. Destructive to protein samples.
Lowry	High ~5 µg	Slow 40–60 min	(1) Biuret reaction (2) Reduction of phosphomolybdate-phosphotungstate by Tyr and Trp	Ammonium sulfate, glycine, zwitterionic buffers, mercaptans	Time-consuming. Color varies with proteins. Critical timing of procedure. Destructive to protein samples.
Bradford	High ~1 µg	Rapid 15 min	λ_{max} of Coomassie dye shifts from 465 nm to 595 nm when protein-bound	Strongly basic buffers, detergents Triton X-100, SDS	Stable color which varies with proteins. Reagents commercially available. Destructive to protein samples. Discoloration of glassware.
BCA	High 1 µg	Slow 60 min	(1) Biuret reaction (2) Copper complex with BCA; $\lambda_{max} =$ 562 nm	EDTA, DTT, ammonium sulfate	Compatible with detergents. Reagents commercially available. Destructive to protein samples.
Spectrophotometric (A_{280})	Moderate 50–1000 µg	Rapid 5–10 min	Absorption of 280-nm light by aromatic residues	Purines, pyrimidines, nucleic acids	Useful for monitoring column eluents. Nucleic acid absorption can be corrected. Nondestructive to protein samples. Varies with proteins.

Cu^{2+}. The amount of product formed depends on the concentration of protein.

In practice, a calibration curve must be prepared by using a standard protein solution. The best protein to use as a standard is a purified preparation of the protein to be assayed. Since this is rarely available, the experimenter must choose another protein as a relative standard. A relative standard must be selected that provides a similar color yield. An aqueous solution of bovine serum albumin is an appropriate standard. Various known amounts of this solution are treated with the biuret reagent and the color is allowed to develop. Measurements of absorbance at 540 nm (A_{540}) are made against a blank containing biuret reagent and buffer or water. The A_{540} data are plotted versus protein concentration (mg/mL). Unknown protein samples are treated with biuret reagent and the A_{540} is measured after color development. The protein concentration is determined from the standard curve. The biuret assay has several advantages including speed, similar color development with different proteins, and few interfering substances. Its primary disadvantage is its lack of sensitivity.

The Lowry protein assay is one of the more sensitive and has been widely used. The Lowry procedure can detect protein levels as low as 5 μg. The principle behind color development is identical to that of the biuret assay except that a second reagent (Folin-Ciocalteu) is added to increase the amount of color development. Two reactions account for the intense blue color that develops: (1) the coordination of peptide bonds with alkaline copper (biuret reaction) and (2) the reduction of the Folin-Ciocalteu reagent (phosphomolybdate-phosphotungstate) by tyrosine and tryptophan residues in the protein. The procedure is similar to that of the biuret assay. A standard curve is prepared with bovine serum albumin or other pure protein, and the concentration of unknown protein solutions is determined from the graph.

The obvious advantage of the Lowry assay is its sensitivity, which is up to 100 times greater than that of the biuret assay; however, more time is required for the Lowry assay. Since proteins have varying contents of tyrosine and tryptophan, the amount of color development changes with different proteins, including the bovine serum albumin standard. Because of this, the Lowry protein assay should be used only for measuring changes in protein concentration, not absolute values of protein concentration.

The Bradford Assay

The many limitations of the biuret and Lowry assays have encouraged researchers to seek better methods for quantitation of protein solutions. The Bradford assay, based on protein binding of a dye (perhaps to arginine residues), provides numerous advantages over other methods.

The binding of Coomassie Brilliant Blue dye (Figure 2.5) to protein in acidic solution causes a shift in wavelengh of maximum absorption (λ_{max}) of the dye from 465 nm to 595 nm. The absorption at 595 nm is directly related to the concentration of protein.

Figure 2.5

Coomassie Brilliant Blue
G-250 dye.

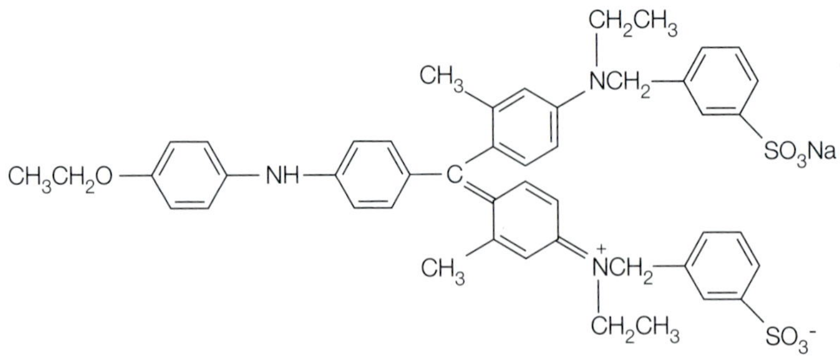

In practice, a calibration curve is prepared using bovine plasma gamma globulin or bovine serum albumin as a standard (Figure 2.6). The assay requires only a single reagent, an acidic solution of Coomassie Brilliant Blue G-250. After addition of dye solution to a protein sample, color development is complete in 2 minutes and the color remains stable for up to 1 hour. The sensitivity of the Bradford assay rivals and may surpass that of the Lowry assay. With a microassay procedure, the Bradford assay can be used to determine proteins in the range of 1 to 20 μg. The Bradford assay shows significant variation with different proteins, but this also occurs with the Lowry assay. The Bradford method not only is rapid but also has very few interferences by nonprotein components. The only known interfering substances are detergents, Triton X-100 and sodium dodecyl sulfate. The many advantages of the Bradford assay have led to its wide adoption in biochemical research laboratories.

The BCA Assay

The BCA protein assay is based on chemical principles similar to those of the biuret and Lowry assays. The protein to be analyzed is reacted with Cu^{2+} (which produces Cu^+) and bicinchoninic acid (BCA). The Cu^+ is chelated by BCA, which converts the apple-green color of the free BCA to the purple color of the copper-BCA complex. Samples of unknown protein and relative standard are treated with the reagent, the absorbance is read in a spectrophotometer at 562 nm, and concentrations are determined from a standard calibration curve. This assay has the same sensitivity level as the Lowry and Bradford assays. Its main advantages are its simplicity and its usefulness in the presence of 1% detergents such as Triton or sodium dodecyl sulfate (SDS).

The Spectrophotometric Assay

Most proteins have relatively intense ultraviolet light absorption centered at 280 nm. This is due to the presence of tyrosine and tryptophan residues in

Figure 2.6

Standard calibration curve
for the Bradford assay using
bovine gamma globulin as
standard protein.

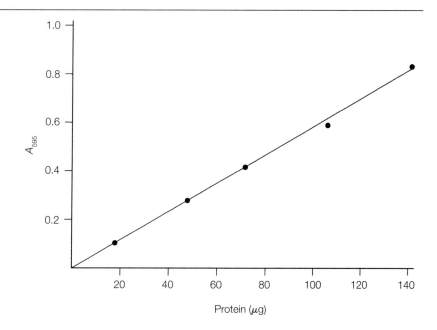

the protein. However, the amount of these amino acid residues varies in different proteins, as was pointed out earlier. If certain precautions are taken, the A_{280} value of a protein solution is proportional to the protein concentration. The procedure is simple and rapid. A protein solution is transferred to a quartz cuvette and the A_{280} is read against a reference cuvette containing the protein solvent only (buffer, water, etc.).

Cellular extracts contain many other compounds that absorb in the vicinity of 280 nm. Nucleic acids, which would be common contaminants in a protein extract, absorb strongly at 280 nm (λ_{max} = 260). Early researchers developed a method to correct for this interference by nucleic acids. Mixtures of pure protein and pure nucleic acid were prepared, and the ratio A_{280}/A_{260} was experimentally determined. The following empirical equation may be used for protein solutions containing up to 20% nucleic acids.

Protein concentration (mg/mL) = $1.55A_{280} - 0.76A_{260}$

Although the spectrophotometric assay of proteins is fast, relatively sensitive, and requires only a small sample size, it is still only an estimate of protein concentration. It has certain advantages over the colorimetric assays in that most buffers and ammonium sulfate do not interfere and the procedure is nondestructive to protein samples. The spectrophotometric assay is particularly suited to the rapid measurement of protein elution from a chromatography column, where only protein concentration changes are required.

C.   MEASUREMENT OF NUCLEIC ACID SOLUTIONS

The Spectrophotometric Assay

Solutions of nucleic acids strongly absorb ultraviolet light with a λ_{max} of 260 nm. The intense absorption is due primarily to the presence of aromatic rings in the purine and pyrimidine bases. The concentration of nucleic acid in a solution can be calculated if one knows the value of A_{260} of the solution. A solution of double-stranded DNA at a concentration of 50 μg/mL in a 1-cm quartz cuvette will give an A_{260} reading of 1.0. Therefore, the concentration (X) of a double-stranded DNA solution with A_{260} of 0.15 is

$$\frac{1.0}{50\ \mu g/mL} = \frac{0.15}{X\ \mu g/mL}, \qquad X = 7.5\ \mu g/mL$$

A solution of single-stranded DNA or RNA that has an A_{260} of 1.0 in a cuvette with a 1-cm path length has a concentration of 40 μg/mL.

The spectrophotometric assay for nucleic acids is simple, rapid, nondestructive, and very sensitive; however, it measures both DNA and RNA. The minimum concentration of nucleic acid that can be accurately determined is 2.5–5.0 μg/mL.

Other Assays for Nucleic Acids

The spectrophotometric assay for nucleic acids, DNA and RNA, is simple and rapid, but it does not have a sufficient level of sensitivity for many applications. Other assays for DNA and RNA depend on the enhanced fluorescence of dyes when bound to nucleic acids. Features of fluorescence assays are compared to the spectrophotometric assay in Table 2.4.

The fluorescence dye bisbenzimidazole, when bound to DNA, emits light at 458 nm that is about five times greater in intensity than the emission by free, unbound dye. The increase in fluorescence is linear with increasing amounts of DNA in the concentration range of 0.01 μg/mL to 5 μg/mL. One special advantage of this assay is that it can be used to quantify DNA in cell suspensions and crude homogenates. It does not respond to RNA or other biomolecules in the extracts. The assay is fast and requires only a fluorimeter, an instrument found in most laboratories.

Another simple and relatively inexpensive fluorescence method uses ethidium bromide–agarose plates. The fluorescence emission of ethidium bromide at 590 nm is enhanced about 25-fold when the dye interacts with DNA (see Experiment 13). DNA samples for measurement are spotted onto 1% agarose containing ethidium bromide in a Petri dish or spot plates. The samples are exposed to UV light, photographed, and the light intensity compared to standard samples of DNA. This method is sensitive (in the 1–10 ng range), but it detects both DNA and RNA.

One new and interesting method for nucleic acid analysis is the DNA Dipstick Kit developed by Invitrogen. The kit is a colorimetric-based

Table 2.4

Measurement of Nucleic Acid Solutions

Method	Sensitivity	Time	Principle	Interferences	Comments
Spectrophotometric (A_{260})	High 2.5 μg	Rapid 5–10 min	Absorption of 260 nm light by aromatic bases	Proteins	Protein absorption may be corrected. A_{280}/A_{260} indicates purity. Nondestructive. Measures both DNA and RNA.
Bisbenzimidazole	Very high 10 ng	Rapid 5–10 min	Enhancement of fluorescence of dye when bound to DNA	None known	Not reliable with plasmids or small DNA. Does not measure RNA. May be used with crude extracts. Nondestructive.
DNA Dipstick	Very high 0.1–10 ng	Moderate 30 min	Proprietary	SDS, agarose	Gives no indication of purity. Measures single and double-stranded DNA and RNA.
EtBr-Agarose	Very high 1–10 ng	Moderate 30–45 min	Fluorescence of bound EtBr	None known	Measures DNA and RNA.

assay where the nucleic acid sample interacts with a proprietary dye solution. The method can detect nucleic acids and oligonucleotides in the range of 0.1–10 ng/μL and requires only about 30 minutes.

D.   TECHNIQUES FOR SAMPLE PREPARATION

Dialysis

One of the oldest procedures applied to the purification and characterization of biomolecules is **dialysis,** an operation used to separate dissolved molecules on the basis of their molecular size. The technique involves sealing an aqueous solution containing both macromolecules and small molecules in a porous membrane. The sealed membrane is placed in a large container of low-ionic-strength buffer. The membrane pores are too small to allow diffusion of macromolecules of molecular weight greater than about 10,000. Smaller molecules diffuse freely through the openings (Figure 2.7). The passage of smaller molecules continues until their concentrations inside the dialysis tubing and outside in the large volume of buffer are equal. Thus, the concentration of small molecules inside the membrane is reduced. Equilibrium is volume dependent and is reached after 4 to 6 hours. Of course, if the outside solution (dialysate) is replaced with fresh buffer after equilibrium is reached, the concentration of small molecules inside the membrane will be further reduced by continued dialysis.

Dialysis membranes are available in a variety of materials and sizes. The most common materials are collodion, cellophane, and cellulose. Recent modifications in membrane construction make a range of pore sizes avail-

Figure 2.7

Diffusion of smaller molecules through dialysis membrane pores.

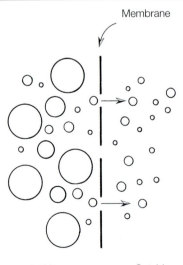

Membrane

Inside   Outside

able. Spectrum Laboratories offers Spectra/Por membrane tubing with complete molecular weight cutoffs ranging from 100 to 300,000.

Dialysis is most commonly used to remove salts and other small molecules from solutions of macromolecules. During the separation and purification of biomolecules, small molecules are added to selectively precipitate or dissolve the desired molecule. For example, proteins are often precipitated by addition of organic solvents or salts such as ammonium or sodium sulfate. Since the presence of organics or salts usually interferes with further purification and characterization of the molecule, they must be removed. Dialysis is a simple, inexpensive, and effective method for removing all small molecules, ionic or nonionic.

Dialysis is also useful for removing small ions and molecules that are weakly bound to biomolecules. Protein cofactors such as NAD, FAD, and metal ions can often be dissociated by dialysis. The removal of metal ions is facilitated by the addition of a chelating agent (EDTA) to the dialysate.

Ultrafiltration

Although dialysis is still used occasionally as a purification tool, it has been largely replaced by ultrafiltration and gel filtration (see Chapter 3). The major disadvantage of dialysis that is overcome by the newer methods is that it may take several days of dialysis to attain a suitable separation. The other methods require 1 to 2 hours or less.

Ultrafiltration involves the separation of molecular species on the basis of size, shape, and/or charge. The solution to be separated is forced through a membrane by an external force. Membranes may be chosen for optimum flow rate, molecular specificity, and molecular weight cutoff. Two applications of membrane filtration are obvious: (1) desalting buffers or other solutions and (2) clarification of turbid solutions by removal of micron- or submicron-sized particles. Other applications are discussed below.

Ultrafiltration membranes have molecular weight cutoffs in the range of 100 to 1,000,000. They are usually composed of two layers: (1) a thin (0.1–0.5 μm), surface, semipermeable membrane made from a variety of materials including cellulose acetate, nylon, and polyvinylidene, and (2) a thicker, inert, support base (Figure 2.8). These filters function by retaining particles on the surfaces, not within the base matrix.

Membrane filters of these materials can be manufactured with a predetermined and accurately controlled pore size. Filters are available with a mean pore size ranging from 0.025 to 15 μm. These filters require suction, pressure, or centrifugal force for liquid flow. A typical flow rate for the commonly used 0.45-μm membrane is 57 mL min^{-1} cm^{-2} at 10 psi.

Ultrafiltration devices are available for macroseparations (up to 50 L) or for microseparations (milli- to microliters). For solutions larger than a few milliliters, gas-pressurized cells or suction-filter devices are used. For concentration and purification of samples in the milli- to microliter range, disposable filters are available. These devices, often called microconcentrators,

Figure 2.8

Electron micrograph of an ultrafiltration membrane showing the two layers. Particles greater than 0.1 μM in diameter are retained on the surface or within pores. *Courtesy of the Millipore Corporation.*

offer the user simplicity, time saving, and high recovery. The sample is placed in a reservoir above the membrane and centrifuged. The time and centrifugal force required depend on the membrane, with spin times varying from 30 minutes to 2 hours and forces from $1000 \times g$ to $7500 \times g$. Figure 2.9 outlines the use of a centrifuge microfilter.

The principles behind ultrafiltration are sometimes misunderstood. The nomenclature implies that separations are the result of physical trapping of the particles and molecules by the filter. With polycarbonate and fiberglass filters, separations are made primarily on the basis of physical size. Other filters (cellulose nitrate, polyvinylidene fluoride, and to a lesser extent cellulose acetate) trap particles that cannot pass through the pores, but also retain macromolecules by adsorption. In particular, these materials have protein and nucleic acid binding properties. Each type of membrane displays a different affinity for various molecules. For protein, the relative binding affinity is polyvinylidene fluoride > cellulose nitrate > cellulose acetate. We can expect to see many applications of the "affinity membranes" in the future as the various membrane surface chemistries are altered and made more specific. Some applications are described in the following pages.

Clarification of Solutions

Because of low solubility and denaturation, solutions of biomolecules or cellular extracts are often turbid. This is a particular disadvantage if spectrophotometric analysis is desired. The transmittance of turbid solutions can be greatly increased by passage through a membrane filter system.

This simple technique may also be applied to the sterilization of nonautoclavable materials such as protein and nucleic acid solutions or heat-labile reagents. Bacterial contamination can be removed from these solutions by passing them through filter systems that have been sterilized by autoclaving.

Figure 2.9

Use of a centrifuge micro-filter. *Courtesy of the Millipore Corporation.*

1. Pipet up to 2 ml of sample into the top reservoir.

2. Cover with the cone-shaped cap and spin in a centrifuge with a fixed-angle rotor.

3. Detach and invert top half.

4. Centrifuge 1 to 2 minutes to spin the concentrate into the cone-shaped cap.

5. Concentrate is now completely accessible for further analysis. Both concentrate and filtrate can be stored in collection cups.

Collection of Precipitates for Analysis

The collection of small amounts of very fine precipitates is the basis for many chemical and biochemical analytical procedures. Membrane filtration is an ideal method for sample collection. This is of great advantage in the collection of radioactive precipitates. Cellulose nitrate and fiberglass filters are often used to collect radioactive samples because they can be analyzed by direct suspension in an appropriate scintillation cocktail (see Chapter 6).

Harvesting of Bacterial Cells from Fermentation Broths

The collection of bacterial cells from nutrient broths is typically done by batch centrifugation. This time-consuming operation can be replaced by membrane filtration. Filtration is faster than centrifugation, and it allows extensive cell washing.

Concentration of Biomolecule Solutions

Protein or nucleic acid solutions obtained from extraction or various purification steps are often too dilute for further investigation. Since they cannot be concentrated by high-temperature evaporation of solvent, gentler methods have been developed. One of the most effective is the use of ultra-filtration pressure cells as shown in Figure 2.10. A membrane filter is placed in the bottom and the solution is poured into the cell. High pressure, exerted by compressed nitrogen (air could cause oxidation and denaturation of biomolecules), forces the flow of small molecules, including solvent, through the filter. Membranes are available in a number of sizes allowing a large variety of molecular weight cutoffs. Larger molecules that cannot pass through the pores are concentrated in the sample chamber. This method of concentration is rapid and gentle and can be performed at cold temperatures to ensure minimal inactivation of the molecules. One major disadvantage is clogging of the pores, which reduces the flow rate through the filter; this is lessened by constant but gentle stirring of the solution.

Lyophilization and Centrifugal Vacuum Concentration

Although ultrafiltration is being used more and more for the concentration of biological solutions, the older technique of lyophilization (freeze-drying) is still used. There are some situations (storing or transporting biological materials) in which lyophilization is preferred. Lyophilization is a drying technique that uses the process of sublimation to change a solvent (water) in the frozen state directly to the vapor state under vacuum. The product after lyophilization is a fluffy matrix that may be reconstituted by the addition of liquid. This is one of the most effective methods for drying or concentrating heat-sensitive materials. In practice, a biological solution to be concentrated is "shell-frozen" on the walls of a round-bottom or freeze-drying flask. Freezing of the solution is accomplished by placing the flask (half full with sample) in a dry ice–acetone bath and slowly rotating it as it is held at a 45° angle. The aqueous solution freezes in layers on the wall of the flask. This provides a large surface area for evaporation of water. The flask is then connected to the lyophilizer, which consists of a refrigeration unit and a

Figure 2.10

Schematic of an ultrafiltration cell. *Courtesy of Amicon Corporation.*

- -

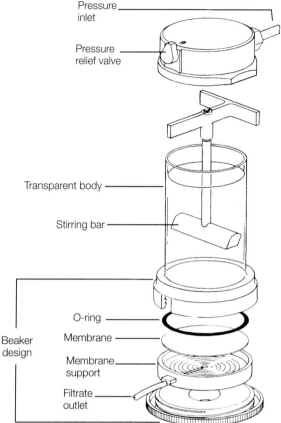

Pressure inlet

Pressure relief valve

Transparent body

Stirring bar

Beaker design

O-ring

Membrane

Membrane support

Filtrate outlet

vacuum pump (see Figure 2.11). The combined unit maintains the sample at −40°C for stability of the biological materials and applies a vacuum of approximately 5 to 25 mm on the sample. Ice formed from the aqueous solution sublimes and is pumped from the sample vial. In fact, all materials that are volatile under these conditions (−40°C, 5 to 25 mm) will be removed, and nonvolatile materials (proteins, buffer salts, nucleic acids, etc.) will be concentrated into a light, sometimes fluffy precipitate. Most freeze-dried biological materials are stable for long periods of time and some remain viable for many years.

As with any laboratory method, there are precautions and limitations of lyophilization that must be understood. Only aqueous solutions should be lyophilized. Organic solvents lower the melting point of aqueous solutions and increase the chances that the sample will melt and become denatured during freeze-drying. There is also the possibility that organic vapors will pass through the cold trap into the vacuum pump, where they may cause damage.

Figure 2.11

A typical freeze dryer.
Courtesy of Savant Instruments (savantec@savec.com).

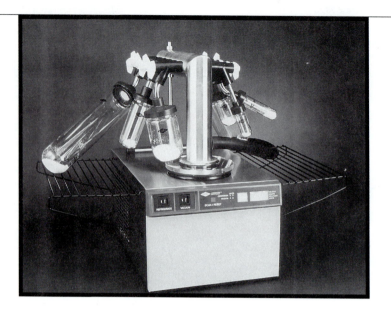

Figure 2.12

The SpeedVac centrifugal vacuum concentrator.
Courtesy of Savant Instruments (savantec@savec.com).

A new and increasingly popular technique for sample concentration and drying is centrifugal vacuum concentration. The method may be used to dry a wide variety of biological samples. The process starts with a sample dissolved in a solvent (water or organic). The solvent is evaporated under vacuum in a centrifuge, thus producing a pellet in the bottom of the container. The most widely used instrument is the SpeedVac available through Savant (Figure 2.12). This method is better than freeze-drying because it is faster, it does not require a prefreezing step, it provides 100% sample recovery, and it may be used with solvents other than water.

Study Problems

> *1.* You need to prepare a buffer for biochemistry lab. The required solution is 0.5 M sodium phosphate, pH 7.0. Use the Henderson-Hasselbalch equation to calculate the number of moles and grams of monobasic sodium phosphate (NaH_2PO_4) and dibasic sodium phosphate (Na_2HPO_4) necessary to make 1 liter of the solution.

> *2.* Design a "shortcut" method for preparing the phosphate buffer in Study Problem 1. Hint: You need only NaH_2PO_4, a solution of NaOH, and a pH meter.

> *3.* Describe how you would prepare a 0.1 M glycine buffer of pH 10.0. You have available isoelectric glycine and sodium glycinate.

> *4.* Describe how you would prepare a 0.20 M Tris-HCl buffer of pH 8.0. The only Tris reagent you have available is Tris base. What other reagent do you need and how would you use it to prepare the solution?

> *5.* Below is a table prepared by a biochemistry student to construct a standard curve for protein analysis. The Bradford assay was used with bovine serum albumin (BSA, 0.1 mg/mL) as standard protein. Complete the table by filling in the weight of BSA in each tube and the approximate A_{595} that will be obtained for each tube. Assume the procedure was conducted correctly.

	Tube No.					
Reagents	1	2	3	4	5	6
H_2O (mL)	1.0	0.9	0.8	0.6	0.2	—
BSA volume (mL)	—	0.1	0.2	0.4	0.8	1.0
BSA weight (μg)	0					
Bradford reagent (mL)	5.0	$\longrightarrow$				
A_{595}	0.00	0.08				

> *6.* Assume that you use the standard curve produced in Study Problem 5 to measure the concentration of an unknown protein. The A_{595} for 1.0 mL of the unknown was 0.52. Prepare a standard curve from the data in

Problem 5 and estimate the concentration of unknown protein in the sample in $\mu g/mL$.

▶ 7. A solution of purified DNA gave in the spectrophotometric assay an A_{260} of 0.35 when measured in a 1-cm quartz cuvette. What is the concentration of the DNA in $\mu g/mL$?

8. Compare the techniques of lyophilization and centrifugal vacuum concentration. Give advantages and disadvantages of each.

▶ 9. What amino acid residues are detected when the spectrophotometric assay is used to quantify proteins? Are those amino acids present in the same quantity in all proteins? Explain how this may affect measurement of proteins by this method.

▶ 10. What biomolecules would interfere with the measurement of nucleic acids using the spectrophotometric (A_{260}) assay?

▶ 11. Select a buffer system that can be used at each of the following pH values.
(a) pH 3.5
(b) pH 6.0
(c) pH 9.5

Further Reading

pH, Buffers, Electrodes, and Biosensors

A. Athani and U. Banakar, *BioPharm.* January 23, pp. 23–30 (1990). "The Development and Potential Use of Biosensors."

K. Barker, *At the Bench: A Laboratory Navigator* (1998), Cold Spring Harbor Laboratory Press (Cold Spring Harbor, NY).

A. Cass, Editor, *Biosensors: A Practical Approach* (1990), IRL Press (Oxford).

B. Eggins and G. McAteer, *Educ. Chem.*, January, p. 20 (1997). "Experimenting with Biosensors."

N. Good and S. Izawa, in *Methods in Enzymology,* Vol. XXIV, A. San Pietro, Editor (1972), Academic Press (New York), pp. 53–68. "Hydrogen Ion Buffers for Photosynthesis Research."

D. E. Gueffroy, Editor, *Buffers—A Guide for the Preparation and Use of Buffers in Biological Systems* (1993), Calbiochem-Novabiochem Corp., PO Box 12087, La Jolla, CA 92039-2087.

R. Hughes et al., *Science* **254,** 74–80 (1991). "Chemical Microsensors."

E. Marques et al., *Biochem. Educ.* **19,** 77–78 (1991). "A Low Cost Oxygen Meter."

C. Mathews, K. vanHolde, and K. Ahern, *Biochemistry* (2000), Benjamin/Cummings (San Francisco), pp. 33–50. Review of pH, buffers, and water chemistry.

J. Risley, *J. Chem. Educ.* **68,** 1054 (1991). "Preparing Solutions in the Biochemistry Lab."

J. Wang and C. Macca, *J. Chem. Educ.* **73,** 797 (1996). "Use of Blood-Glucose Strips for Introducing Enzyme Electrodes and Modern Biosensors."

Measurement of Proteins and Nucleic Acids

M. Bradford, *Anal. Biochem.* **72,** 248–254 (1976). "A Rapid and Sensitive Method for the Quantitation of Microgram Quantities of Protein Utilizing the Principle of Ligand-Dye Binding."

P. Hengen, *Trends Biochem. Sci.* **19,** 93–94 (1994). "Methods and Reagents: Determining DNA Concentrations and Rescuing PCR Primers."

C. Labarca and K. Paigen, *Anal. Biochem.* **102,** 344–352 (1980). "A Simple, Rapid, and Sensitive DNA Assay Procedure."

P. Smith et al., *Anal. Biochem.* **150,** 76–85 (1985). "Measurement of Protein Using Bicinchoninic Acid."

On the Web

http://ntri. tamuk. edu/fplc/fplc1.html
> Click on Buffers. Review Preparation of Buffers, Definitions of pH, Henderson-Hasselbalch Equation and Buffer Calculator.

http://www.turnerdesigns.com/applications/998_2600.htm
> Measurement of DNA solution using bisbenzimidazole dye (Hoechst 33258).

http://www.mamgen.tubitak.gov.tr/taylan/protocols/spot~~00.txt
> EtBr spot test for DNA and RNA analysis.

http://www.swmed.edu/home_pages/personal/davie/Protein.html
> "Bradford plate assay" for protein.

http://nmr.bmb.colostate.edu/bmb/instructions/farrell/bc352/spectro.htm
> Review principles of spectrophotometry and the Bradford protein assay.

http://www.ruf.rice.edu/~bioslabs/methods/protein/protein.html
> Review methods for protein assay including absorbance and colorimetric (Lowry, Biuret, Bradford, BCA).

3

PURIFICATION AND IDENTIFICATION OF BIOMOLECULES BY CHROMATOGRAPHY

The molecular details of a biochemical process cannot be fully elucidated until the reacting molecules have been isolated and characterized. Therefore, our understanding of biochemical principles has increased at about the same pace as the development of techniques for the separation and identification of biomolecules. Chromatography has been and will continue to be the most effective technique for isolating and purifying all types of biomolecules. In addition, it is widely used as an analytical tool to measure quantitative properties.

A.   INTRODUCTION TO CHROMATOGRAPHY

All types of chromatography are based on a very simple principle. The sample to be examined (called the **solute)** is allowed to interact with two physically distinct entities–a **mobile phase** and a **stationary phase** (see Figure 3.1). The mobile phase, which may be a gas or liquid, moves the sample through a region containing the solid or liquid stationary phase called the **sorbent.** The stationary phase will not be described in detail at this time, since it varies from one chromatographic method to another. However, it may be considered as having the ability to "bind" some types of solutes. The sample, which may contain one or many molecular components, comes into contact with the stationary phase. The components distribute themselves between the mobile and stationary phases. If some of the sample components are preferentially bound by the stationary phase, they spend more time in the stationary phase and, hence, are retarded in their movement through the chromatography system. Molecules that show weak affinity for the stationary phase spend more time with the mobile phase and

Figure 3.1

A representation of the principles of chromatography.

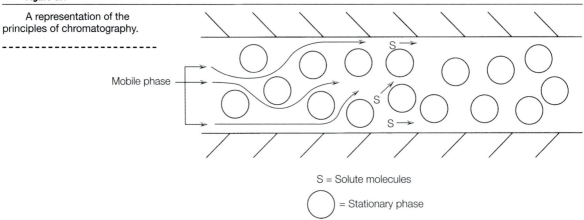

Mobile phase

S = Solute molecules

= Stationary phase

are more rapidly removed or **eluted** from the system. The many interactions that occur between solute molecules and the stationary phase bring about a separation of molecules because of different affinities for the stationary phase. The general process of moving a solute mixture through a chromatographic system is called **development.**

The mobile phase can be collected as a function of time at the end of the chromatographic system. The mobile phase, now called the **effluent,** contains the solute molecules. If the chromatographic process has been effective, fractions or "cuts" that are collected at different times will contain the different components of the original sample. In summary, molecules are separated because they differ in the extent to which they are distributed between the mobile phase and the stationary phase.

Throughout this chapter and others, biochemical techniques will be designated as **preparative** or **analytical,** or both. A preparative procedure is one that can be applied to the purification of a relatively large amount of a biological material. The purpose of such an experiment would be to obtain purified material for further characterization and study. Analytical procedures are used most often to determine the purity of a biological sample; however, they may be used to evaluate any physical, chemical, or biological characteristic of a biomolecule or biological system.

Partition versus Adsorption Chromatography

Chromatographic methods are divided into two types according to how solute molecules bind to or interact with the stationary phase. **Partition** chromatography is the distribution of a solute between two liquid phases. This may involve direct extraction using two liquids, or it may use a liquid immobilized on a solid support as in the case of paper, thin-layer, and gas-liquid chromatography. For partition chromatography, the stationary phase in Figure 3.1 consists of inert solid particles coated with liquid adsorbent. The distribution of solutes between the two phases is based primarily on solubility differences. The distribution may be quantified by using the **partition coefficient,** K_D (Equation 3.1).

>> $$K_D = \frac{\text{concentration of solute in stationary phase}}{\text{concentration of solute in mobile phase}}$$ *Equation 3.1*

Adsorption chromatography refers to the use of a stationary phase or support, such as an ion-exchange resin, that has a finite number of relatively specific binding sites for solute molecules. There is not a clear distinction between the processes of partition and adsorption. All chromatographic separations rely, to some extent, on adsorptive processes. However, in some methods (paper, thin-layer, and gas chromatography) these specific adsorptive effects are minimal and the separation is based primarily on nonspecific solubility factors. Adsorption chromatography relies on relatively specific interactions between the solute molecules and binding sites on the surface of the stationary phase. The attractive forces between solute and support may be ionic, hydrogen bonding, or hydrophobic interactions. Binding of solute is, of course, reversible.

Because of the different interactions involved in partition and adsorption processes, they may be applied to different separation problems. Partition processes are the most effective for the separation of small molecules, especially those in homologous series. Partition chromatography has been widely used for the separation and identification of amino acids, carbohydrates, and fatty acids. Adsorption techniques, represented by ion-exchange chromatography, are most effective when applied to the separation of macromolecules including proteins and nucleic acids.

In the rest of the chapter, various chromatographic methods will be discussed. You should recognize that no single chromatographic technique relies solely on adsorption or partition effects. Therefore, little emphasis will be placed on a classification of the techniques; instead, theoretical and practical aspects will be discussed.

B.   PLANAR CHROMATOGRAPHY (PAPER AND THIN-LAYER)

Because of the similarities in the theory and practice of these two procedures, they will be considered together. Both are examples of partition chromatography. In paper chromatography, the cellulose support is extensively hydrated, so distribution of the solutes occurs between the immobilized water (stationary phase) and the mobile developing solvent. The initial stationary liquid phase in thin-layer chromatography (TLC) is the solvent used to prepare the thin layer of adsorbent. However, as developing solvent molecules move through the stationary phase, polar solvent molecules may bind to the immobilized support and become the stationary phase.

Preparation of the Stationary Support

The support medium may be a sheet of cellulose or a glass or plastic plate covered with a thin coating of silica gel, alumina, or cellulose. Large sheets of cellulose chromatography paper are available in different porosities. These may be cut to the appropriate size and used without further treatment. The paper should never be handled with bare fingers. Although thin-layer plates can easily be prepared, it is much more convenient to purchase ready-made plates. These are available in a variety of sizes, materials, and thicknesses of stationary support. They are relatively inexpensive and have a more uniform support thickness than hand-made plates.

Figure 3.2 outlines the application procedure. The sample to be analyzed is usually dissolved in a volatile solvent. A very small drop of solution is spotted onto the plate with a disposable microcapillary pipet and allowed

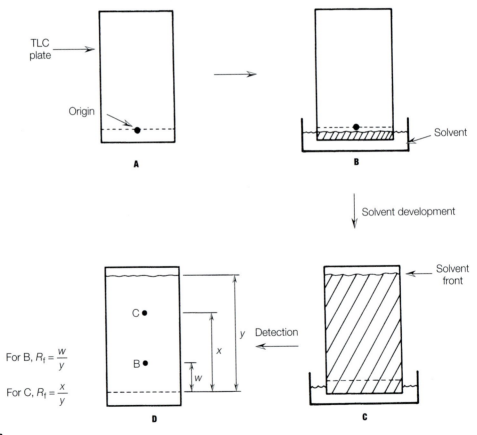

For B, $R_f = \dfrac{w}{y}$

For C, $R_f = \dfrac{x}{y}$

Figure 3.2

The procedure of paper and thin-layer chromatography.   **A** Application of the sample.   **B** Setting plate in solvent chamber.   **C** Movement of solvent by capillary action.   **D** Detection of separated components and calculation of R_f.

to dry; then the spotting process is repeated by superimposing more drops on the original spot. The exact amount of sample applied is critical. There must be enough sample so the developed spots can be detected, but overloading will lead to "tailing" and lack of resolution. Finding the proper sample size is a matter of trial and error. It is usually recommended that two or three spots of different concentrations be applied for each sample tested. Spots should be applied along a very faint line drawn with a pencil and ruler. TLC plates should not be heavily scratched or marked. Identifying marks may be made on the top of the chromatogram.

Solvent Development of the Support

A wide selection of solvent systems is available in the biochemical literature. If a new solvent system must be developed, a preliminary analysis must be done on the sample with a series of solvents. Solvents can be rapidly screened by developing several small chromatograms (2 × 6 cm) in small sealed bottles containing the solvents. For the actual analysis, the sample should be run on a larger plate with appropriate standards in a development chamber (Figure 3.3). The chamber must be airtight and saturated with solvent vapors. Filter paper on two sides of the chamber, as shown in Figure 3.3, enhances vaporization of the solvent.

Paper chromatograms may be developed in either of two types of arrangements—ascending or descending solvent flow. Descending solvent flow leads to faster development because of assistance by gravity, and it can offer better resolution for compounds with small R_f values because the solvent can be allowed to run off the paper. R_f values cannot be determined under these conditions, but it is useful for qualitative separations.

Figure 3.3

A typical chamber for paper and thin-layer chromatography.

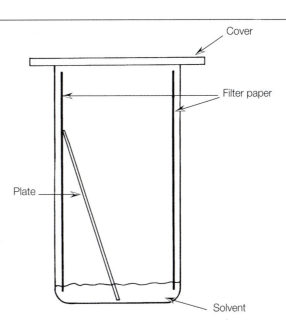

Two-dimensional chromatography is used for especially difficult separations. The chromatogram is developed in one direction by a solvent system, air dried, turned 90°, and developed in a second solvent system.

Detection and Measurement of Components

Unless the components in the sample are colored, their location on a chromatogram will not be obvious after solvent development. Several methods can be used to locate the spots, including fluorescence, radioactivity, and treatment with chemicals that develop colors. Substances that are highly conjugated may be detected by fluorescence under a UV lamp. Chromatograms may be treated with different types of reagents to develop a color. **Universal reagents** produce a colored spot with any organic compound. When a solvent-developed plate is sprayed with concentrated H_2SO_4 and heated at 100°C for a few minutes, all organic substances appear as black spots. A more convenient universal reagent is I_2. The solvent-developed chromatogram is placed in an enclosed chamber containing a few crystals of I_2. The I_2 vapor reacts with most organic substances on the plate to produce brown spots. The spots are more intense with unsaturated compounds.

Specific reagents react with a particular class of compound, For example, rhodamine B is often used for visualization of lipids, ninhydrin for amino acids, and aniline phthalate for carbohydrates.

The position of each component of a mixture is quantified by calculating the distance traveled by the component relative to the distance traveled by the solvent. This is called **relative mobility** and symbolized by R_f. In Figure 3.2D, the R_f values for components B and C are calculated. The R_f for a substance is a constant for a certain set of experimental conditions. However, it varies with solvent, type of stationary support (paper, alumina, silica gel), temperature, humidity, and other environmental factors. R_f values are always reported along with solvent and temperature.

Applications of Planar Chromatography

Thin-layer chromatography is now more widely used than paper chromatography. In addition to its greater resolving power, TLC is faster and plates are available with several sorbents (cellulose, alumina, silica gel).

Partition chromatography as described in this section may be applied to two major types of problems: (1) identification of unknown samples and (2) isolation of the components of a mixture. The first application is, by far, the more widely used. Paper chromatography and TLC require only a minute sample size, the analysis is fast and inexpensive, and detection is straightforward. Unknown samples are applied to a plate along with appropriate standards, and the chromatogram is developed as a single experiment. In this way any changes in experimental conditions (temperature, humidity, etc.) affect standards and unknowns to the same extent. It is then possible to compare the R_f values directly.

Purified substances can be isolated from developed chromatograms; however, only tiny amounts are present. In paper chromatography, the spot may be cut out with a scissors and the piece of paper extracted with an appropriate solvent. Isolation of a substance from a TLC plate is accomplished by scraping the solid support from the region of the spot with a knife edge or razor blade and extracting the sorbent with a solvent. "Preparative" thin-layer plates with a thick coating of sorbent (up to 2 mm) are especially useful because they have higher sample capacity.

C.   GAS CHROMATOGRAPHY (GC)

When the mobile phase of a chromatographic system is gaseous and the stationary phase is a liquid coated on inert solid particles, the technique is **gas-liquid chromatography,** or simply gas chromatography. Separation by GC methods is based primarily on partitioning processes. The stationary phase, inert particles coated with a thin layer of liquid, is confined to a long stainless steel or glass tube, called the **column,** which is maintained at a suitable (usually elevated) temperature. A gaseous mobile phase under high pressure is continuously swept through the column. The sample to be analyzed is vaporized, introduced into the warm gaseous phase, and swept through the stationary phase. The vaporous chemical constituents in the sample then distribute themselves between the mobile phase and the stationary liquid film on the solid support. Components of the sample mixture that have affinity for the stationary phase are retarded in their movement through the column. Ideally, each component will have a different partition coefficient, K_D (see Equation 3.1), and each will pass through the column at a different rate.

Instrumentation

The essential components of a gas chromatography system are shown in Figure 3.4. The mobile phase (called the carrier gas) is inert, usually helium, nitrogen, or argon. The gas is directed past an injection port, the entry point of the sample. The sample, dissolved in a solvent, is injected with a syringe through a rubber septum into the injection port. The column, injection port, and detector are in individual ovens maintained at elevated temperatures so that the sample components remain vaporized throughout their residence time in the system.

Two types of columns are used. A **packed column** is one filled with inert, solid particles coated with a liquid stationary phase. Standard tubing is about 0.5 cm in diameter, with lengths ranging from 1 m to 20 m; however, columns for large-scale preparative work may be up to 5 cm in diameter and several meters long. Commonly used solid supports are diatomaceous earth, Teflon powder, and glass beads. The stationary liquid must be chosen on the basis of the compounds to be analyzed. A more recently developed and more widely used type of column is the **open-tubular** or **capillary column.** This is prepared by coating the inner wall of the column with the stationary liquid phase. The inside diameter of a typical capillary tube is 0.25 mm, and

Figure 3.4

The essential components of a gas chromatography system.

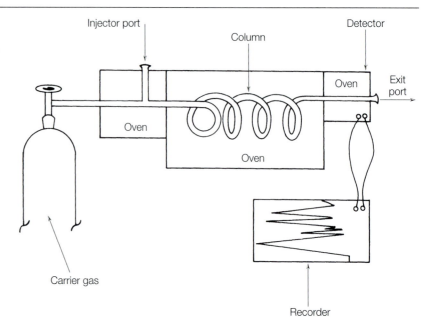

the length ranges from 10 to 100 meters. The length of a column depends on the degree of resolution required. Clearly, the longer the column, the better the separation; however, a long column increases the pressure differential from the beginning to the end of the column. A large pressure change hampers analysis by increasing retention time and causing peak broadening. Most columns used today are made of fused silica glass.

After passage through the column, the carrier gas and separated components of the mixture are directed through a detector. Several types of detectors are available, but the two most commonly used for bioanalytical purposes are the **thermal conductivity cell (TC cell)** and the **flame ionization detector (FID).** The TC cell functions by measuring the temperature-dependent electrical resistance of a hot wire. The detector unit consists of a platinum or platinum-alloy wire through which an electric current is passed. The hot wire is cooled as the carrier gas passes over it. The extent of cooling depends on the gas flow rate and the thermal conductivity of the gas. Since the rate of carrier gas flow remains constant during a typical GC analysis, only changes in the thermal conductivity of the vapor will change the temperature and, therefore, the resistance of the wire. Organic vapors usually have lower thermal conductivities than the carrier gas. When carrier gas containing an organic vapor exits the column and passes over the wire, the electrical resistance of the wire decreases. The changing current in the wire is amplified and monitored by a recorder. The extent of resistance change depends on the amount of organic vapor in the carrier gas. Therefore, the size of the recorder signal is a measure of the amount of that chemical constituent in the sample. The thermal conductivity cell has a poor level of sensitivity (about 5 μg), but it is widely used because it responds to all organic compounds. Since it does not destroy the sample, a TC cell can be used when samples are to be collected from the column.

In contrast to the TC detector, the FID is sensitive to about $10^{-5}\,\mu$g. A flame supported by hydrogen gas and air is used to burn organic vapors as they leave the column. Electrons and ionic fragments are produced upon combustion of the organics. A wire loop (called the ion collector) collects the charged particles and produces an electrical current, which is fed into a printer. The FID is sensitive only to oxidizable compounds and does not respond to water vapor or CO_2. Since the detector response is proportional to the amount of organic material in the carrier gas, quantitative analysis can be done. Samples cannot be collected after passage through the FID; however, it is possible to split the carrier gas flow before it enters the detector. Only a small fraction of the carrier gas with organic vapor is directed through the detector, and the rest is collected for further analysis.

The electrical signal from a detector is amplified and fed into a recorder or computer for analysis. A typical recorder trace is shown in Figure 3.5. Each peak represents a component in the original mixture. A peak is identified by a **retention time,** the time lapse between injection of the sample and the maximum signal from the recorder. This number is a constant for a particular compound under specified conditions of the carrier gas flow rate; temperature of the injector, column, and detector; and type of column. Retention time in GC analysis is analogous to the R_f value in thin-layer or paper chromatography.

Figure 3.5

A recorder trace from gas chromatography.

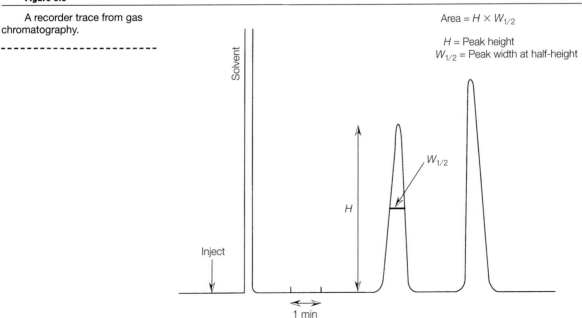

Area = $H \times W_{1/2}$

H = Peak height
$W_{1/2}$ = Peak width at half-height

Selection of Operating Conditions

Each type of gas chromatograph has its own set of operating instructions, but general experimental conditions are appropriate for all instruments. Three important factors must always be considered when a GC analysis is to be completed: (1) selection of the proper column; (2) choice of temperatures for injector, oven, and detector; and (3) adjustment of gas flow. Because hundreds of stationary phases are available, it is impossible to outline the characteristics of each. Selecting the stationary phase requires some knowledge of the nature of the sample to be analyzed.

The operating temperature of the column oven is critical, as it greatly affects the resolving ability of the column. If the temperature is too high, low-boiling components are swept rapidly through the column without equilibration with the stationary phase. A temperature that is too low causes condensation of some organic compounds in the liquid stationary phase, resulting in very extended retention times. A column temperature set near the boiling point of the primary component in the sample is often the best. If a sample containing a mixture of compounds with a wide range of boiling points is to be analyzed, a temperature gradient may be required. Research-grade chromatographs are equipped with temperature programmers that gradually increase the column temperature at a preselected rate.

The detector and injector temperatures should be maintained about 10° above the column temperature. This ensures against condensation of vapor and causes rapid vaporization of the sample upon injection.

The separation of a mixture by GC also depends on the flow rate of the carrier gas. A low rate allows extensive equilibration of the sample between the stationary phase and the gaseous phase and leads to better separation; however, the recorder peaks become very broad. A high flow rate greatly decreases column efficiency by decreasing equilibration time. A suitable flow rate for a 0.5–cm packed column is between 50 and 120 mL/min.

Analysis of GC Data

Gas chromatography can be applied to both qualitative and quantitative analyses. There are two general methods for qualitative identification of eluted compounds. A compound may be identified by **retention time** or by **peak enhancement.** If you have some idea of the identity of an unknown peak or if you know it is one of three or four compounds, then you can inject the known compounds into the GC under conditions identical to those for the unknown sample. Compounds with the same retention times (± 2 or 3%) can generally be considered identical. Alternatively, you can add pure sample of a suspected component to the unknown sample and inject the mixture into the GC. If a recorder peak is increased in size, this also provides evidence for the identity of an unknown. The latter technique is called **peak enhancement** or **spiking.**

Identification of unknown compounds by retention times or peak enhancement is not conclusive or absolute proof of identity. It is possible for two different substances to have identical retention times under the same experimental conditions. For positive identification, the sample must be collected at the exit port and characterized by mass spectrometry, infrared, nuclear magnetic resonance, or chemical analysis.

The response from most recorders and detectors is proportional to the amount of compound in the original sample. In other words, the area of each recorder peak is a relative measure of the concentration of that substance in the sample. GC then becomes an important tool for quantitative analysis. Methods for measuring peak area range from primitive to sophisticated. The most ideal method is to use an electronic integrator, which automatically measures peak area and can be programmed to yield % composition of each component. Alternatively, one can use the method of triangulation. Here the peaks are assumed to be triangles with an area calculated according to Equation 3.2, where H is height and $W_{1/2}$ is the width at one-half height.

>> $$Area = HW_{1/2}$$ **Equation 3.2**

The % composition of the sample can be estimated using Equation 3.3.

>> $$\%x = \frac{Area_x}{Area_a + Area_b + Area_x} \times 100$$ **Equation 3.3**

The percent of x in the sample is equal to the area represented by component x divided by the sum of all peak areas. $Area_a$ and $Area_b$ refer to other components in the mixture.

Advantages and Limitations of GC

Gas chromatography has many advantages as an analytical technique. It provides excellent separation of most small biomolecules; it is simple, versatile, and rapid; it is highly sensitive; and, by most standards, it is relatively inexpensive.

The major limitation of GC is the requirement for heat stability and volatility of the sample. Obviously, compounds that decompose at elevated temperatures (below 250°C) cannot normally be subjected to GC analysis. Many compounds of biochemical interest are not volatile in the useful temperature range of GC (up to about 200–250°C). Such compounds can often be converted to volatile derivatives. Hydroxyl groups in alcohols, carbohydrates, and sterols are converted to derivatives by trimethylsilylation or acetylation. Amino groups can also be converted to volatile derivatives by acetylation and silylation. Fatty acids are transformed to methyl esters for GC analysis, as described in Experiment 6.

D. COLUMN CHROMATOGRAPHY

Adsorption chromatography in biochemical applications usually consists of a solid stationary phase and a liquid mobile phase. The most useful technique is column chromatography, in which the stationary phase is confined to a glass or plastic tube and the mobile phase (a solvent or buffer) is allowed to flow through the solid adsorbent. A small amount of the sample to be analyzed is layered on top of the column. The sample mixture enters the column of adsorbing material and is distributed between the mobile phase and the stationary phase. The various components in the sample have different affinities for the two phases and move through the column at different rates. Collection of the liquid phase emerging from the column yields separate fractions containing the individual components in the sample.

Specific terminology is used to describe various aspects of column chromatography. When the actual adsorbing material is made into a column, it is said to be **poured** or **packed.** Application of the sample to the top of the column is **loading** the column. Movement of solvent through the loaded column is called **developing** or **eluting** the column. The **bed volume** is the total volume of solvent and adsorbing material taken up by the column. The volume taken up by the liquid phase in the column is the **void volume.** The **elution volume** is the amount of solvent required to remove a particular solute from the column. This is analogous to R_f values in thin-layer or paper chromatography or to retention time in GC.

In adsorption chromatography, solute molecules take part in specific interactions with the stationary phase. Herein lies the great versatility of adsorption chromatography. Many varieties of adsorbing materials are available, so a specific sorbent can be chosen that will effectively separate a mixture. There is still an element of trial and error in the selection of an effective stationary phase. However, experiences of many investigators are recorded in the literature and are of great help in choosing the proper system. Table 3.1 lists the most common stationary phases employed in adsorption column chromatography.

Adsorbing materials come in various forms and sizes. The most suitable forms are dry powders or a slurry form of the material in an aqueous buffer or organic solvent. Alumina, silica gel, and fluorisil do not normally need special pretreatment. The size of particles in an adsorbing material is de-

---

Table 3.1

Adsorbents Useful in Biochemical Applications

Adsorbing Material	Uses
Alumina	Small organics, lipids
Silica gel	Amino acids, lipids, carbohydrates
Fluorisil (magnesium silicate)	Neutral lipids
Calcium phosphate (hydroxyapatite)	Proteins, polynucleotides, nucleic acids

Table 3.2

Mesh Sizes of Adsorbents and Typical Applications

Mesh Size	Applications
20–50	Crude preparative work, very high flow rate
50–100	Preparative applications, high flow rate
100–200	Analytical separations, medium flow rate
200–400	High-resolution analytical separations, low flow rate

fined by **mesh size.** This refers to a standard sieve through which the particles can pass. A 100 mesh sieve has 100 small openings per square inch. Adsorbing material with high mesh size (400 and greater) is extremely fine and is most useful for very high resolution chromatography. Table 3.2 lists standard mesh sizes and the most appropriate application. For most biochemical applications, 100 to 200 mesh size is suitable.

Operation of a Chromatographic Column

A typical column setup is shown in Figure 3.6. The heart of the system is, of course, the column of adsorbent. In general, the longer the column, the better the resolution of components. However, a compromise must be made because flow rate decreases with increasing column length. The actual size of a column depends on the nature of the adsorbing material and the amount of chemical sample to be separated. For preparative purposes, column heights of 20 to 50 cm are usually sufficient to achieve acceptable resolution. Column inside diameters may vary from 0.5 to 5 cm.

Packing the Column

Once the adsorbing material and column size have been selected, the column is poured. If the tube does not have a fritted disc in the bottom, a small piece of glass wool or cotton should be used to support the column. Most columns are packed by pouring a slurry of the sorbent into the tube and allowing it to settle by gravity into a tight bed. The slurry is prepared with the solvent or buffer that will be used as the initial developing solvent. Pouring of the slurry must be continuous to avoid formation of sorbent layers. Excess solvent is eluted from the bottom of the column while the sorbent is settling. The column must never run dry. Additional slurry is added until the column bed reaches the desired height. The top of the settled adsorbent is then covered with a small circle of filter paper or glass wool to protect the surface while the column is loaded with sample or the eluting solvent is changed.

Sometimes it is necessary to pack a column under pressure (5 to 10 psi). This leads to a tightly packed bed that yields more reproducible results, especially with gradient elution (see below).

CHAPTER 3

Figure 3.6

Setup for the operation of a chromatography column.

- -

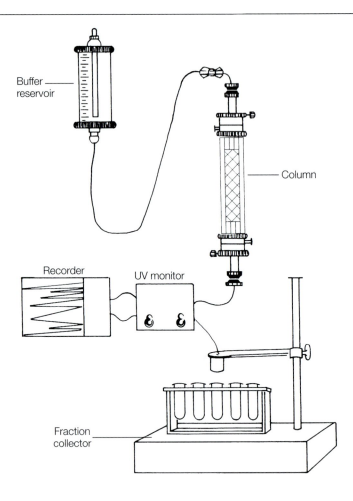

Buffer reservoir

Column

Recorder

UV monitor

Fraction collector

Loading the Column

The sample to be analyzed by chromatography should be applied to the top of the column in a concentrated form. If the sample is solid, it is dissolved in a minimum amount of solvent; if already in solution, it may be concentrated by ultrafiltration as described in Chapter 2. After the sample is loaded onto the column with a graduated or disposable pipet, it is allowed to percolate into the adsorbent. A few milliliters of solvent are then carefully added to wash the sample into the column material. The column is then filled with eluting solvent.

Eluting the Column

The chromatography column is developed by continuous flow of a solvent. Maintaining the appropriate flow rate is important for effective separation.

If the flow rate is set too high, there is not sufficient time for complete equilibration of the sample components with the two phases. Too low a flow rate allows diffusion of solutes, which leads to poor resolution and broad elution peaks. It is difficult to give guidelines for the proper flow rate of a column, but, in general, a column should be adjusted to a rate slightly less than "free flow." Sometimes it is necessary to find the proper flow rate by trial and error. One problem encountered during column development is a changing flow rate. As the solvent height above the column bed is reduced, there is less of a "pressure head" on the column, so the flow rate decreases. This can be avoided by storing the developing solvent in a large reservoir and allowing it to enter the column at the same rate as it is emerging from the column (see Figure 3.6).

Adsorption columns are eluted in one of three ways. All components may be eluted by a single solvent or buffer. This is referred to as **continual elution.** In contrast, **stepwise elution** refers to an incremental change of solvent to aid development. The column is first eluted with a volume of one solvent and then with a second solvent. This may continue with as many solvents or solvent mixtures as desired. In general, the first solvent should be the least polar of any used in the analysis, and each additional solvent should be of greater polarity or ionic strength. Finally, adsorption columns may be developed by **gradient elution** brought about by a gradual change in solvent composition. The composition of the eluting solvent can be changed by the continuous mixing of two different solvents to gradually change the ratio of the two solvents. Alternatively, the concentration of a component in the solvent can be gradually increased. This is most often done by addition of a salt (KCl, NaCl, etc.). Devices are commercially available to prepare predetermined, reproducible gradients.

Collecting the Eluent

The separated components emerging from the column in the eluent are usually collected as discrete fractions. This may be done manually by collecting specified volumes of eluent in Erlenmeyer flasks or test tubes. Alternatively, if many fractions are to be collected, a mechanical fraction collector is convenient and even essential. An automatic fraction collector (see Figure 3.6) directs the eluent into a single tube until a predetermined volume has been collected or until a preselected time period has elapsed; then the collector advances another tube for collection. Specified volumes are collected by a drop counter activated by a photocell, whereas a timer can be set to collect a fraction over a specific period.

Detection of Eluting Components

The completion of a chromatographic experiment calls for a means to detect the presence of solutes in the collected fractions. The detection method used will depend on the nature of the solutes. Smaller molecules such as lipids, amino acids, and carbohydrates can be detected by spotting fractions on a thin-layer plate or a piece of filter paper and treating them with a chemical reagent that produces a color. The same reagents that are used to visualize spots on a thin-layer or paper chromatogram are useful for this. Proteins and nucleic acids are conveniently detected by spectroscopic absorption measurements at 280 and 260 nm, respectively. Enzymes can be detected by measurements of catalytic activity associated with each fraction. Research-grade chromatographic systems are equipped with detectors that continuously monitor some physical property of the eluent and display the separation results on a computer screen (see Figure 3.6). The newest advance in detectors is the diode array (see Chapter 5). Most often the eluent is directed through a flow cell where absorbance or fluorescence characteristics can be measured. The detector is connected to a recorder or computer for a permanent record of spectroscopic changes. When the location of the various solutes is determined, fractions containing identical components are pooled and stored for later use.

E. ION-EXCHANGE CHROMATOGRAPHY

Ion-exchange chromatography is a form of adsorbtion chromatography in which ionic solutes display reversible electrostatic interactions with a charged stationary phase. The chromatographic setup is identical to that described in the last section and Figure 3.6. The column is packed with a stationary phase consisting of a synthetic resin that is tagged with ionic functional groups. The steps involved in ion-exchange chromatography are outlined in Figure 3.7. In stage 1, the insoluble resin material (positively charged) in the column is surrounded by buffer counterions. Loading of the column in stage 2 brings solute molecules of different charge into the ion-exchange medium. Solutes entering the column may be negatively charged, positively charged, or neutral under the experimental conditions. Solute molecules that have a charge opposite to that of the resin bind tightly but reversibly to the stationary

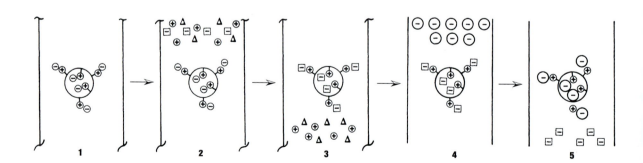

Figure 3.7

Illustration of the principles of ion-exchange chromatography. See text for explanation.

phase (stage 3). The strength of binding depends on the size of the charge and the charge density (amount of charge per unit volume of molecule) of the solute. The greater the charge or the charge density, the stronger the interaction. Neutral solute molecules (Δ) or those with a charge identical to that of the resin show little or no affinity for the stationary phase and move with the eluting buffer. The bound solute molecules can be released by eluting the column with a buffer of increased ionic strength or pH (stage 4). An increase in buffer ionic strength releases bound solute molecules by displacement. Increasing the buffer pH decreases the strength of the interaction by reducing the charge on the solute or on the resin (stage 5).

The following sections will focus on the properties of ion-exchange resins, selection of experimental conditions, and applications of ion-exchange chromatography.

Ion-Exchange Resins

Ion exchangers are made up of two parts—an insoluble, three-dimensional matrix and chemically bonded charged groups within and on the surface of the matrix. The resins are prepared from a variety of materials, including polystyrene, acrylic resins, polysaccharides (dextrans), and celluloses. An ion exchanger is classified as **cationic** or **anionic** depending on whether it exchanges cations or anions. A resin that has negatively charged functional groups exchanges positive ions and is a **cation exchanger.** Each type of exchanger is also classified as **strong** or **weak** according to the ionizing strength of the functional group. An exchanger with a quaternary amino group is, therefore, a **strongly basic anion exchanger,** whereas primary or secondary aromatic or aliphatic amino groups would lead to a **weakly basic anion exchanger.** A **strongly acidic cation exchanger** contains the sulfonic acid group. Table 3.3 lists the more common ion exchangers according to each of these classifications.

The ability of an ion exchanger to adsorb counterions is defined quantitatively by *capacity.* The *total capacity* of an ion exchanger is the quantity of charged and potentially charged groups per unit weight of dry exchanger. It is usually expressed as milliequivalents of ionizable groups per milligram of dry weight, and it can be experimentally determined by titration. The capacity of an ion exchanger is a function of the porosity of the resin. The resin matrix contains covalent cross-linking that creates a "molecular sieve." Ionized functional groups within the matrix are not readily accessible to large molecules that cannot fit into the pores. Only surface charges would be available to these molecules for exchange. The purely synthetic resins (polystyrene and acrylic) have cross-linking ranging from 2 to 16%, with 8% being the best for general purposes.

With so many different experimental options and resin properties to consider, it is difficult to select the proper conditions for a particular separation. The next section will outline the choices and offer guidelines for proper experimental design.

Table 3.3

Ion-Exchange Resins

Name	Functional Group	Matrix	Class
Anion Exchangers			
AG I	Tetramethylammonium	Polystyrene	Strong
AG 3	Tertiary amine	Polystyrene	Weak
DEAE-Sephacel	Diethylaminoethyl	Sephacel	Weak
PEI-cellulose	Polyethyleneimine	Cellulose	Weak
DEAE-Sephadex	Diethylaminoethyl	Dextran	Weak
QAE-Sephadex	Diethyl-(2–hydroxyl-propyl)-aminoethyl	Dextran	Strong
Cation Exchangers			
AG 50	Sulfonic acid	Polystyrene	Strong
Bio-Rex 70	Carboxylic acid	Acrylic	Weak
CM-Sephacel	Carboxymethyl	Sephacel	Weak
P-Cellulose	Phosphate	Cellulose	Intermediate
CM-Sephadex	Carboxymethyl	Dextran	Weak
SP-Sephadex	Sulfopropyl	Dextran	Strong

Selection of the Ion Exchanger

Before a proper choice of ion exchanger can be made, the nature of the solute molecules to be separated must be considered. For relatively small, stable molecules (amino acids, lipids, nucleotides, carbohydrates, pigments, etc.) the synthetic resins based on polystyrene are most effective. They have relatively high capacity for small molecules because the extensive cross-linking still allows access to the interior of the resin beads. For separations of peptides, proteins, nucleic acids, polysaccharides, and other large biomolecules, one must consider the use of fibrous cellulosic ion exchangers and low-percent cross-linked dextran or acrylic exchangers. The immobilized functional groups in these resins are readily available for exchange even to larger molecules.

The choice of ion exchanger has now been narrowed considerably. The next decision is whether to use a cationic or anionic exchanger. If the solute molecule has only one type of charged group, the choice is simple. A solute that has a positive charge will bind to a cationic exchanger and vice versa. However, many biomolecules have more than one type of ionizing group and may have both negatively and positively charged groups (they are amphoteric). The net charge on such molecules depends on pH. At the isoelectric point, the substance has no net charge and would not bind to any type of ion exchanger.

In principle, amphoteric molecules should bind to both anionic and cationic exchangers. However, when one is dealing with large biomolecules, the pH "range of stability" must also be evaluated. The range of stability refers to the pH range in which the biomolecule is not denatured. Figure 3.8

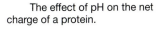

Figure 3.8

The effect of pH on the net charge of a protein.

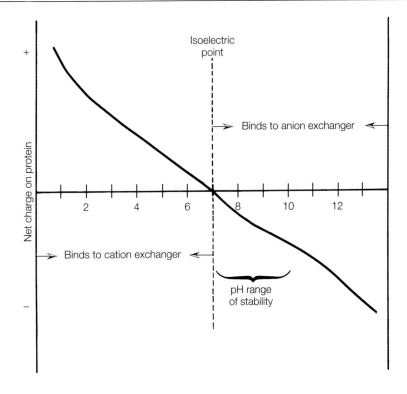

shows how the net charge of a hypothetical protein changes as a function of pH. Below the isoelectric point, the molecule has a net positive charge and would be bound to a cation exchanger. Above the isoelectric point, the net charge is negative, and the protein would bind to an anion exchanger. Superimposed on this graph is the pH range of stability for the hypothetical protein. Because it is stable in the range of pH 7.0–9.0, the ion exchanger of choice is an anionic exchanger. In most cases, the isoelectric point of the protein is not known. The type of ion exchanger must be chosen by trial and error as follows. Small samples of the solute mixture in buffer are equilibrated for 10 to 15 minutes in separate test tubes, one with each type of ion exchanger. The tubes are then centrifuged or let stand to sediment the ion exchanger. Check each supernatant for the presence of solute (A_{260} for nucleic acids, A_{280} for proteins, catalytic activity for enzymes, etc.). If a supernatant has a relatively low level of added solute, that ion exchanger would be suitable for use. This simple test can also be extended to find conditions for elution of the desired macromolecule from the ion exchanger. The ion exchanger charged with the macromolecule is treated with buffers of increasing ionic strength or changing pH. The supernatant after each treatment is analyzed as before for release of the macromolecule.

Choice of Buffer

This decision includes not just the buffer substance but also the pH and the ionic strength. Buffer ions will, of course, interact with ion-exchange resins. Buffer ions with a charge opposite to that on the ion exchanger compete with solute for binding sites and greatly reduce the capacity of the column. Cationic buffers should be used with anionic exchangers; anionic buffers should be used with cationic exchangers.

The pH chosen for the buffer depends first of all on the range of stability of the macromolecule to be separated (see Figure 3.8). Second, the buffer pH should be chosen so that the desired macromolecule will bind to the ion exchanger. In addition, the ionic strength should be relatively low to avoid "damping" of the interaction between solute and ion exchanger. Buffer concentrations in the range 0.05 to 0.1 M are recommended.

Preparation of the Ion Exchanger

The commercial suppliers of ion exchangers provide detailed instructions for the preparation of the adsorbents. Failure to pretreat ion exchangers will greatly reduce the capacity and resolution of a column. Most new ion-exchange resins are commercially available in slurry form and are ready to use with a minimum number of pretreatment steps.

Sometimes pretreatment steps do not remove the small particles that are present in most ion-exchange materials. If left in suspension, these particles, called fines, result in decreased resolution and low column flow rates. The fines are removed from an exchanger by suspending the swollen adsorbent in a large volume of water in a graduated cylinder and allowing at least 90% of the exchanger to settle. The cloudy supernatant containing the fines is decanted. This process is repeated until the supernatant is completely clear. The number of washings necessary to remove most fines is variable, but for a typical ion exchanger 8 to 10 times is probably sufficient.

Using the Ion-Exchange Resin

Ion exchangers are most commonly used in a column form. The column method discussed earlier in this chapter can be directly applied to ion-exchange chromatography.

An alternative method of ion exchange is **batch separation.** This involves mixing and stirring equilibrated exchanger directly with the solute mixture to be separated. After an equilibration time of approximately 1 hour, the slurry is filtered and washed with buffer. The ion exchanger can be chosen so that the desired solute is adsorbed onto the exchanger or remains unbound in solution. If the latter is the case, the desired material is in the filtrate. If the desired solute is bound to the exchanger, it can be removed by suspending the exchanger in a buffer of greater ionic strength or different pH. Batch processes have some advantages over column methods. They are rapid, and the problems of packing, channeling, and dry columns are avoided.

Another development in ion-exchange column chromatography allows the separation of proteins according to their isoelectric points. This technique, **chromatofocusing,** involves the formation of a pH gradient on an ion-exchange column. If a buffer of a specified pH is passed through an ion-exchange column that was equilibrated at a second pH, a pH gradient is formed on the column. Proteins bound to the ion exchanger are eluted in the order of their isoelectric points. In addition, protein band concentration (focusing) takes place during elution. Chromatofocusing is similar to isoelectric focusing, introduced in Chapter 4, in which a column pH gradient is produced by an electric current.

Storage of Resins

Most ion exchangers in the dry form are stable for many years. Aqueous slurried ion exchangers are still useful after several months. One major storage problem with a wet exchanger is microbial growth. This is especially true for the cellulose and dextran exchangers. If it is necessary to store pretreated exchangers, an antimicrobial agent must be added to the slurry. Sodium azide (0.02%) is suitable for cation exchangers and phenylmercuric salts (0.001%) are effective for anion exchangers. Since these preservative reagents are toxic, they must be used with caution.

F.   GEL EXCLUSION CHROMATOGRAPHY

The chromatographic methods discussed up to this point allow the separation of molecules according to polarity, volatility, and charge. The method of **gel exclusion chromatography** (also called gel filtration, molecular sieve chromatography, or gel permeation chromatography) exploits the physical property of molecular size to achieve separation. The molecules of nature range in molecular weight from less than 100 to as large as several million. It should be obvious that a technique capable of separating molecules of molecular weight 10,000 from those of 100,000 would be very popular among research biochemists. Gel filtration chromatography has been of major importance in the purification of thousands of proteins, nucleic acids, enzymes, polysaccharides, and other biomolecules. In addition, the technique may be applied to molecular weight determination and quantitative analysis of molecular interactions. In this section the theory and practice of gel filtration will be introduced and applied to several biochemical problems.

Theory of Gel Filtration

The operation of a gel filtration column is illustrated in Figure 3.9. The stationary phase consists of inert particles that contain small pores of a controlled size. Microscopic examination of a particle reveals an interior resembling a sponge. A solution containing solutes of various molecular sizes is allowed to pass through the column under the influence of continuous solvent flow. Solute molecules larger than the pores cannot enter the interior

Figure 3.9

Separation of molecules by gel filtration. **A** Application of sample containing large and small molecules. **B** Large molecules cannot enter gel matrix, so they move more rapidly through the column. **C** Elution of the large molecules.

A   B   C

of the gel beads, so they are limited to the space between the beads. The volume of the column accessible to very large molecules is, therefore, greatly reduced. As a result, they are not slowed in their progress through the column and elute rapidly in a single zone. Small molecules capable of diffusing in and out of the beads have a much larger volume available to them. Therefore, they are delayed in their journey through the column bed. Molecules of intermediate size migrate through the column at a rate somewhere between those for large and small molecules. Therefore, the order of elution of the various solute molecules is directly related to their molecular dimensions.

Physical Characterization of Gel Chromatography

Several physical properties must be introduced to define the performance of a gel and solute behavior. Some important properties are:

1. Exclusion Limit   This is defined as the molecular mass of the smallest molecule that cannot diffuse into the inner volume of the gel matrix. All molecules above this limit elute rapidly in a single zone. The exclusion limit of a typical gel, Sephadex G-50, is 30,000 daltons. All solute molecules having a molecular size greater than this value would pass directly through the column bed without entering the gel pores.

2. Fractionation Range   Sephadex G-50 has a fractionation range of 1500 to 30,000 daltons. Solute molecules within this range would be separated in a somewhat linear fashion.

3. Water Regain and Bed Volume   Gel chromatography media are often supplied in dehydrated form and must be swollen in a solvent, usually water, before use. The weight of water taken up by 1 g of dry gel is known as the water regain. For G-50, this value is 5.0 ± 0.3 g. This value does not include the water surrounding the gel particles, so it cannot be used as an estimate of the final volume of a packed gel column. Most commercial suppliers of gel materials provide, in addition to water regain, a bed volume value. This is the final volume taken up by 1 g of dry gel when swollen in water. For G-50, bed volume is 9 to 11 mL/g dry gel.

4. Gel Particle Shape and Size   Ideally, gel particles should be spherical to provide a uniform bed with a high density of pores. Particle size is defined either by mesh size or bead diameter (μm). The degree of resolution afforded by a column and the flow rate both depend on particle size. Larger particle sizes (50 to 100 mesh, 100 to 300 μm) offer high flow rates but poor chromatographic separation. The opposite is true for very small particle sizes ("superfine," 400 mesh, 10 to 40 μm). The most useful particle size, which represents a compromise between resolution and flow rate, is 100 to 200 mesh (50 to 150 μm).

5. Void Volume   This is the total space surrounding the gel particles in a packed column. This value is determined by measuring the volume of solvent required to elute a solute that is completely excluded from the gel matrix. Most columns can be calibrated for void volume with a dye, blue dextran, which has an average molecular mass of 2,000,000 daltons.

6. Elution Volume   This is the volume of eluting buffer necessary to remove a particular solute from a packed column.

Chemical Properties of Gels

Four basic types of gels are available: **dextran, polyacrylamide, agarose, and combined polyacrylamide-dextran.** The first gels to be developed were those based on a natural polysaccharide, dextran. These are supplied by Pharmacia under the trade name Sephadex. Table 3.4 gives the physical properties of the various sizes of Sephadex. The number given each gel refers to the water regain multiplied by 10. Sephadex is available in various particle sizes labeled coarse, medium, fine, and superfine. Dextran-based gels cannot be manufactured with an exclusion limit greater than 600,000 daltons because the small extent of cross-linking is not sufficient to prevent collapse of the particles. If the dextran is cross-linked with N,N'- methylenebisacrylamide, gels for use in higher fractionation ranges are possible. Table 3.4 lists these gels, which are called Sephacryl and are also supplied by Pharmacia.

Polyacrylamide gels are produced by the copolymerization of acrylamide and the cross-linking agent N,N'- methylenebisacrylamide. These are supplied by Bio-Rad Laboratories (Bio-Gel P). The Bio-Gel media are available in 10 sizes with exclusion limits ranging from 1800 to 400,000 daltons. Table 3.4 lists the acrylamide gels and their physical properties.

The agarose gels have the advantage of very high exclusion limits. Agarose, the neutral polysaccharide component of agar, is composed of alternating galactose and anhydrogalactose units. The gel structure is stabilized by hydrogen bonds rather than by covalent cross-linking. Agarose gels, supplied by Bio-Rad Laboratories (Bio-Gel A) and by Pharmacia (Sepharose and Superose), are listed in Table 3.4.

The combined polyacrylamide-agarose gels are commercially available under the trade name Ultragel. These consist of cross-linked polyacrylamide with agarose trapped within the gel network. The polyacrylamide gel allows a high degree of separation and the agarose maintains gel rigidity, so high flow rates may be used.

Table 3.4

Properties of Gel Filtration Media

Name	Fractionation Range for Proteins (daltons)	Water Regain (mL/g dry gel)	Bed Volume (mL/g dry gel)
Dextran (Sephadex)[1]			
G-10	0–700	1.0 ± 0.1	2–3
G-15	0–1500	1.5 ± 0.2	2.5–3.5
G-25	1000–5000	2.5 ± 0.2	4–6
G-50	1500–30,000	5.0 ± 0.3	9–11
G-75	3000–80,000	7.5 ± 0.5	12–15
G-100	4000–150,000	10 ± 1.0	15–20
G-150	5000–300,000	15 ± 1.5	20–30
G-200	5000–600,000	20 ± 2.0	30–40
Polyacrylamide (Bio-Gels)[2]			
P-2	100–1800	1.5	3.0
P-4	800–4000	2.4	4.8
P-6	1000–6000	3.7	7.4
P-10	1500–20,000	4.5	9.0
P-30	2500–40,000	5.7	11.4
P-60	3000–60,000	7.2	14.4
P-100	5000–100,000	7.5	15.0
P-150	15,000–150,000	9.2	18.4
P-200	30,000–200,000	14.7	29.4
P-300	60,000–400,000	18.0	36.0
Dextran-polyacrylamide (Sephacryl)[1]			
S-100 HR	—	—	—
S-200 HR	5000–250,000	—	—
S-300 HR	10,000–1,500,000	—	—
S-400 HR	20,000–8,000,000	—	—
Agarose			
Sepharose[1] 6B	10,000–4,000,000	—	—
Sepharose 4B	60,000–20,000,000	—	—
Sepharose 2B	70,000–40,000,000	—	—
Superose[1] 12 HR	1000–300,000	—	—
Superose 6 HR	5000–5,000,000	—	—
Bio-Gel[2] A-0.5	10,000–500,000	—	—
Bio-Gel A-1.5	10,000–1,500,000	—	—
Bio-Gel A-5	10,000–5,000,000	—	—
Bio-Gel A-15	40,000–15,000,000	—	—
Bio-Gel A-50	100,000–50,000,000	—	—
Bio-Gel A-150	1,000,000–150,000,000	—	—
Vinyl (Fractogel TSK)[3]			
HW-40	100–10,000	—	—
HW-55	1000–700,000	—	—
HW-65	50,000–5,000,000	—	—
HW-75	500,000–50,000,000	—	—

[1] *Pharmacia Biotechnology.*

[2] *Bio-Rad Laboratories.*

[3] *Pierce Chemical Co.*

Selecting a Gel

The selection of the proper gel is a critical stage in successful gel chromatography. Most gel chromatographic experiments can be classified as either **group separations** or **fractionations.** Group separations involve dividing a solute sample into two groups, a fraction of relatively low-molecular-weight solutes and a fraction of relatively high-molecular-weight solutes. Specific examples of this are desalting a protein solution or removing small contaminating molecules from protein or nucleic acid extracts. For group separations, a gel should be chosen that allows complete exclusion of the high-molecular-weight molecules in the void volume. Sephadex G-25, Bio-Gel P-6, and Sephacryl S-100HR are recommended for most group separations. The particle size recommended is 100 to 200 mesh or 50 to 150 μm diameter.

Gel fractionation involves separation of groups of solutes of similar molecular weights in a multicomponent mixture. In this case, the gel should be chosen so that the fractionation range includes the molecular weights of the desired solutes. If the solute mixture contains macromolecules up to 120,000 in molecular weight, then Bio-Gel P-150, Sephacryl S-200HR, or Sephadex G-150 would be most appropriate. If P-100, G-100, or Sephacryl S-100HR were used, some of the higher-molecular-weight proteins in the sample would elute in the void volume. On the other hand, if P-200, P-300, or G-200 were used, there would be a decrease in both resolution and flow rate. If the molecular weight range of the solute mixture is unknown, empirical selection is necessary. The recommended gel grade for most fractionations is 100–200 or 200–400 mesh (20–80 μm or 10–40 μm). The finest grade that allows a suitable flow rate should be selected. For very critical separations, superfine grades offer the best resolution but with very low flow rates.

Gel Preparation and Storage

The dextran and acrylamide gel products are sometimes supplied in dehydrated form and must be allowed to swell in water before use. The swelling time required differs for each gel, but the extremes are 3 to 4 hours at 20°C for highly cross-linked gels and up to 72 hours at 20°C for P-300 or G-200. The swelling time can be shortened if a boiling-water bath is used. Agarose gels and combined polyacrylamide-agarose gels are supplied in a hydrated state, so there is no need for swelling.

Before a gel slurry is packed into the column, it should be defined and deaerated. Defining is necessary to remove very fine particles, which would reduce flow rates. To define, pour the gel slurry into a graduated cylinder and add water equivalent to two times the gel volume. Invert the cylinder several times and allow the gel to settle. After 90 to 95% of the gel has settled, decant the supernatant, add water, and repeat the settling process. Two or three defining operations are usually sufficient to remove most small particles.

Deaerating (removing dissolved gases) should be done on the gel slurry and all eluting buffers. Gel particles that have not been deaerated tend to float and form bubbles in the column bed. Dissolved gases are removed by placing the gel slurry in a side-arm vacuum flask and applying a vacuum from a water aspirator. The degassing process is complete when no more small air bubbles are released from the gel (usually 1 to 2 hours).

Antimicrobial agents must be added to stored, hydrated gels. One of the best agents is sodium azide (0.02%).

Operation of a Gel Column

The procedure for gel column chromatography is very similar to the general description given earlier. The same precautions must be considered in packing, loading, and eluting the column. A brief outline of important considerations follows.

Column Size

For fractionation purposes, it is usually not necessary to use columns greater than 100 cm in length. The ratio of bed length to width should be between 25 and 100. For group separations, columns less than 50 cm long are sufficient, and appropriate ratios of bed length to width are between 5 and 10.

Eluting Buffer

There are fewer restrictions on buffer choice in gel chromatography than in ion-exchange chromatography. Dextran and polyacrylamide gels are stable in the pH range 1.0 to 10.0, whereas agarose gels are limited to pH 4 to 10. Since there is such a wide range of stability of the gels, the buffer pH should be chosen on the basis of the range of stability of the macromolecules to be separated.

Sample Volume

The sample volume is a critical factor in planning a gel chromatography experiment. If too much sample is applied to a column, resolution is decreased; if the sample size is too small, the solutes are greatly diluted. For group separations, a sample volume of 10 to 25% of the column total volume is suitable. The sample volume for fractionation procedures should be between 1 and 5% of the total volume. Column total volume is determined by measuring the volume of water in the glass column that is equivalent to the height of the packed bed.

Column Flow Rate

The flow rate of a gel column depends on many factors, including length of column and type and size of the gel. It is generally safe to elute a gel column at a rate slightly less than free flow. A high flow rate reduces sample diffusion or zone broadening but may not allow complete equilibration of solute molecules with the gel matrix.

A specific flow rate cannot be recommended, since each type of gel requires a different range. The average flow rate given in literature references for small-pore-size gels is 8 to 12 mL/cm² of cross-sectional bed area per hour (15 to 25 mL/hr). For large-pore-size gels, a value of 2 to 5 mL/cm² of cross-sectional bed area per hour (5 to 10 mL/hr) is average.

Eluent can be made to flow through a column by either of two methods, gravity or pump elution. Gravity elution is most often used because no special equipment is required. It is quite acceptable for developing a column used for group separations and fractionations when small-pore-sized gels are used. However, if the flow rate must be maintained at a constant value throughout an experiment or if large-pore gels are used, pump elution is recommended.

One variation of gel chromatography is ascending eluent flow. Some investigators report more reproducible results, better resolution, and a more constant flow rate if the eluting buffer is pumped backward through the gel. This type of experiment requires special equipment, including a specialized column and a peristaltic pump.

Applications of Gel Exclusion Chromatography

Several experimental applications of gel chromatography have already been mentioned, but more detail will be given here.

Desalting

Inorganic salts, organic solvents, and other small molecules are used extensively for the purification of macromolecules. Gel chromatography provides an inexpensive, simple, and rapid method for removal of these small molecules. One especially attractive method for desalting very small samples (0.1 mL or less) of proteins or nucleic acid solutions is to use spin columns. These are prepacked columns of polyacrylamide exclusion gels. Spin columns are used in a similar fashion to microfiltration centrifuge tubes (Chapter 2, pp. 49–51). The sample is placed on top of the gel column and spun in a centrifuge. Large molecules are eluted from the column and collected in a reservoir. The small molecules to be removed remain in the gel.

Purification of Biomolecules

This is probably the most popular use of gel chromatography. Because of the ability of a gel to fractionate molecules on the basis of size, gel filtration complements other purification techniques that separate molecules on the basis of polarity and charge.

Estimation of Molecular Weight

The elution volume for a particular solute is proportional to the molecular size of that solute. This indicates that it is possible to estimate the molecular weight of a solute on the basis of its elution characteristics on a gel column. An elution curve for several standard proteins on Sephadex G-100 is shown in Figure 3.10. This curve, a plot of protein concentration (A_{280}) vs. volume collected, is representative of data obtained from a gel filtration experiment. The elution volume, V_e, for each protein can be estimated as shown in the figure. A plot of log molecular mass vs. elution volume for the proteins is shown in Figure 3.11. Note that the linear portion of the curve in Figure 3.11 covers the molecular mass range of 10,000 to 100,000 daltons. Molecules below 10,000 daltons are not eluted in an elution volume proportional to size. Since they readily diffuse into the gel particles, they are retarded. Molecules larger than 100,000 daltons are all excluded from the gel in the void volume. A solution of the unknown protein is chromatographed through the calibrated column under conditions identical to those for the standards and the elution volume is measured. The unknown molecular size is then read directly from the graph. This method of molecular weight estimation is widely used because it is simple, accurate, inexpensive, and fast. It can be used with highly purified or impure samples. There are, of course, some limitations to consider. The gel must be chosen so that the molecular weight of the unknown is within the linear section of the curve. The method assumes that only steric and partition effects influence the elution of the standards and unknown. If a protein interacts with the gel by adsorptive or ionic processes, the estimate of molecular weight is lower than the true value. The assumption is also made that the unknown molecules have a general spherical shape, not an elongated or rod shape.

Figure 3.10

Elution curve for a mixture of several proteins using gel filtration chromatography.
A = hemoglobin,
B = egg albumin,
C = chymotrypsinogen,
D = myoglobin,
E = cytochrome *c*.

- -

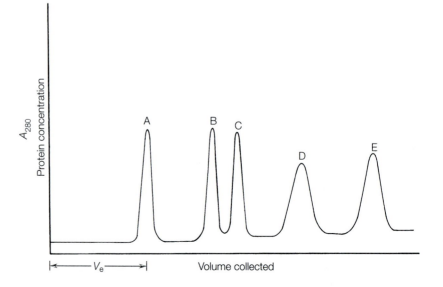

Figure 3.11

A plot of log molecular mass vs. elution volume for the proteins in Figure 3.10.

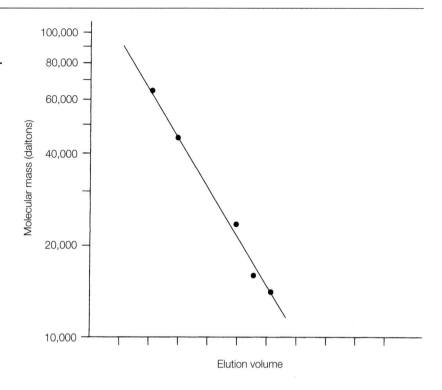

Gel Chromatography In Organic Solvents

The gels discussed so far in this section are hydrophilic, and the inner matrix retains its integrity only in aqueous solvents. Because there is a need for gel chromatography of nonhydrophilic solutes, gels have been produced that can be used with organic solvents. The trade names of some of these products are Sepharose CL and Sephadex LH (Pharmacia), Bio Beads S (Bio-Rad), and Styragel (Dow Chemical). Organic solvents that are of value in gel chromatography are ethanol, acetone, dimethylsulfoxide, dimethylformamide, tetrahydrofuran, chlorinated hydrocarbons, and acetonitrile.

G.   HIGH-PERFORMANCE LIQUID CHROMATOGRAPHY (HPLC)

The previous discussions on the theory and practice of the various chromatographic methods should convince you of the tremendous influence chromatography has had on our biochemical understanding. It is tempting, but unfair, to make comparisons about the relative importance of the methods; however, each serves a specific purpose. There will, for example, always be a need for fast, inexpensive, and qualitative analyses as afforded by planar chromatography. Traditional column chromatography will probably always be preferred in large-scale protein purification.

However, during the past three decades, an analytical method has been developed that currently rivals and may soon surpass the traditional liquid chromatographic techniques in importance for analytical separations. This technique, high-performance liquid chromatography (HPLC), is ideally suited for the separation and identification of amino acids, carbohydrates, lipids, nucleic acids, proteins, pigments, steroids, pharmaceuticals, and many other biologically active molecules.

The future promise of HPLC is indicated by its classification as "modern liquid chromatography" when compared to other forms of column-liquid chromatography, now referred to as "classical" or "traditional." Compared to the classical forms of liquid chromatography (paper, TLC, column), HPLC has several advantages:

1. Resolution and speed of analysis far exceed the classical methods.

2. HPLC columns can be reused without repacking or regeneration.

3. Reproducibility is greatly improved because the parameters affecting the efficiency of the separation can be closely controlled.

4. Instrument operation and data analysis are easily automated.

5. HPLC is adaptable to large-scale, preparative procedures.

Gas chromatography will continue to be used in biochemical research; however, its major disadvantage is that many biochemical samples either are not volatile or are thermally unstable. HPLC, which can be operated at ambient temperatures, does not have this limitation.

The advantages of HPLC are the result of two major advances: (1) the development of stationary supports with very small particle sizes and large surface areas, and (2) the improvement of elution rates by applying high pressure to the solvent flow.

The great versatility of HPLC is evidenced by the fact that all chromatographic modes, including partition, adsorption, ion exchange, chromatofocusing, and gel exclusion, are possible. In a sense, HPLC can be considered as automated liquid chromatography. The theory of each of these chromatographic modes has been discussed and needs no modification for application to HPLC. However, there are unique theoretical and practical characteristics of HPLC that should be introduced.

The **retention time** of a solute in HPLC (t_R) is defined as the time necessary for maximum elution of the particular solute. This is analogous to retention time measurements in GC. **Retention volume** (V_R) of a solute is the solvent volume required to elute the solute and is defined by Equation 3.4, where F is the flow rate of the solvent.

$$V_R = Ft_R$$

Equation 3.4

In all forms of chromatography, a measure of column efficiency is **resolution,** *R.* Resolution indicates how well solutes are separated; it is defined by Equation 3.5, where t_R and t_R' are the retention times of two solutes and w and w' are the base peak widths of the same two solutes.

>> $$R = 2\frac{t_R - t_R'}{w + w'}$$ **Equation 3.5**

Instrumentation

The increased resolution achieved in HPLC compared to classical column chromatography is primarily the result of adsorbents of very small particle sizes (less than 20 μm) and large surface areas. The smallest gel beads used in gel exclusion chromatography are "superfine" grade with diameters of 20 to 50 μm. Recall that the smaller the particle size, the lower the flow rate; therefore, it is not feasible to use very small gel beads in liquid column chromatography because low flow rates lead to solute diffusion and the time necessary for completion of an analysis would be impractical. In HPLC, increased flow rates are obtained by applying a pressure differential across the column. A combination of high pressure and adsorbents of small particle size leads to the high resolving power and short analysis times characteristic of HPLC.

A schematic diagram of a typical high-pressure liquid chromatograph is shown in Figure 3.12. The basic components are a solvent reservoir, high-pressure pump, packed column, detector, and recorder. A computer is used to control the process and to collect and analyze data. The similarities between a gas chromatograph and an HPLC are obvious. The tank of carrier gas in GC is replaced by the solvent reservoir and high-pressure pump in HPLC.

Solvent Reservoir

The solvent chamber should have a capacity of at least 500 mL for analytical applications, but larger reservoirs are required for preparative work. In order to avoid bubbles in the column and detector, the solvent must be degassed. Several methods may be used to remove unwanted gases, including refluxing, filtration through a vacuum filter, ultrasonic vibration, and purging with an inert gas. The solvent should also be filtered to remove particulate matter that would be drawn into the pump and column.

Pumping Systems

The purpose of the pump is to provide a constant, reproducible flow of solvent through the column. Two types of pumps are available—constant pressure and constant volume. Typical requirements for a pump are:

1.   It must be capable of pressure outputs of at least 500 psi and preferably up to 5000 psi.

2.   It should have a controlled, reproducible flow delivery of about 1 mL/min for analytical applications and up to 100 mL/min for preparative applications.

Figure 3.12

A schematic diagram of a high-performance liquid chromatograph.

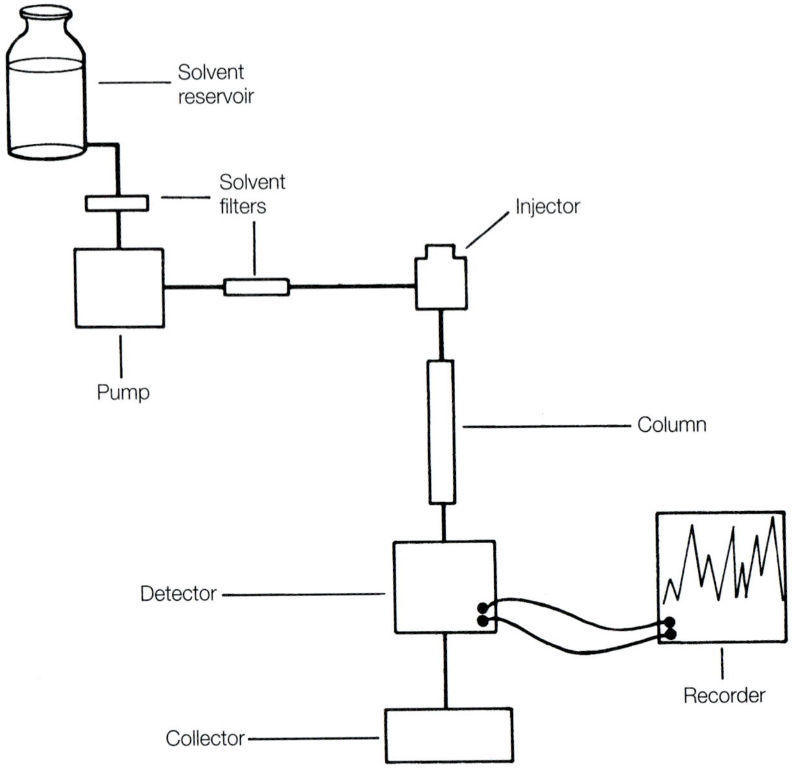

- -

3. It should yield pulse-free solvent flow.

4. It should have a small holdup volume.

Although neither type of pump meets all these criteria, constant-volume pumps maintain a more accurate flow rate and a more precise analysis is obtained.

Injection Port

A sample must be introduced onto the column in an efficient and reproducible manner. One of the most popular injectors is the syringe injector. The sample, in a microliter syringe, is injected through a neoprene/Teflon septum. This type of injection can be used at pressures up to 3000 psi.

Columns

HPLC columns are prepared from stainless steel or glass-Teflon tubing. Typical column inside diameters are 2.1, 3.2, or 4.5 mm for analytical separations and up to 30 mm for preparative applications. The length of the column can range from 5 to 100 cm, but 10 to 20 cm columns are common.

Detector

Liquid chromatographs are equipped with a means to continuously monitor the column effluent and recognize the presence of solute. Only small sample sizes are used with most HPLC columns, so a detector must have high sensitivity. The type of detector that has the most universal application is the **differential refractometer.** This device continuously monitors the refractive index difference between the mobile phase (pure solvent) and the mobile phase containing sample (column effluent). The sensitivity of this detector is on the order of 0.1 μg, which, compared to other detectors, is only moderately sensitive. The major advantage of the refractometer detector is its versatility; its main limitation is that there must be at least 10^{-7} refractive index units between the mobile phase and sample.

The most widely used HPLC detectors are the **photometric detectors.** These detectors measure the extent of absorption of ultraviolet or visible radiation by a sample. Since few compounds are colored, visible detectors are of limited value. Ultraviolet detectors are the most widely used in HPLC. The typical UV detector functions by focusing radiation from a low-pressure mercury lamp on a flow cell that contains column effluent. The mercury lamp provides a primary radiation at 254 nm. The use of filters or other lamps provides radiation at 220, 280, 313, 334, and 365 nm. Many compounds absorb strongly in this wavelength range, and sensitivities on the order of 1 ng are possible. Most biochemicals are detected, including proteins, nucleic acids, pigments, vitamins, some steroids, and aromatic amino acids. Aliphatic amino acids, carbohydrates, lipids, and other biochemicals that do not absorb UV can be detected by chemical derivatization with UV-absorbing functional groups. UV detectors have many positive characteristics, including high sensitivity, small sample volumes, linearity over wide concentration ranges, nondestructiveness to sample, and suitability for gradient elution.

A third type of detector that has only limited use is the **fluorescence detector.** This type of detector is extremely sensitive: its use is limited to samples containing trace quantities of biological materials. Its response is not linear over a wide range of concentrations, but it may be up to 100 times more sensitive than the UV detector.

Collection of Eluent

All of the detectors described here are nondestructive to the samples, so column effluent can be collected for further chemical and physical analysis.

Analysis of HPLC Data

Most HPLC instruments are on line with an integrator and a computer for data handling. For quantitative analysis of HPLC data, operating parameters such as rate of solvent flow must be controlled. In modern instruments, the whole system (including the pump, injector, detector, and data system) is under the control of a computer.

Figure 3.13 illustrates the separation by HPLC of several phenylhydantoin derivatives of amino acids.

Stationary Phases in HPLC

The adsorbents in HPLC are typically small-diameter. porous materials. Two types of stationary phases are available. **Porous layer beads** (Figure 3.14A) have an inert solid core with a thin porous outer shell of silica, alumina, or ion-exchange resin. The average diameter of the beads ranges from 20 to 45 μm. They are especially useful for analytical applications, but, because of their short pores, their capacities are too low for preparative applications.

Microporous particles are available in two sizes: 20 to 40 μm diameter with longer pores and 5 to 10 μm with short pores (see Figure 3.14B). These are now more widely used than the porous layer beads because they offer greater resolution and faster separations with lower pressures. The microporous beads are prepared from alumina, silica, ion-exchanger resins, and chemically bonded phases (see next section).

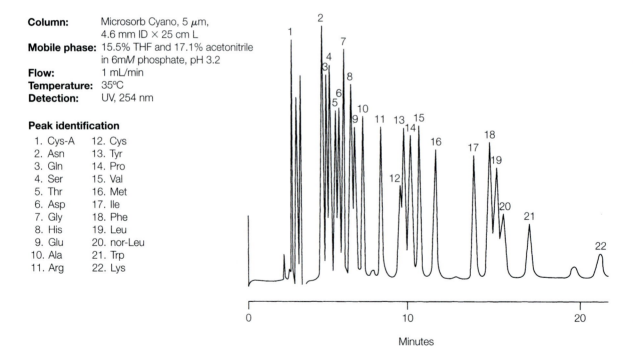

Column: Microsorb Cyano, 5 μm, 4.6 mm ID × 25 cm L
Mobile phase: 15.5% THF and 17.1% acetonitrile in 6mM phosphate, pH 3.2
Flow: 1 mL/min
Temperature: 35°C
Detection: UV, 254 nm

Peak identification

1. Cys-A	12. Cys
2. Asn	13. Tyr
3. Gln	14. Pro
4. Ser	15. Val
5. Thr	16. Met
6. Asp	17. Ile
7. Gly	18. Phe
8. His	19. Leu
9. Glu	20. nor-Leu
10. Ala	21. Trp
11. Arg	22. Lys

Figure 3.13

The separation of several amino acid phenylhydantoins by HPLC. *Courtesy of Rainin Instrument Co., Woburn, MA.*

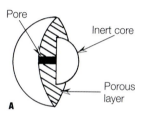

Figure 3.14

Adsorbents used in HPLC.
A Porous layer with short pores.
B Microporous particle with longer pores.

The HPLC can function in several chromatographic modes. Each type of chromatography will be discussed below, with information about stationary phases.

Liquid-Solid (Adsorption) Chromatography

HPLC in the adsorption mode can be carried out with silica or alumina porous-layer-bead columns. Small glass beads are often used for the inert core. Some of the more widely used packings are μ Porasil (Waters Associates), BioSilA (Bio-Rad Laboratories), LiChrosorb Si-100 Partisil, Vydac, ALOX 60D (several suppliers), and Supelcosil (Supelco).

In high-pressure adsorption chromatography, solutes adsorb with different affinities to binding sites in the solid stationary phase. Separation of solutes in a sample mixture occurs because polar solutes adsorb more strongly than nonpolar solutes. Therefore, the various components in a sample are eluted with different retention times from the column. This form of HPLC is usually called **normal phase** (polar stationary phase and a nonpolar mobile phase).

Liquid-Liquid (Partition) Chromatography

In the early days of HPLC (1970–78), solid supports were coated with a liquid stationary phase as in gas chromatography. Columns with these packings had short lifetimes and a gradual decrease in resolution because there was continuous loss of the liquid stationary phase with use of the column.

This problem was remedied by the discovery of methods for chemically bonding the active stationary phase to the inert support. Most chemically bonded stationary phases are produced by covalent modification of the surface silica. Three modification processes are shown in Equations 3.6–3.8.

>> Silicate esters   *Equation 3.6*

$$-Si-OH + ROH \longrightarrow -Si-O-R$$

>> Silica-carbon   *Equation 3.7*

$$-Si-Cl + \begin{array}{c} RMgBr \\ or \\ RLi \end{array} \longrightarrow -Si-R$$

>> Siloxanes   *Equation 3.8*

$$-Si-OH + \begin{array}{c} ClSiR_3 \\ or \\ ROSiR_3 \end{array} \longrightarrow -Si-O-SiR_3$$

The major advantage of a bonded stationary phase is stability. Since it is chemically bonded, there is very little loss of stationary phase with column use. The siloxanes are the most widely used silica supports. Functional groups that can be attached as siloxanes are alkylnitriles ($-Si-CH_2CH_2-CN$), phenyl ($-Si-C_6H_5$), alkylamines ($-Si(CH_2)_n-NH_2$), and alkyl side chains ($-Si-C_8H_{17}$; $-Si-C_{18}H_{37}$).

The use of nonpolar chemically bonded stationary phases with a polar mobile phase is referred to as **reverse-phase HPLC.** This technique separates sample components according to hydrophobicity. It is widely used for the separation of all types of biomolecules, including peptides, nucleotides, carbohydrates, and derivatives of amino acids. Typical solvent systems are water-methanol, water-acetonitrile, and water-tetrahydrofuran mixtures. Figure 3.15 shows the results of protein separation on a silica-based reverse-phase column.

Ion-Exchange Chromatography

Ion-exchange HPLC uses column packings with charged functional groups. Structures of typical ion exchangers are shown in Figure 3.16. They are prepared by chemically bonding the ionic groups to the support via silicon atoms or by using polystyrene-divinylbenzene resins. These stationary phases may be used for the separation of proteins, peptides, and other

---

Figure 3.15

Reverse-phase chromatography of a mixture of standard proteins. *Separation courtesy of Bio-Rad Laboratories, Chemical Division, Richmond, CA. Used by permission.*

- -

Conditions
column: Hi-Pore RP-304 column,
                  250 × 4.6 mm
Sample: Protein standard
Buffers: A. H_2O, 0.1% B
                   B. 95% Acetonitrile,
                   5% H_2O, 0.1% TFA
Gradient: 25%–100% B in 30 minutes
Flow rate: 1.5 mL/min
Temperature: Ambient
Detection: UV @ 280 nm; conductivity

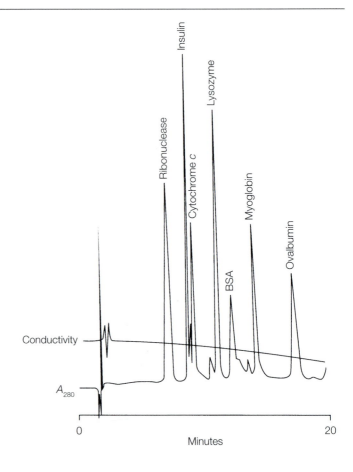

Figure 3.16

Ion exchangers used in HPLC.

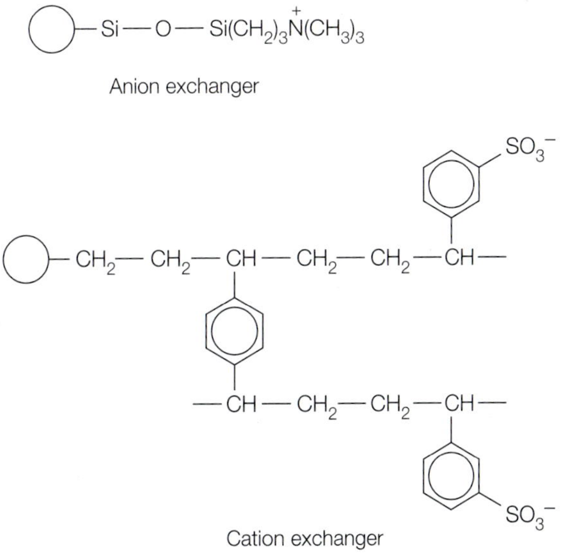

Anion exchanger

Cation exchanger

charged biomolecules. Two widely used phases are DEAE-silica and CM-silica. Both of these columns can be used with aqueous buffers in the pH range 2.5 to 7.5.

Gel Exclusion Chromatography

The combination of HPLC and gel exclusion chromatography is used extensively for the separation of large biomolecules, especially proteins and nucleic acids.

The exclusion gels discussed in Section F are not appropriate for HPLC use because they are soft and, except for small-pore beads (G-25 and less), collapse under high-pressure conditions. Semirigid gels based on cross-linked styrene-divinylbenzene, polyacrylamide, and vinyl-acetate copolymer are available with various fractionation ranges useful for the separation of molecules up to 10,000,000 daltons.

Rigid packings for HPLC gel exclusion are prepared from porous glass or silica. They have several advantages over the semirigid gels, including several fractionation ranges, ease of packing, and compatibility with water and organic solvents.

The Mobile Phase

Selection of a column packing that is appropriate for a given analysis does not ensure a successful HPLC separation. A suitable solvent system must also be chosen. Several critical solvent properties will be considered here.

Purity   Very-high-purity solvents with no particulate matter are required. Many laboratory workers do not purchase expensive prepurified solvents, but rather they purify lesser grade solvents by microfiltration through a Millipore system or distillation in glass.

Reactivity   The mobile phase must not react with the analytical sample or column packing. This does not present a major limitation since many relatively unreactive hydrocarbons, alkyl halides, and alcohols are suitable.

Detector Compatibility   A solvent must be carefully chosen to avoid interference with the detector. Most UV detectors monitor the column effluent at 254 nm. Any UV-absorbing solvent, such as benzene or olefins, would be unacceptable because of high background. Since refractometer detectors monitor the difference in refractive index between solvent and column effluent, a greater difference leads to greater ability to detect the solute.

Solvents for HPLC Operation

It has long been recognized that the eluting power of a solvent is related to its polarity. Chromatographic solvents have been organized into a list according to their ability to displace adsorbed solutes (eluting power, $\varepsilon°$). This list of solvents, called an **eluotropic series,** is shown in Table 3.5. It should be noted that $\varepsilon°$ increases with an increase in polarity. Using the eluotropic series makes solvent choice less a matter of trial and error.

Occasionally, a single solvent does not provide suitable resolution of solutes. Solvent **binary mixtures** can be prepared with eluent strengths intermediate between the $\varepsilon°$ values for the individual solvents.

Table 3.5

Eluotropic Series of HPLC Solvents

Solvent	$\varepsilon°[1]$	n_0^2	UV Cutoff (nm)
n-Pentane	0.00	1.358	210
n-Hexane	0.01	1.375	210
Cyclohexane	0.04	1.427	210
Carbon tetrachloride	0.18	1.466	265
2–Chloropropane	0.29	1.378	225
Toluene	0.29	1.496	285
Ethyl ether	0.38	1.353	220
Chloroform	0.40	1.443	245
Tetrahydrofuran	0.45	1.408	230
Acetone	0.56	1.359	330
Ethyl acetate	0.58	1.370	260
Dimethylsulfoxide	0.62	1.478	270
Acetonitrile	0.65	1.344	210
2–Propanol	0.82	1.380	210
Ethanol	0.88	1.361	210
Methanol	0.95	1.329	210
Ethylene glycol	1.11	1.427	210

[1] *Eluent strength.*

[2] *Refractive index.*

Gradient Elution in HPLC

Figure 3.17A illustrates the separation of a multicomponent sample using a single solvent and a silica column. Note that some earlier components (1 to 5) are poorly resolved and later peaks (9 to 11) display extensive broadening. This common difficulty encountered in the analysis of multicomponent samples is referred to as the **general elution problem.** It is due to the fact that the components have a wide range of K_D values and no single solvent system is equally effective in displacing all components from the column.

The general elution problem is solved by the use of **gradient elution.** This is achieved by varying the composition of the mobile phase during elution. The similarity between gradient elution and temperature programming in gas chromatography should be readily apparent. In practice, gradient elution is performed by beginning with a weakly eluting solvent (low $\varepsilon°$) and gradually increasing the concentration of a more strongly eluting solvent

Figure 3.17

The general elution problem.
A Normal elution.   **B** Gradient
elution. *From L. R. Snyder, J.
Chromatogr. Sci.* **8,** *692 (1970)
Reproduced by permission of
Preston Publications.*

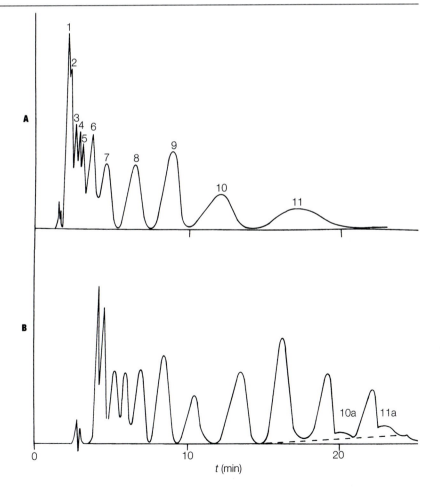

(higher $\varepsilon°$). The weaker solvent is able to improve resolution of components having low K_D values (peaks 1 to 5, Figure 3.17B). In addition, the gradual increase in $\varepsilon°$ of the solvent mixture decreases line broadening of components 9 to 11 and provides a more effective separation.

The choice of solvents for gradient elution is still somewhat empirical; however, using the data from Table 3.5 narrows the choices. Modern HPLC instruments are equipped with **solvent programming units** that control gradient elution in a stepwise or continuous manner.

Sample Preparation and Selection of HPLC Operating Conditions

During the initial stages of biochemical sample preparation, the sample is often quite crude; it may contain hundreds of components in addition to the desired biomolecules. Most samples must be pretreated before optimum HPLC results can be expected. The following procedures may be needed in order to convert a crude sample into a clean one: desalting, removal of anions and cations, removal of metal ions, concentration of the desired macromolecules, removal of detergent, and particulate removal. Sample preparation techniques used to achieve these results are gel exclusion chromatography, ion-exchange chromatography, microfiltration, and metal affinity chromatography. These procedures may be completed by commercially available prefilters or precolumns.

Each type of HPLC instrument has its own characteristics and operating directions. It is not feasible to describe those here. However, it is appropriate to outline the general approach taken when an HPLC analysis is desired. The following items must be considered:

1. Chemical nature and proper preparation of the sample.
2. Selection of type of chromatography (partition, adsorption, ion exchange, gel exclusion).
3. Choice of solvent system and mode of elution.
4. Selection of column packing.
5. Choice of equipment (type of detector).

FPLC—A Modification of HPLC

In 1982 Pharmacia introduced an innovative chromatographic method that provides a link between classical column chromatography and HPLC. This technique, called **fast protein liquid chromatography (FPLC),** uses experimental conditions intermediate between those of column chromatography and HPLC. The typical FPLC system requires a pump that will deliver solvent to the column in the flow rate range 1–499 mL/hr with operating pressures of 0–40 bar. (HPLC pumps deliver solvent in a flow rate range of 0.010–10 mL/min with operating pressures of 1–400 bar. Classical chromatography columns are operated at atmosphere pressure.) Also required for FPLC are a controller, detector, and fraction collector. Since lower pressures are used in FPLC than in HPLC, a wider range of column supports is possible. Chromatographic techniques incorporated in an FPLC system are gel filtration, ion exchange, affinity (see next section), hydrophobic interaction, reversed phase, and chromatofocusing.

H.   AFFINITY CHROMATOGRAPHY AND IMMUNOADSORPTION

The more conventional chromatographic procedures that we have studied up to this point rely on rather nonspecific physicochemical interactions between a stationary support and solute. The molecular characteristics of net charge, size, and polarity do not provide a basis for high selectivity in the separation and isolation of biomolecules. The desire for more specificity in chromatographic separations has led to the development of **affinity chromatography.** This technique offers the ultimate in specificity—separation on the basis of biological interactions. The biological function displayed by most macromolecules (antibodies, transport proteins, enzymes, nucleic acids, polysaccharides, receptor proteins, etc.) is a result of recognition of and interaction with specific molecules called **ligands.** This is illustrated by Equation 3.9, where A represents a macromolecule and B a smaller molecule or ligand. The two molecules interact in a specific manner to form a complex, A:B.

$$\blacktriangleright\blacktriangleright \qquad A + B \rightleftharpoons A:B \longrightarrow \text{biological response} \qquad \textit{Equation 3.9}$$

In a biological system, the formation of the complex often triggers some response such as immunological action, control of a metabolic process, hormone action, catalytic breakdown of a substrate, or membrane transport. The biological response depends on proper molecular recognition and binding as shown in the reaction. The most common example of Equation 3.9 is the interaction that occurs between an enzyme molecule, E, and a substrate, S, with reversible formation of an ES complex. The biological event resulting from this interaction is the transformation of S to a metabolic product, P. Only the first step in Equation 3.9, formation of the complex, is of concern in affinity chromatography.

In practice, affinity chromatography requires the preparation of an insoluble stationary phase, to which appropriate ligand molecules (B) are covalently affixed. Thus, ligand molecules are immobilized on the stationary support. The affinity support is packed into a column through which a mixture containing the desired macromolecule, A, is allowed to percolate. There are many types of molecules in the mixture, especially if it is a crude cell extract, but only macromolecules that recognize and bind to immobilized B are retarded in their movement through the column. After the nonbinding molecules have washed through the column, the desired macromolecules, B,

Figure 3.18

The steps of affinity chromatography.

- -

1. Attach ligand B to gel:

Gel   Modified gel

2. Pack modified gel into column and adsorb sample containing a mixture of components A, C, and D:

3. Dissociate complex with Y and elute A:

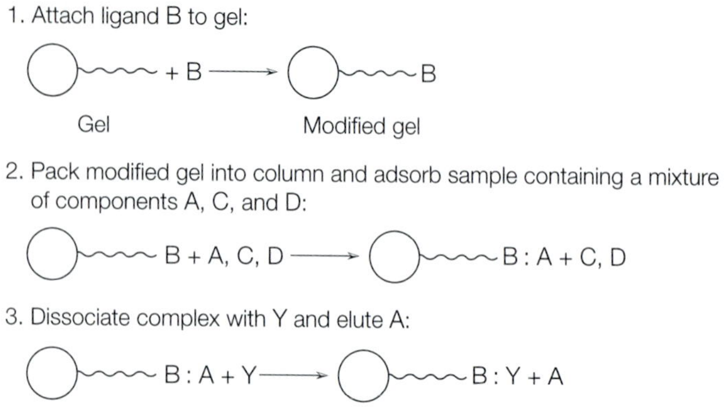

are eluted by gentle disruption of the A:B complex. Study Figure 3.18 for an illustration of affinity chromatography.

Affinity chromatography can be applied to the isolation and purification of virtually all biological macromolecules. It has been used to purify nucleic acids, enzymes, transport proteins, antibodies, hormone receptor proteins, drug-binding proteins, neurotransmitter proteins, and many others.

Successful application of affinity chromatography requires careful design of experimental conditions. The essential components, which are outlined below, are (1) creation and preparation of a stationary matrix with immobilized ligand and (2) design of column development and eluting conditions.

Chromatographic Media

Selection of the matrix used to immobilize a ligand requires consideration of several properties. The stationary supports used in gel exclusion chromatography are found to be quite suitable for affinity chromatography because (1) they are physically and chemically stable under most experimental conditions, (2) they are relatively free of nonspecific adsorption effects, (3) they have satisfactory flow characteristics, (4) they are available with very large pore sizes, and (5) they have reactive functional groups to which an appropriate ligand may be attached.

Four types of media possess most of these desirable characteristics: agarose, polyvinyl, polyacrylamide, and controlled-porosity glass (CPG) beads. Highly porous agarose beads such as Sepharose 4B (Pharmacia) and Bio-Gel A-150 m (Bio-Rad Laboratories) have virtually all of these characteristics and are the most widely used matrices. Polyacrylamide gels such as Bio-Gel P-300 (Bio-Rad) display many of the recommended features; however, the porosity is not especially high.

The Immobilized Ligand

The ligand (B in Equation 3.9 and Figure 3.18) can be selected only after the nature of the macromolecule to be isolated is known. When a hormone receptor protein is to be purified by affinity chromatography, the hormone itself is an ideal candidate for the ligand. For antibody isolation, an antigen or hapten may be used as ligand. If an enzyme is to be purified, a substrate analog, inhibitor, cofactor, or effector may be used as the immobilized ligand. The actual substrate molecule may be used as a ligand, but only if column conditions can be modified to avoid catalytic transformation of the bound substrate.

In addition to the foregoing requirements, the ligand must display a strong, specific, but reversible interaction with the desired macromolecule and it must have a reactive functional group for attachment to the matrix. It should be recognized that several types of ligand may be used for affinity purification of a particular macromolecule. Of course, some ligands will work better than others, and empirical binding studies can be performed to select an effective ligand.

Attachment of Ligand to Matrix

Several procedures have been developed for the covalent attachment of the ligand to the stationary support. All procedures for gel modification proceed in two separate chemical steps: (1) activation of the functional groups on the matrix and (2) joining of the ligand to the functional group on the matrix.

A wide variety of activated gels is now commercially available. The most widely used are described as follows:

Cyanogen Bromide-Activated Agarose

This gel is especially versatile because all ligands containing primary amino groups are easily attached to the agarose. It is available under the trade name CNBr-activated Sepharose 4B (Pharmacia). Since the gel is extremely reactive, very gentle conditions may be used to couple the ligand. One disadvantage of CNBr activation is that small ligands are coupled very closely to the matrix surface; macromolecules, because of steric repulsion, may not be able to interact fully with the ligand. The procedure for CNBr activation and ligand coupling is outlined in Figure 3.19A.

6–Aminohexanoic Acid (CH)-Agarose and 1,6–Diaminohexane (AH)-Agarose

These activated gels overcome the steric interference problems stated above by positioning a six-carbon spacer arm between the ligand and the matrix. Ligands with free primary amino groups can be covalently attached to CH-agarose, whereas ligands with free carboxyl groups can be coupled to

AH-agarose. The attachment of ligands to AH and CH gels is outlined in Figure 3.19B,C.

Carbonyldiimidazole (CDI)-Activated Supports

Reaction with CDI produces gels that contain uncharged N-alkylcarbamate groups (see Figure 3.19D). CDI-activated agarose, dextran, and polyvinyl acetate are sold by Pierce Chemical Co. under the trade name Reacti-Gel.

Epoxy-Activated Agarose

The structure of this gel is shown in Figure 3.19E. It provides for the attachment of ligands containing hydroxyl, thiol, or amino groups. The hydroxyl groups of mono-, oligo-, and polysaccharides can readily be attached to the gel. Epoxy-activated Sepharose 6B is available from Pharmacia.

Group-Specific Adsorbents

The affinity materials described up to this point are modified with a ligand having specificity for a particular macromolecule. Therefore, each time a biomolecule is to be isolated by affinity chromatography, a new adsorbent

---

Figure 3.19

Attachment of specific ligands to activated gels. R = ligand.

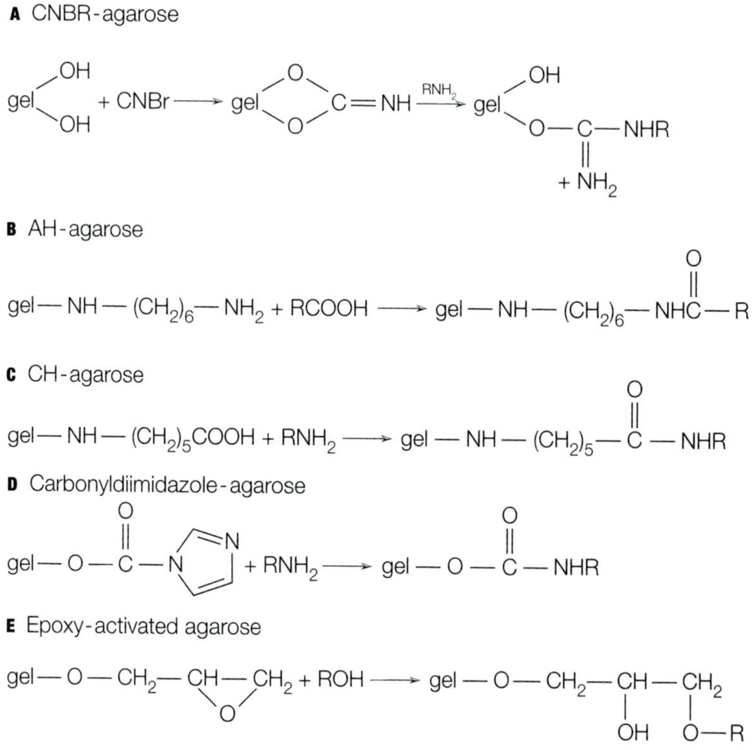

- - - - - - - - - - - - - - - - -

Table 3.6

Group-Specific Adsorbents Useful in Biochemical Applications

Group-Specific Adsorbent	Group Specificity
5'-AMP-agarose	Enzymes that have NAD$^+$ cofactor; ATP-dependent kinases
Benzamidine-Sepharose	Serine proteases
Boronic acid–agarose	Compounds with *cis*-diol groups; sugars, catecholamines, ribonucleotides, glycoproteins
Cibracron blue–agarose	Enzymes with nucleotide cofactors (dehydrogenases, kinases, DNA polymerases); serum albumin
Concanavalin A–agarose	Glycoproteins and glycolipids
Heparin-Sepharose	Nucleic acid–binding proteins, restriction endonucleases, lipoproteins
Iminodiacetate–agarose	Proteins with affinity for metal ions, serum proteins, interferons
Lentil lectin–Sepharose	Detergent–soluble membrane proteins
Lysine-Sepharose	Nucleic acids
Octyl-Sepharose	Weakly hydrophobic proteins, membrane proteins
Phenyl-Sepharose	Strongly hydrophobic proteins
Poly(A)-agarose	Nucleic acids containing poly(U) sequences, mRNA-binding proteins
Poly(U)-agarose	Nucleic acids containing poly(A) sequences, poly(U)-binding proteins
Protein A–agarose	IgG-type antibodies
Thiopropyl-Sepharose	— SH containing proteins

must be designed and prepared. Ligands of this type are called substance specific. In contrast, group-specific adsorbents contain ligands that have affinity for a class of biochemically related substances. Table 3.6 shows several commercially available group-specific adsorbents and their specificities. The principles behind binding of nucleic acids and proteins to group-specific adsorbents depend on the actual affinity adsorbent. In most cases, the immobilized ligand and macromolecule (protein or nucleic acid) interact through one or more of the following forces: hydrogen bonding, hydrophobic interactions, and/or covalent interactions. Some group-specific adsorbents deserve special attention. Phenyl- and octyl-Sepharose are gels used for **hydrophobic interaction chromatography.** These adsorbent materials separate proteins on the basis of their hydrophobic character. Because most proteins contain hydrophobic amino acid side chains, this method is widely used. Octyl-Sepharose is strongly hydrophobic; hence it binds strongly to nonpolar proteins. Phenyl-Sepharose is more weakly hydrophobic; therefore, it is more likely to reversibly bind strongly hydrophobic proteins.

The use of thiopropyl-Sepharose and boronic acid–agarose is an example of **covalent chromatography,** since relatively strong but reversible covalent bonds are formed between the affinity gel and specific macromolecules.

Metal affinity chromatography is a relatively new method that separates proteins on the basis of metal binding. This technique is used in Experiment 4 to isolate α-lactalbumin from milk.

The availability of a great variety of group-specific adsorbents in prepacked columns makes possible the combination of FPLC and affinity chromatography for the separation and purification of proteins.

One of the most specific modifications of affinity chromatography is **immunoaffinity.** The unique high specificity of antibodies for their antigens is valuable for the purification of antigens. In practice, the antibody is immobilized on a column support. When a mixture containing several other proteins along with the protein antigen for the antibody is passed through the column, only the antigen binds; the other proteins, which have no affinity for the antibody, wash off the column. Protein A–agarose in Table 3.6 is an example of immunoaffinity; however, this adsorbent does not recognize specific antibodies but, rather, the general family of immunoglobulin G antibodies.

Experimental Procedure for Affinity Chromatography

Although the procedure is different for each type of substance isolated, a general experimental plan is outlined here. Figure 3.20 provides a step-by-step plan in flowchart form. Many types of matrix-ligand systems are commercially available and the costs are reasonable, so it is not always necessary to spend valuable laboratory time for affinity gel preparation. Even if a specific gel is not available, time can be saved by purchasing preactivated gels for direct attachment of the desired ligand. Once the gel is prepared, the procedure is similar to that described earlier. The major difference is the use of shorter columns. Most affinity gels have high capacities and column beds less than 10 cm in length. A second difference is the mode of elution. Ligand-macromolecule complexes immobilized on the column are held together by hydrogen bonding, ionic interactions, and hydrophobic effects. Any agent that diminishes these forces causes the release and elution of the macromolecule from the column. The common methods of elution are change of buffer pH, increase of buffer ionic strength, affinity elution, and chaotropic agents. The choice of elution method depends on many factors, including the types of forces responsible for complex formation and the stability of the ligand matrix and isolated macromolecule.

Buffer pH or Ionic Strength

If ionic interactions are important for complex formation, a change in pH or ionic strength weakens the interaction by altering the extent of ionization of ligand and macromolecule. In practice, either a decrease in pH or a gradual increase in ionic strength (continual or stepwise gradient) is used.

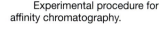

Experimental procedure for
affinity chromatography.

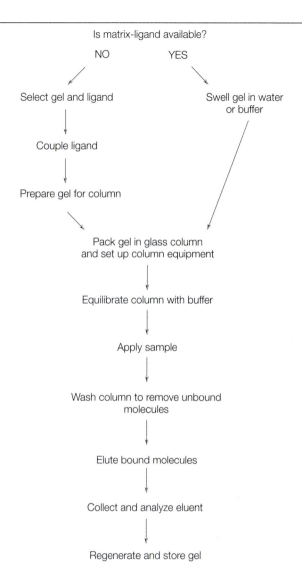

Affinity Elution

In this method of elution, a selective substance added to the eluting
buffer competes for binding to the ligand or for binding to the adsorbed
macromolecule.

Chaotropic Agents

If gentle and selective elution methods do not release the bound macromolecule, then mild denaturing agents can be added to the buffer. These

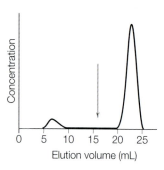

Figure 3.21

Purification of α-chymotrypsin by affinity chromatography on immobilized D-tryptophan methyl ester. *From Affinity Chromatography: Principles and Methods, Pharmacia, Uppsala, Sweden.*

substances deform protein and nucleic acid structure and decrease the stability of the complex formed on the affinity gel. The most useful agents are urea, guanidine · HCl, CNS^-, ClO_4^-, and CCl_3COO^-. These substances should be used with care, because they may cause irreversible structural changes in the isolated macromolecule.

The application of affinity chromatography is limited only by the imagination of the investigator. Every year literally hundreds of research papers appear with new and creative applications of affinity chromatography. Figure 3.21 illustrates the purification of α-chymotrypsin by affinity chromatography on immobilized D-tryptophan methyl ester. α-Chymotrypsin can recognize and bind, but not chemically transform, D-tryptophan methyl ester. The enzyme catalyzes the hydrolysis of L-tryptophan methyl ester. The impure α-chymotrypsin mixture was applied to the gel, D-tryptophan methyl ester coupled to CH-Sepharose 4B, and the column washed with Tris buffer. At the point shown by the arrow, the eluent was changed to 0.1 M acetic acid. The decrease in pH caused release of α-chymotrypsin from the column.

I. PERFUSION CHROMATOGRAPHY

A separation method that improves resolution and decreases the time required for analysis of biomolecules has recently been introduced. This method, called **perfusion chromatography,** relies on a type of particle support called POROS, which may be used in low-pressure and high-pressure liquid chromatography applications. In conventional chromatographic separations, some biomolecules in the sample move rapidly around and past the media particles while other molecules diffuse slowly through the particles (Figure 3.22A). The result is loss of resolution because some biomolecules exit the column before others. To improve resolution, the researcher with conventional media found it necessary to reduce the flow rate to allow for diffusion processes, increasing the time required for analysis. In other words, before the development of perfusion chromatography, the researcher had to choose between high speed–low resolution and low speed–high resolution. POROS particles have two types of pores—**through pores** (6000–8000 Å in diameter), which provide channels through the particles, and connected **diffusion pores** (800–1500 Å in diameter), which line the through pores and have very short diffusion path lengths (Figure 3.22B). This combination pore system increases the porosity and the effective surface area of the particles and results in improved resolution and shorter analysis times (30 seconds to 3 minutes for POROS versus 30 minutes to several hours for conventional media).

POROS media, made by copolymerization of styrene and divinylbenzene, have high mechanical strength and are resistant to many solvents and chemicals. The functional surface chemistry of the particles can be modified to provide supports for many types of chromatography, including ion exchange, hydrophobic interaction, immobilized metal affinity, reversed

Figure 3.22

Transport of biomolecules through chromatographic media. **A** Conventional support particles. **B** POROS particles for perfusion chromatography. *Courtesy of PerSeptive Biosystems, Cambridge, MA.*

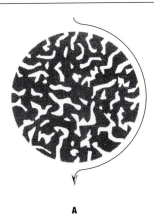
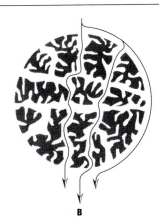

A   B

phase, group-selective affinity, and conventional bioaffinity. Perfusion chromatography has been applied with success to the separation of peptides, proteins, and polynucleotides on both preparative and analytical scales.

In addition to high resolution and short analysis times, perfusion chromatography has the advantage of improved recovery of biological activity because active biomolecules spend less time on the column, where denaturing conditions may exist.

Study Problems

▶ *1.* Amino acid analyzers are instruments that automatically separate amino acids by cation-exchange chromatography. Predict the order of elution (first to last) for each of the following sets of amino acids at pH = 4.
(a) Gly, Asp, His
(b) Arg, Glu, Ala
(c) Phe, His, Glu

▶ *2.* Predict the relative order of paper chromatography R_f values for the amino acids in the following mixture: Ser, Lys, Leu, Val, and Ala. Assume that the developing solvent is *n*-butanol, water, and acetic acid.

▶ *3.* In what order would the following proteins be eluted from a DEAE-cellulose ion exchanger by an increasing salt gradient. The pH_I is listed for each protein.

Egg albumin, 4.6   Cytochrome *c*, 10.6
Pepsin, 1.0   Myoglobin, 6.8
Serum albumin, 4.9   Hemoglobin, 6.8

▶ *4.* Draw the elution curve (A_{280} vs. fraction number) obtained by passing a mixture of the following proteins through a column of Sephadex G-100. The molecular weight is given for each protein.

Myoglobin, 16,900   Myosin, 524,000

Catalase, 222,000   Serum albumin, 68,500

Cytochrome c, 13,370   Chymotrypsinogen, 23,240

▶ 5. Amino acids and fatty acids do not readily elute from gas chromatography columns even at temperatures above 200°C. What can be done to these biomolecules to allow gas chromatographic analysis?

▶ 6. Describe the various detection methods that can be used in HPLC. What types of biomolecules are detected by each method?

▶ 7. Name three enzymes that you predict will bind to the affinity support, 5'-AMP-agarose.

8. Briefly describe how you would experimentally measure the exclusion limit for a Sephadex gel whose bottle has lost its label.

▶ 9. Describe how you would use "affinity elution" to remove the enzyme alcohol dehydrogenase bound to a Cibracron blue–agarose column.

10. Explain the elution order of amino acids in Figure 3.13.

Further Reading

H. Barth and B. Boyes, *Anal. Chem.* **62**, 382R–394R (1990). "Size Exclusion Chromatography."

R. Boyer, *Concepts in Biochemistry* (1999), Brooks/Cole (Pacific Grove, CA), pp. 102–104. An introduction to chromatography.

M. Chakravarthy, L. Snyder, T. Vanyo, J. Holbrook, and H. Jakubowski, *J. Chem. Educ.* **73**, 268–272 (1996). Protein structure and chromatographic behavior.

S. Fulton and D. Vanderburgh, *The Busy Researcher's Guide to Biomolecule Chromatography* (1996), PerSeptive Biosystems (Framingham, MA). Practical guide to chromatography of biomolecules with emphasis on perfusion methods.

R. Garrett and C. Grisham, *Biochemistry,* 2nd ed. (1999), Saunders (Orlando, FL), pp. 128–130. An introduction to chromatography.

D. Grant, *Capillary Gas Chromatography* (1995), John Wiley & Sons (United Kingdom). An introduction to specialized GC.

N. Grinberg, Editor, *Modern Thin-Layer Chromatography* (1990), Marcel Dekker (New York). Up-to-date developments in TLC.

T. Hanai, *HPLC: A Practical Guide* (1999), Springer-Verlag (New York). HPLC introduction and applications.

E. Katz, Editor, *High Performance Liquid Chromatography* (1995), John Wiley & Sons (United Kingdom). A modern introduction to HPLC.

A. Lehninger, D. Nelson, and M. Cox, *Principles of Biochemistry,* 3rd ed. (1999), Worth Publishers (New York), pp. 130–133, 384–386. Introduction to chromatography.

P. Matejtschuk, Editor, *Affinity Separations: A Practical Approach* (1997), IRL Press (Oxford). Principles and applications.

C. Mathews, K. van Holde, and K. Ahern, *Biochemistry,* 3rd ed. (2000), Benjamin/Cummings (San Francisco), pp. 150–152. Introduction to purification by chromatography.

M. McMaster, *HPLC: A Practical User's Guide* (1994), John Wiley & Sons (New York). Very useful in the laboratory.

V. Meyer, *Practical High-Performance Liquid Chromatography,* 3rd ed. (1999), John Wiley & Sons (New York). Practical applications.

V. Meyer, *Pitfalls and Errors of HPLC in Pictures* (1998), John Wiley & Sons (New York). A useful book for the beginner.

C. Morgan and N. Moir, *J. Chem. Educ.* **73,** 1040–1041 (1996). "Rapid Microscale Isolation and Purification of Yeast Alcohol Dehydrogenase Using Cibacron Blue Affinity Chromatography."

A. Niederwieser, in *Methods in Enzymology,* Vol. XXV, C. Hirs and S. Timasheff, Editors (1972), Academic Press (New York), pp. 60–99. Thin-layer chromatography of amino acids.

Pharmacia Biotechnology, *Gel Filtration: Principles and Methods,* 5th ed. (1991), Pharmacia Biotechnology, S-751 82 Uppsala, Sweden, or 800 Centennial Avenue, Piscataway, NJ 08854. An excellent guide on theory, practice, and applications of gel filtration.

F. Regnier, *Nature* **350,** 634 (1991). "Perfusion Chromatography."

R. Scopes, *Protein Purification: Principles and Practice,* 3rd ed. (1993), Springer-Verlag (Berlin). Introduction to all methods of chromatography.

J. Sherma, *Anal. Chem.* **62,** 371R–381R (1990). "Planar Chromatography."

L. Stryer, *Biochemistry,* 4th ed. (1995), Freeman (New York), pp. 48–50. Affinity, ion-exchange, and gel-filtration chromatography.

J. Touchstone, Editor, *Planar Chromatography in the Life Sciences* (1990), John Wiley & Sons (New York). Modern aspects of TLC.

D. Voet and J. Voet, *Biochemistry,* 2nd ed. (1995), John Wiley & Sons (New York), pp. 78–89. Use of chromatography in protein purification.

D. Voet, J. Voet, and C. Pratt, *Fundamentals of Biochemistry* (1999), John Wiley & Sons (New York), pp. 99–104. Use of chromatography in protein purification.

P. Williams and M. Hudson, Editors, *Recent Developments in Ion Exchange* (1990), Elsevier Applied Science (Amsterdam). Modern discussion of ion-exchange chromatography.

Chromatography on the Web

http://www.bio.mtu.edu/campbell/482w91a.htm
    Graphical presentation of the steps in affinity chromatography.

http://www.affinity-chrom.com/
    Introduction and applications of affinity chromatography.

http://kerouac.pharm.uky.edu/ASRG/HPLC/hplcmytry.html
    A users' guide to HPLC.

http://www.md.huji.ac.il/spectroscopy/gc.htm
  Principles of chromatography.
  Select GC, HPLC, LC, TLC

http://www.eng.rpi.edu/dept/chem-eng/Biotech-Environ/CHROMO/chromtypes.html
  Descriptions of GC, LC, Ion Exchange, and Affinity Chromatography.

http://ntri.tamuk.edu/fplc/affin.html
  Discussion of affinity chromatography.

http://ntri.tamuk.edu/fplc/fplc1.html
  Introduction to fast performance liquid chromatography. Review Buffer preparation, Definitions of pH, Henderson-Hasselbalch equation, and Buffer calculations.

http://ultranet.com/~jkimball/BiologyPages/A/AffinityChrom.html
  Application of Affinity Chromatography to purification of antibodies. Also a link to Exclusion Chromatography.

CHARACTERIZATION OF PROTEINS AND NUCLEIC ACIDS BY ELECTROPHORESIS

Electrophoresis is an analytical tool by which biochemists can examine the movement of charged molecules in an electric field. Modern electrophoretic techniques use a polymerized gel-like matrix as a support medium. The sample to be analyzed is applied to the medium as a spot or thin band, hence the term "zonal" electrophoresis. The migration of molecules is influenced by the applied electric field; the rigid, mazelike matrix of the gel support; and the size, shape, charge, and chemical composition of the molecules to be separated. Electrophoresis, which is a relatively rapid and convenient technique, is capable of analyzing and purifying several different types of biomolecules, but especially proteins and nucleic acids. Although it is difficult to provide an accurate theoretical description of the electrophoretic movement of molecules in a gel support, zonal electrophoresis has gained widespread use in purifying and identifying biochemicals and in determining their molecular sizes.

A. THEORY OF ELECTROPHORESIS

The movement of a charged molecule subjected to an electric field is represented by Equation 4.1.

>>
$$v = \frac{Eq}{f}$$
Equation 4.1

where

E = the electric field in volts/cm

q = the net charge on the molecule

111

f = frictional coefficient, which depends on the mass and shape of the molecule

v = the velocity of the molecule

The charged particle moves at a velocity that depends directly on the electrical field (E) and charge (q), but inversely on a counteracting force generated by the viscous drag (f). The applied voltage represented by E in Equation 4.1 is usually held constant during electrophoresis, although some experiments are run under conditions of constant current (where the voltage changes with resistance) or constant power (the product of voltage and current). Under constant-voltage conditions, Equation 4.1 shows that the movement of a charged molecule depends only on the ratio q/f. For molecules of similar conformation (for example, a collection of linear DNA fragments or spherical proteins), f varies with size but not shape; therefore, the only remaining variables in Equation 4.1 are the charge (q) and mass dependence of f, meaning that under such conditions molecules migrate in an electric field at a rate proportional to their charge-to-mass ratio.

The movement of a charged particle in an electric field is often defined in terms of mobility, μ, the velocity per unit of electric field (Equation 4.2).

>> $$\mu = \frac{v}{E}$$

Equation 4.2

This equation can be modified using Equation 4.1.

>> $$\mu = \frac{Eq}{Ef} = \frac{q}{f}$$

Equation 4.3

In theory, if the net charge, q, on a molecule is known, it should be possible to measure f and obtain information about the hydrodynamic size and shape of that molecule by investigating its mobility in an electric field. Attempts to define f by electrophoresis have not been successful, primarily because Equation 4.3 does not adequately describe the electrophoretic process. Important factors that are not accounted for in the equation are interaction of migrating molecules with the support medium and shielding of the molecules by buffer ions. This means that electrophoresis is not useful for describing specific details about the shape of a molecule. Instead, it has been applied to the analysis of purity and size of macromolecules. Each molecule in a mixture is expected to have a unique charge and size, and its mobility in an electric field will therefore be unique. This expectation forms the basis for analysis and separation by all electrophoretic methods. The technique is especially useful for the analysis of amino acids, peptides, proteins, nucleotides, nucleic acids, and other charged molecules.

B.   METHODS OF ELECTROPHORESIS

All types of electrophoresis are based on the principles just outlined. The major difference between methods is the type of support medium, which can be either cellulose or thin gels. Cellulose is used as a support medium for low-molecular-weight biochemicals such as amino acids and carbohydrates, and polyacrylamide and agarose gels are widely used as support media for larger molecules. Geometries (vertical and horizontal), buffers, and electrophoretic conditions for these two types of gels provide several different experimental arrangements, as described below.

Polyacrylamide Gel Electrophoresis (PAGE)

Gels formed by polymerization of acrylamide have several positive features in electrophoresis: (1) high resolving power for small and moderately sized proteins and nucleic acids (up to approximately 1×10^6 daltons), (2) acceptance of relatively large sample sizes, (3) minimal interactions of the migrating molecules with the matrix, and (4) physical stability of the matrix. Recall from the earlier discussion of gel filtration (Chapter 3) that gels can be prepared with different pore sizes by changing the concentration of cross-linking agents. Electrophoresis through polyacrylamide gels leads to enhanced resolution of sample components because the separation is based on both molecular sieving and electrophoretic mobility. The order of molecular movement in gel filtration and PAGE is very different, however. In gel filtration (Chapter 3), large molecules migrate through the matrix faster than small molecules. The opposite is the case for gel electrophoresis, where there is no void volume in the matrix, only a continuous network of pores throughout the gel. The electrophoresis gel is comparable to a single bead in gel filtration. Therefore, large molecules do not move easily through the medium, and the rate of movement is small molecules followed by large molecules.

Polyacrylamide gels are prepared by the free radical polymerization of acrylamide and the cross-linking agent N,N'- methylene-bis-acrylamide (Figure 4.1). Chemical polymerization is controlled by an initiator-catalyst system, ammonium persulfate–N,N,N',N'-tetramethylethylenediamine (TEMED). Photochemical polymerization may be initiated by riboflavin in the presence of ultraviolet (UV) radiation. A standard gel for protein separation is 7.5% polyacrylamide. It can be used over the molecular size range of 10,000 to 1,000,000 daltons; however, the best resolution is obtained in the range of 30,000 to 300,000 daltons. The resolving power and molecular size range of a gel depend on the concentrations of acrylamide and bis-acrylamide. Lower concentrations give gels with larger pores, allowing analysis of higher-molecular-weight biomolecules. In contrast, higher concentrations of acrylamide give gels with smaller pores, allowing analysis of lower-molecular-weight biomolecules (Table 4.1).

Figure 4.1

Chemical reactions illustrating the copolymerization of acrylamide and N,N'-methylene-bis-acrylamide. See text for details.

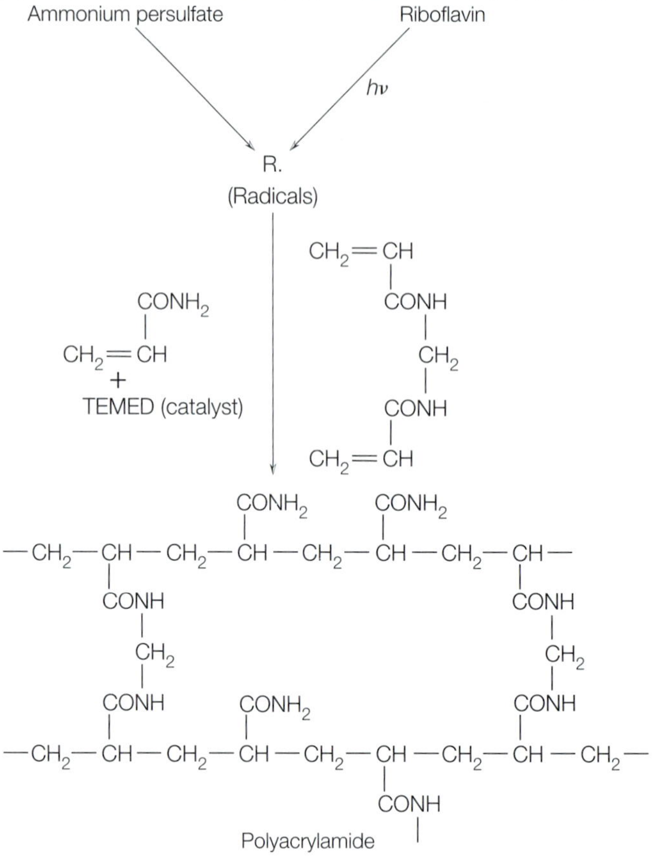

Table 4.1

Effective Range of Separation of DNA by PAGE

Acrylamide[1] (% w/v)	Range of Separation (bp)	Bromophenol Blue[2]	Xylene Cyanol[2]
3.5	1000–2000	100	450
5.0	80–500	65	250
8.0	60–400	50	150
12.0	40–200	20	75
20.0	5–100	10	50

[1] Ratio of acrylamide to bis-acrylamide, 20: 1.

[2] The numbers (in bp) represent the size of DNA fragment with the same mobility as the dye.

Polyacrylamide electrophoresis can be done using either of two arrangements, column or slab. Figure 4.2 shows the typical arrangement for a column gel. Glass tubes (10 cm $\times$ 6 mm i.d.) are filled with a mixture of acrylamide, N,N'-methylene-bis-acrylamide, buffer, and free radical initiator-catalyst. Polymerization occurs in 30 to 40 minutes. The gel column is inserted between two separate buffer reservoirs. The upper reservoir usually contains the cathode and the lower the anode. Gel electrophoresis is usually carried out at basic pH, where most biological polymers are anionic; hence, they move down toward the anode. The sample to be analyzed is layered on top of the gel and voltage is applied to the system. A "tracking dye" is also applied, which moves more rapidly through the gel than the sample components. When the dye band has moved to the opposite end of the column, the voltage is turned off and the gel is removed from the column and stained with a dye. Chambers for column gel electrophoresis are commercially available or can be constructed from inexpensive materials.

Slab gels are now more widely used than column gels. A slab gel on which several samples may be analyzed is more convenient to make and use than several individual column gels. Slab gels also offer the advantage that all samples are analyzed in a matrix environment that is identical in composition. A typical vertical slab gel apparatus is shown in Figure 4.3. The polyacrylamide slab is prepared between two glass plates that are separated by

Figure 4.2

A column gel for polyacrylamide electrophoresis.

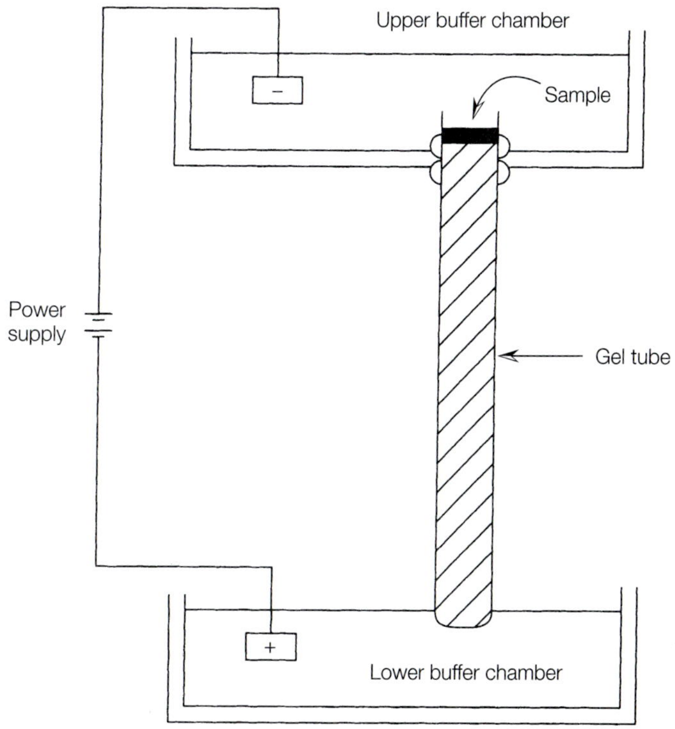

Figure 4.3

Figure 4.3

A vertical electrophoresis apparatus for a slab gel. *Courtesy of Hoefer Pharmacia Biotech, Inc., San Francisco.*

spacers (Figure 4.4). The spacers allow a uniform slab thickness of 0.5 to 2.0 mm, which is appropriate for analytical procedures. Slab gels are usually 8×10 cm or 10×10 cm, but for nucleotide sequencing, slab gels as large as 20×40 cm are often required.

A plastic "comb" inserted into the top of the slab gel during polymerization forms indentations in the gel that serve as sample wells. Up to 20 sample wells may be formed. After polymerization, the comb is carefully removed and the wells are rinsed thoroughly with buffer to remove salts and any unpolymerized acrylamide. The gel plate is clamped into place between two buffer reservoirs, a sample is loaded into each well, and voltage is applied. For visualization, the slab is removed and stained with an appropriate dye.

Perhaps the most difficult and inconvenient aspect of polyacrylamide gel electrophoresis is the preparation of gels. The monomer, acrylamide, is a neurotoxin and a cancer suspect agent; hence, special handling is required. Other necessary reagents including catalysts and initiators also require special handling and are unstable. In addition, it is difficult to make gels that have reproducible thicknesses and compositions. Many researchers are now

Figure 4.4

Arrangement of two glass plates with spacers to form a slab gel. The comb is used to prepare wells for placement of samples.

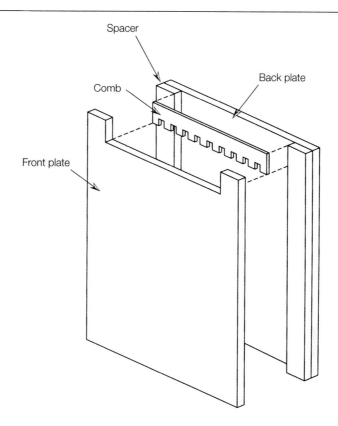

turning to the use of precast polyacrylamide gels. Several manufacturers now offer gels precast in glass or plastic cassettes. Gels for all experimental operations are available including single percentage (between 3 and 27%) or gradient gel concentrations and a variety of sample well configurations and buffer chemistries. More details on precast gels will be given in Section C, Practical Aspects of Electrophoresis.

Several modifications of PAGE have greatly increased its versatility and usefulness as an analytical tool.

Discontinuous Gel Electrophoresis

The experimental arrangement for "disc" gel electrophoresis is shown in Figure 4.5. Three significant characteristics of this method are that (1) there are two gel layers, a lower or **resolving gel** and an upper or **stacking gel;** (2) the buffers used to prepare the two gel layers are of different ionic strengths and pH; and (3) the stacking gel has a lower acrylamide concentration, so its pore sizes are larger. These three changes in the experimental conditions cause the formation of highly concentrated bands of sample in the stacking gel and greater resolution of the sample components in the lower gel.

Figure 4.5

The process of disc gel electrophoresis. **A** Before electrophoresis. **B** Movement of chloride, glycinate, and protein through the stacking gel. **C** Separation of protein samples by the resolving gel.

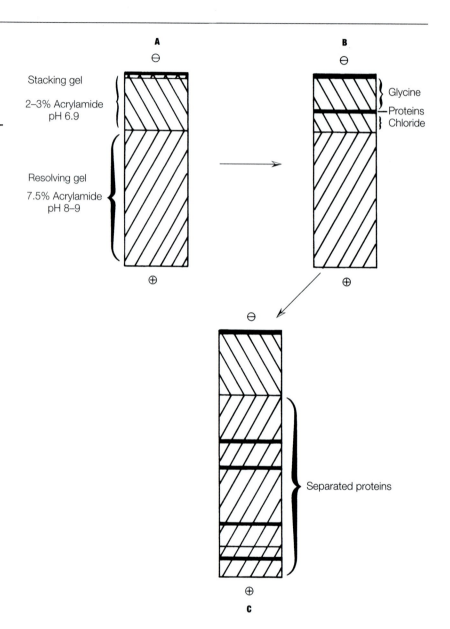

Sample concentration in the upper gel occurs in the following manner. The sample is usually dissolved in glycine-chloride buffer, pH 8 to 9, before loading on the gel. Glycine exists primarily in two forms at this pH, a zwitterion and an anion (Equation 4.4).

$$H_3\overset{+}{N}CH_2COO^- \rightleftharpoons H_2NCH_2COO^- + H^+$$

Equation 4.4

The average charge on glycine anions at pH 8.5 is about -0.2. When the voltage is turned on, buffer ions (glycinate and chloride) and protein or nucleic acid sample move into the stacking gel, which has a pH of 6.9. Upon entry into the upper gel, the equilibrium of Equation 4.4 shifts toward the left, increasing the concentration of glycine zwitterion, which has no net charge and hence no electrophoretic mobility. In order to maintain a constant current in the electrophoresis system, a flow of anions must be maintained. Since most proteins and nucleic acid samples are still anionic at pH 6.9, they replace glycinate as mobile ions. Therefore, the relative ion mobilities in the stacking gel are chloride > protein or nucleic acid sample > glycinate. The sample will tend to accumulate and form a thin, concentrated band sandwiched between the chloride and glycinate as they move through the upper gel. Since the acrylamide concentration in the stacking gel is low (2 to 3%), there is little impediment to the mobility of the large sample molecules.

Now, when the ionic front reaches the lower gel with pH 8 to 9 buffer, the glycinate concentration increases and anionic glycine and chloride carry most of the current. The protein or nucleic acid sample molecules, now in a narrow band, encounter both an increase in pH and a decrease in pore size. The increase in pH would, of course, tend to increase electrophoretic mobility, but the smaller pores decrease mobility. The relative rate of movement of anions in the lower gel is chloride > glycinate > protein or nucleic acid sample. The separation of sample components in the resolving gel occurs as described in an earlier section on gel electrophoresis. Each component has a unique charge/mass ratio and a discrete size and shape, which directly influence its mobility.

Disc gel electrophoresis yields excellent resolution and is the method of choice for analysis of proteins and nucleic acid fragments. Protein or nucleic acid bands containing as little as 1 or 2 μg can be detected by staining the gels after electrophoresis.

Sodium Dodecyl Sulfate–Polyacrylamide Gel Electrophoresis (SDS-PAGE)

The electrophoretic techniques previously discussed are not applicable to the measurement of the molecular weights of biological molecules because mobility is influenced by both charge and size. If protein samples are treated so that they have a uniform charge, electrophoretic mobility then depends primarily on size (see Equation 4.3). The molecular weights of proteins may be estimated if they are subjected to electrophoresis in the presence of a detergent, sodium dodecyl sulfate (SDS), and a disulfide bond reducing agent, mercaptoethanol. This method is often called "denaturing electrophoresis."

When protein molecules are treated with SDS, the detergent disrupts the secondary, tertiary, and quaternary structure to produce linear polypeptide chains coated with negatively charged SDS molecules. The presence of mercaptoethanol assists in protein denaturation by reducing all disulfide bonds. The detergent binds to hydrophobic regions of the denatured protein chain in a constant ratio of about 1.4 g of SDS per gram of protein. The bound detergent molecules carrying negative charges mask the native charge of the protein. In essence, polypeptide chains of a constant charge/mass ratio and uniform shape are produced. The electrophoretic mobility of the SDS-protein complexes is influenced primarily by molecular size: the larger molecules are retarded by the molecular sieving effect of the gel, and the smaller molecules have greater mobility. Empirical measurements have shown a linear relationship between the log molecular weight and the electrophoretic mobility (Figure 4.6).

In practice, a protein of unknown molecular weight and subunit structure is treated with 1% SDS and 0.1 M mercaptoethanol in electrophoresis buffer. A standard mixture of proteins with known molecular weights must also be subjected to electrophoresis under the same conditions. Two sets of standards are commercially available, one for low-molecular-weight proteins (molecular weight range 14,000 to 100,000) and one for high-molecular-weight proteins (45,000 to 200,000). Figure 4.7 shows a stained gel after electrophoresis of a standard protein mixture. After electrophoresis and dye staining, mobilities are measured and molecular weights determined graphically.

SDS-PAGE is valuable for estimating the molecular weight of protein subunits. This modification of gel electrophoresis finds its greatest use in characterizing the sizes and different types of subunits in oligomeric pro-

---

Figure 4.6

Graph illustrating the linear relationship between electrophoretic mobility of a protein and the log of its molecular weight. Thirty-seven different polypeptide chains with a molecular weight range of 11,000 to 70,000 are shown. Gels were run in the presence of SDS (denaturing conditions). *From K. Weber and M. Osborn, J. Biol. Chem. **244**, 4406 (1969). By permission of the copyright owner, the American Society for Biochemistry and Molecular Biology, Inc.*

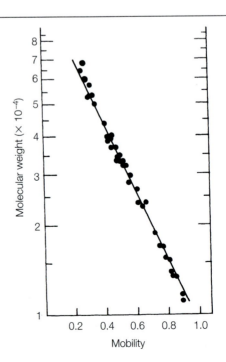

Figure 4.7

A silver-stained gel obtained by electrophoresis of a standard protein mixture under denaturing conditions. Samples were run on 12% polyacrylamide gels, 0.75 mm. Lane 1: Bovine brain homogenate soluble fraction, 20 μL. Lane 2: Bovine brain homogenate soluble fraction, 10 μL. Lanes 3, 4, 5: Bio-Rad SDS-PAGE low-molecular-weight standards, three different dilutions. *Courtesy of Bio-Rad Laboratories, Richmond, CA.*

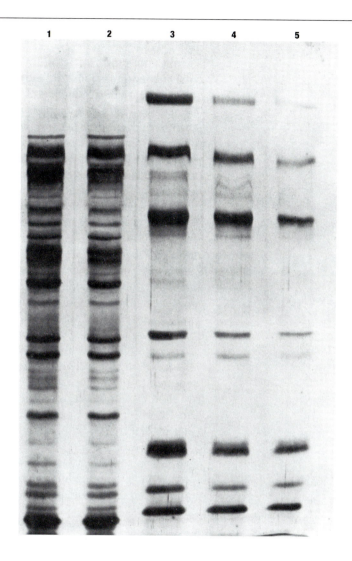

teins. SDS-PAGE is limited to a molecular weight range of 10,000 to 200,000. Gels of less than 2.5% acrylamide must be used for determining molecular weights above 200,000, but these gels do not set well and are very fragile because of minimal cross-linking. A modification using gels of agarose-acrylamide mixtures allows the measurement of molecular weights above 200,000.

Nucleic Acid Sequencing Gels

The amino acid sequence of a protein is determined by identifying amino acid residues as they are sequentially cleaved from the intact protein (see Experiment 2). Sequence analysis of nucleic acids is based on the generation of sets of DNA or RNA fragments with common ends and the separation of these oligonucleotide fragments by polyacrylamide electrophoresis. Two methods have been developed for sequencing nucleic acids: (1) the partial chemical degradation method of Maxam and Gilbert, which uses four specific chemical reactions to modify bases and cleave phosphodiester bonds, and (2) the chain termination method developed by Sanger, which requires a single-stranded DNA template and chain extension processes, followed by chain termination caused by the presence of dideoxynucleoside triphosphates. Both sequencing methods result in nested sets of DNA or RNA fragments that have one common end and chains varying in length. The smallest possible size difference of nucleic acid fragments is one nucleotide. Separation of the nucleic acid fragments by polyacrylamide electrophoresis allows one to "read" the sequence of nucleotides from the gel.

The experimental arrangement is the same as that previously described for PAGE; however, the gel is prepared with many sample wells to accommodate a large number of samples. Sequence gels of 6, 8, 12, and 20% polyacrylamide are routinely used. Gels of 20% may be used to sequence the first 50 to 100 nucleotides of a nucleic acid, and lower percentage gels allow sequencing out to 250 nucleotides. Sequencing gels are large (up to 40 cm) and power supplies must provide more power than for conventional methods. Precast sequencing gels are now commercially supplied by Stratagene. They have a gel concentration of 5.5%, have 32 sample wells, and will sequence up to 500 nucleotides. Denaturants such as urea and formamide are required to prevent renaturing of the nucleic acid fragments during electrophoresis. For detection, nucleic acid chains for sequencing must be end labeled with ^{32}P, ^{35}S or a fluorescent tag. ^{32}P and ^{35}S-labeled nucleic acids on gels are detected by autoradiography (see later). Nucleic acids end labeled with fluorescent molecules are detected by fluorimeter scanning of the gels. Many researchers working on the large and expensive human genome project[1] are generating huge amounts of DNA sequence data. Much of this information is stored in computer data banks for use by researchers around the world.

Agarose Gel Electrophoresis

The electrophoretic techniques discussed up to this point are useful for analyzing proteins and small fragments of nucleic acids up to 350,000 daltons (500 bp) in molecular size; however, the small pore sizes in the gel are not appropriate for analysis of large nucleic acid fragments or intact DNA molecules. The standard method used to characterize RNA and DNA in the range 200 to 50,000 base pairs (50 kilobases) is electrophoresis with agarose as the support medium.

Agarose, a product extracted from seaweed, is a linear polymer of galactopyranose derivatives. Gels are prepared by dissolving agarose in warm elec-

---

[1] *The human genome project is a federal government-sponsored program to sequence all DNA in human chromosomes.*

trophoresis buffer. After cooling the gel mixture to 50°C, the agarose solution is poured between glass plates as described for polyacrylamide. Gels with less than 0.5% agarose are rather fragile and must be used in a horizontal arrangement (Figure 4.8). The sample to be separated is placed in a sample well made with a comb, and voltage is applied until separation is complete. Precast agarose gels of all shapes, sizes, and percent composition are commercially available.

Nucleic acids can be visualized on the slab gel after separation by soaking in a solution of ethidium bromide, a dye that displays enhanced fluorescence when intercalated between stacked nucleic acid bases. Ethidium bromide may be added directly to the agarose solution before gel formation. This method allows monitoring of nucleic acids during electrophoresis. Irradiation of ethidium bromide–treated gels by UV light results in orange-red bands where nucleic acids are present.

The mobility of nucleic acids in agarose gels is influenced by the agarose concentration and the molecular size and molecular conformation of the nucleic acid. Agarose concentrations of 0.3 to 2.0% are most effective for nucleic acid separation (Table 4.2). Experiments 14 and 15 illustrate the separation of DNA fragments on agarose gels. Like proteins, nucleic acids migrate at a rate that is inversely proportional to the logarithm of their molecular weights; hence, molecular weights can be estimated from electrophoresis results using standard nucleic acids or DNA fragments of known molecular weight. The DNA conformations most frequently encountered are superhelical circular (form I), nicked circular (form II), and linear (form III). The small, compact, supercoiled form I molecules usually have the greatest mobility, followed by the rodlike, linear form III molecules. The extended, circular form II molecules migrate more slowly. The relative electrophoretic mobility of the three forms of DNA, however, depends on experimental conditions such as agarose concentration and ionic strength.

Figure 4.8

An apparatus for horizontal slab gel electrophoresis. *Courtesy of Hoefer Pharmacia Biotech Inc., San Francisco.*

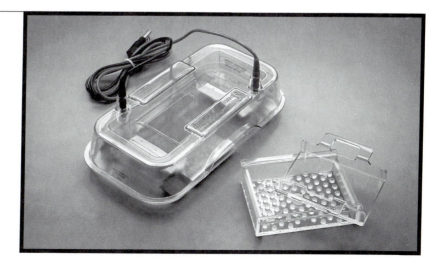

Table 4.2

Effective Range of Separation of DNA by Agarose

Agarose (% w/v)	Effective Range (kb)
0.3	5–50
0.5	2–25
0.7	0.8–10
1.2	0.4–5
1.5	0.2–3
2.0	0.1–2

The versatility of agarose gels is obvious when one reviews their many applications in nucleic acid analysis. The rapid advances in our understanding of nucleic acid structure and function in recent years are due primarily to the development of agarose gel electrophoresis as an analytical tool. Two of the many applications of agarose gel electrophoresis will be described here.

Analysis of DNA Fragments after Digestion by Restriction Endonucleases

As described in Experiment 15, restriction endonucleases recognize a specific base sequence in double-stranded DNA and catalyze cleavage (hydrolysis of phosphodiester bonds) in or near that specific region. Many viral, bacterial, or animal DNA molecules are substrates for the enzymes. When each type of DNA is treated with a restriction endonuclease, a specific number of DNA fragments is produced. The base sequence recognized by the enzyme occurs only a few times in any particular DNA molecule; therefore, the smaller the DNA molecule, the fewer specific cleavage sites there are. Viral or phage DNA, for example, is cleaved into about 50 fragments depending on the enzyme used, whereas larger bacterial or animal DNA may be cleaved into hundreds or thousands of fragments. Smaller DNA molecules, upon cleavage with a particular enzyme, will produce a limited set of fragments. It is unlikely that this set of fragments will be the same for any two different DNA molecules, so the fragmentation pattern can be considered a "fingerprint" of the DNA substrate. **The restriction pattern** is produced by electrophoresis of the cleavage reaction mixture through agarose gels, followed by staining with ethidium bromide (Figure 4.9). The separation of the fragments is based on molecular size, with large fragments remaining near the origin and smaller fragments migrating farther down the gel. In addition to characterization of DNA structure, endonuclease digestion coupled with agarose gel electrophoresis is a valuable tool for plasmid mapping (Experiment 15) and DNA recombination experiments.

Characterization of Superhelical Structure of DNA

The structure of plasmid, viral, and bacterial DNA is often closed circular with negative superhelical turns. It is possible under various experimental

Figure 4.9

Restriction patterns produced by agarose electrophoresis of DNA fragments after restriction endonuclease action. *Courtesy of Bio-Rad Laboratories, Richmond, CA.*

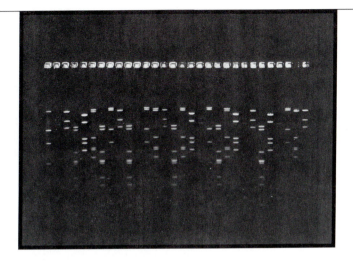

conditions to induce reversible changes in the conformation of DNA. The intercalating dye ethidium bromide causes an unwinding of supercoiled DNA that affects its electrophoretic mobility. Electrophoresis of DNA on agarose in the presence of increasing concentrations of ethidium bromide provides an unambiguous method for distinguishing between closed circular and other DNA conformations.

Closed circular, negatively supercoiled DNA (form I) usually has the greatest electrophoretic mobility of all DNA forms because supercoiled DNA molecules tend to be compact. If ethidium bromide is added to form I DNA, the dye intercalates between the stacked DNA bases, causing unwinding of some of the negative supercoils. As the concentration of ethidium bromide is increased, more and more of the negative supercoils are removed until no more are present in the DNA. The conformational change of the DNA supercoil can be monitored by electrophoresis because the mobility decreases with each unwinding step. With increasing concentration of ethidium bromide, the negative supercoils are progressively unwound and the electrophoretic mobility decreases to a minimum. This minimum represents the free dye concentration necessary to remove all negative supercoils. (The free dye concentration at this minimum has been shown to be related to the superhelix density, which is a measure of the extent of supercoiling in a DNA molecule.) The circular DNA at this point is equivalent to the "relaxed" form. If more ethidium bromide is added to the relaxed DNA, positive superhelical turns are induced in the structure and the electrophoretic mobility increases. Forms II and III DNA, under the same conditions of increasing ethidium bromide concentration, show a gradual decrease in electrophoretic mobility throughout the entire concentration range.

Agarose gel electrophoresis is able to resolve topoisomers of native, covalently closed, circular DNA that differ only in their degree of supercoiling. This technique has proved useful in the analysis and characterization of enzymes that catalyze changes in the conformation or topology of native DNA. These enzymes, called topoisomerases, have been isolated from bacterial and mammalian cells. They change DNA conformations by catalyzing nicking and closing of phosphodiester bonds in circular duplex DNA. Agarose gel electrophoresis is an ideal method for identifying and assaying topoisomerases because the intermediate DNA molecules can be resolved on the basis of the extent of supercoiling. Topoisomerases may be assayed by incubating native DNA with an enzyme preparation, removing aliquots after various periods of time, and subjecting them to electrophoresis on an agarose gel with standard supercoiled and relaxed DNA.

Pulsed Field Gel Electrophoresis (PFGE)

Conventional agarose gel electrophoresis is limited in use for the separation of nucleic acid fragments smaller than 50,000 bp (50 kb). In practice, that limit is closer to 20,000 to 30,000 bp if high resolution is desired. Since chromosomal DNA from most organisms contains thousands and even millions of base pairs, the DNA must be cleaved by restriction enzymes before analysis by standard electrophoresis. In the early 1980s it was discovered by Schwartz and Cantor at Columbia University that large molecules of DNA (yeast chromosomes, 200–3000 kb) could be separated by **pulsed field gel electrophoresis (PFGE).** There is one major distinction between standard gel electrophoresis and PFGE. In PFGE, the electric field is not constant as in the standard method but is changed repeatedly (pulsed) in direction and strength during the separation (Figure 4.10). The physical mechanism for

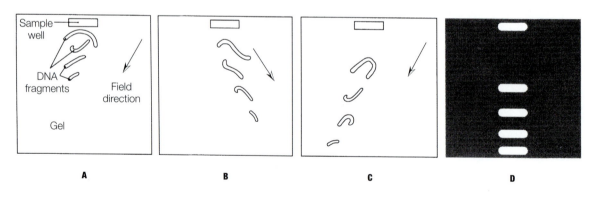

A   B   C   D

Figure 4.10

    Pulsed field gel electrophoresis. Generalized PFGE separation of four DNA fragments of different sizes in one lane. **A** DNA molecules of various shapes and configurations move toward the positive field. **B** The new field orientation pulls the DNA in a different direction, realigning the molecules. **C** The field returns to the original configuration. **D** The bands show the final position of a large collection of the molecules. *Reprinted with permission from the* Journal of NIH Research *1, 115 (Nov.–Dec. 1989). Illustration by Terese Winslow.*

separation of the large DNA molecules as they move through the gel under these conditions is not yet well understood. An early explanation was that the electrical pulses abruptly perturbed the conformation of the DNA molecules. They would be oriented by the influence of the electric field coming from one direction and then reoriented as a new electric field at a different angle to the first was turned on. According to this explanation, it takes longer for larger molecules to reorient, so smaller fragments respond faster to the new pulse and move faster. More recent experiments on dyed DNA moving in gels have shown that conformational changes of the DNA are not abrupt but more gradual in response to the electrical pulse, and DNA molecules tend to "slither" through the gel matrix. In addition, it has been discovered that the gel becomes more fluid during electrical pulsing.

Even though our theoretical understanding of PFGE is lacking, practical applications and experimental advances are expanding rapidly. The availability of PFGE has sparked changes in DNA research. New methods for isolating intact DNA molecules have been developed. Because of mechanical breakage, the average size of DNA isolated from cells in the presence of lysozyme, detergent, and EDTA (Experiment 13) is about 400–500 kb. Intact chromosomal DNA can be isolated by embedding cells in an agarose matrix and disrupting the cells with detergents and enzymes. Slices or "plugs" of the agarose with intact DNA are then placed on the gel for PFGE analysis. Newly discovered restriction endonucleases that cut DNA only rarely can now be used to subdivide chromosome-sized DNA. Two important endonucleases with eight-base recognition sites are *Not* I and *Sfi* I.

There are also many instrumental advances that allow changes in the experimental design of PFGE. Some of the variables that can be changed for each experiment are voltage, pulse length, number of electrodes, relative angle of electrodes, gel box design, temperature, agarose concentration, buffer pH, and time of electrophoresis.

Like all laboratory techniques, PFGE has its disadvantages and problems. Long periods of electrophoresis are often required for good resolution, and migration of fragments is extremely dependent on experimental conditions. Therefore it is difficult to compare gels even when they are run under similar conditions. In spite of these shortcomings, PFGE will continue to advance as a significant tool for the characterization of very large molecules. The technique, which is being widely used in the human genome project, will greatly increase our understanding of chromosome structure and function.

Isoelectric Focusing of Proteins

Another important and effective use of electrophoresis for the analysis of proteins is **isoelectric focusing (IEF),** which examines electrophoretic mobility as a function of pH. The net charge on a protein is pH dependent. Proteins below their isoelectric pH (pH_I, or the pH at which they have zero net charge) are positively charged and migrate in a medium of fixed pH toward the negatively charged cathode. At a pH above its isoelectric point, a protein is deprotonated and negatively charged and migrates toward the anode. If the pH of the electrophoretic medium is identical to the pH_I of a protein, the protein has a net charge of zero and does not migrate toward either electrode. Theoretically, it should be possible to separate protein molecules and to estimate the pH_I of a protein by investigating the electrophoretic mobility in a series of separate experiments in which the pH of the medium is changed. The pH at which there is no protein migration should coincide with the pH_I of the protein. Because such a repetitive series of electrophoresis runs is a rather tedious and time-consuming way to determine the pH_I, IEF has evolved as an alternative method for performing a single electrophoresis run in a medium of gradually changing pH (i.e., a pH gradient).

Figure 4.11 illustrates the construction and operation of an IEF pH gradient. An acid, usually phosphoric, is placed at the cathode; a base, such as triethanolamine, is placed at the anode. Between the electrodes is a medium in which the pH gradually increases from 2 to 10. The pH gradient can be formed before electrophoresis is conducted or formed during the course of electrophoresis. The pH gradient can be either broad (pH 2–10) for separating several proteins of widely ranging pH_I values or narrow (pH 7–8) for precise determination of the pH_I of a single protein. P in Figure 4.11 represents different molecules of the same protein in two different regions of the pH gradient. Assuming that the pH in region 1 is less than the pH_I of the protein and the pH in region 2 is greater than the pH_I of the protein, molecules of P in region 1 will be positively charged and will migrate in an applied electric field toward the cathode. As P migrates, it will encounter an increasing pH, which will influence its net charge. As it migrates *up* the pH gradient, P will become increasingly deprotonated and its net charge will decrease toward zero. When P reaches a region where its net charge is zero (region 3), it will stop migrating. The pH in this region of the electrophoretic medium will coincide with the pH_I of the protein and can be measured with

Figure 4.11

Illustration of isoelectric focusing. See text for details.

- -

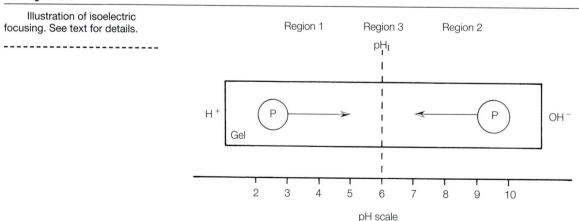

a surface microelectrode, or the position of the protein can be compared to that of a calibration set of proteins of known pH_I values. P molecules in region 2 will be negatively charged and will migrate toward the anode. In this case, the net charge on P molecules will gradually decrease to zero as P moves *down* the pH gradient, and P molecules originally in region 2 will approach region 3 and come to rest. The P molecules move in opposite directions, but the final outcome of IEF is that P molecules located anywhere in the gradient will migrate toward the region corresponding to their isoelectric point and will eventually come to rest in a sharp band; that is, they will "focus" at a point corresponding to their pH_I.

Since different protein molecules in mixtures have different pH_I values, it is possible to use IEF to separate proteins. In addition, the pH_I of each protein in the mixture can be determined by measuring the pH of the region where the protein is focused.

The pH gradient is prepared in a horizontal glass tube or slab. Special precautions must be taken so that the pH gradient remains stable and is not disrupted by diffusion or convective mixing during the electrophoresis experiment. The most common stabilizing technique is to form the gradient in a polyacrylamide, agarose, or dextran gel. The pH gradient is formed in the gel by electrophoresis of synthetic polyelectrolytes, called **ampholytes,** which migrate to the region of their pH_I values just as proteins do and establish a pH gradient that is stable for the duration of the IEF run. Ampholytes are low-molecular-weight polymers that have a wide range of isoelectric points because of their numerous amino and carboxyl or sulfonic acid groups. The polymer mixtures are available in specific pH ranges (pH 5–7, 6–8, 3.5–10, etc.). It is critical to select the appropriate pH range for the ampholyte so that the proteins to be studied have pH_I values in that range. The best resolution is, of course, achieved with an ampholyte mixture over a small pH range (about two units) encompassing the pH_I of the sample proteins. If the pH_I values for the proteins under study are unknown, an ampholyte of wide pH range (pH 3–10) should be used first and then a narrower pH range selected for use.

The gel medium is prepared as previously described except that the appropriate ampholyte is mixed prior to polymerization. The gel mixture is poured into the desired form (column tubes, horizontal slabs, etc.) and allowed to set. Immediately after casting of the gel, the pH is constant throughout the medium, but application of voltage will induce migration of ampholyte molecules to form the pH gradient. The standard gel for proteins with molecular sizes up to 100,000 daltons is 7.5% polyacrylamide; however, if larger proteins are of interest, gels with larger pore sizes must be prepared. Such gels can be prepared with a lower concentration of acrylamide (about 2%) and 0.5 to 1% agarose to add strength. Precast gels for isoelectric focusing are also commercially available.

The protein sample can be loaded on the gel in either of two ways. A concentrated, salt-free sample can be layered on top of the gel as previously described for ordinary gel electrophoresis. Alternatively, the protein can be added directly to the gel preparation, resulting in an even distribution of protein throughout the medium. The protein molecules move more slowly than the low-molecular-weight ampholyte molecules, so the pH gradient is established before significant migration of the proteins occurs. Very small protein samples can be separated by IEF. For analytical purposes, 10 to 50 μg is a typical sample size. Larger sample sizes (up to 20 mg) can be used for preparative purposes.

Two-Dimensional Electrophoresis of Proteins

The separation of proteins by IEF is based on charge, whereas SDS-PAGE separates molecules based on molecular size. A combination of the two methods leads to enhanced resolution of complex protein mixtures. Such an experiment was first reported by O'Farrell (1975), and the combined method has since become a routine and powerful separatory technique. Figure 4.12 shows the results of O'Farrell's analysis of total *Escherichia coli* protein. The sample was first separated in one dimension by IEF. The sample gel was then transferred to an SDS-PAGE slab and electrophoresis was continued in the second dimension. At least 1000 discrete protein spots are visible. This technique is becoming increasingly valuable in developmental biochemistry, where the increase or decrease in intensity of a spot representing a specific protein can be monitored as a function of cell growth. In addition, two-dimensional electrophoresis is a standard method for judging protein purity.

Capillary Electrophoresis (CE)

Capillary electrophoresis is a new technique that combines the high resolving power of electrophoresis with the speed, versatility, and automation of

Figure 4.12

SDS–isoelectric focusing gel electrophoresis of total *E. coli* protein. *Photo courtesy of Dr. P. O'Farrell.*

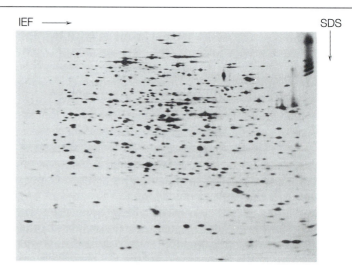

high-performance liquid chromatography (HPLC). It offers the ability to analyze very small samples (5–10 nL) utilizing up to 1 million theoretical plates to achieve high resolution and sensitivity to the attomole level (10^{-18} mole). It will become a widely used technique in the analysis of amino acids (see Experiment 2), peptides, proteins, nucleic acids, and pharmaceuticals.

A general experimental design is diagrammed in Figure 4.13. The equipment consists of a power supply, two buffer reservoirs, a buffer-filled capillary tube, and an on-line detector. Platinum electrodes connected to the power supply are immersed in each buffer reservoir. A high voltage is applied along the capillary and a small plug of sample solution is injected into one end of the capillary. Components in the solution migrate along the length of the capillary under the influence of the electric field. Molecules are detected as they exit from the opposite end of the capillary. The detection method used depends on the type of molecules separated, but the most common are UV-VIS fixed-wavelength detectors and diode-array detectors (see Chapter 5). The capillaries used are flexible, fused, silica tubes of 50–100 μm i.d. and 25–100 cm length that may or may not be filled with chromatographic matrix.

A major advantage of capillary electrophoresis is that many analytical experimental designs are possible, just as in the case of HPLC. In HPLC, a wide range of molecules can be separated by changing the column support (see Chapter 3). In CE, the capillary tube may be coated or filled with a variety of materials. For separation of small, charged molecules, bare silica or polyimide-coated capillaries are often used. If separation by molecular sieving is desired, the tube is filled with polyacrylamide or SDS-polyacrylamide. If the capillary is filled with electrolyte and an ampholyte pH gradient, isoelectric focusing experiments on proteins may be done. We can expect to see numerous applications of CE in all aspects of biochemistry and molecular biology. New applications will include DNA sequencing, analysis of single cells, and separations of neutral molecules.

Figure 4.13

Experimental setup for capillary electrophoresis. *Courtesy of Bio-Rad Laboratories, Life Science Group, Hercules, CA.*

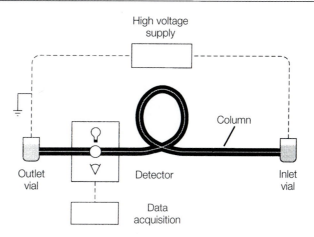

Immunoelectrophoresis (IE)

In immunoelectrophoresis two sequential procedures are applied to the analysis of complex protein mixtures: (1) separation of the protein mixture by agarose gel electrophoresis, followed by (2) interaction with specific antibodies to examine the antigenic properties of the separated proteins.

The technique of IE was first reported by Grabar and Williams in 1953 for the separation and immunoanalysis of serum proteins, but it can be applied to the analysis of any purified protein or complex mixture of proteins. In practice (Figure 4.14), a protein mixture is separated by standard electrophoresis in an agarose gel prepared on a small glass plate. This is followed by exposing the separated proteins to a specific antibody preparation. The antibody is added to a trough cut into the gel, as shown in Figure 4.14, and is allowed to diffuse through the gel toward the separated proteins. If the antibody has a specific affinity for one of the proteins, a visible precipitin arc forms. This is an insoluble complex formed at the boundary of antibody and antigen protein. The technique is most useful for the analysis of protein purity, composition, and antigenic properties. The basic IE technique described here allows only qualitative examination of antigenic proteins. If quantitative results in the form of protein antigen concentration are required, the advanced modifications, rocket immunoelectrophoresis and two-dimensional (crossed) immunoelectrophoresis, may be used. The basic principles of IE are applied in the technique of Western blotting in Experiment 7.

Figure 4.14

Immunoelectrophoresis. **A** Antigen is placed in sample wells. **B** Electrophoresis. **C** Antiserum containing antibody is placed in trough. **D** Insoluble antigen-antibody complexes form precipitin arcs.

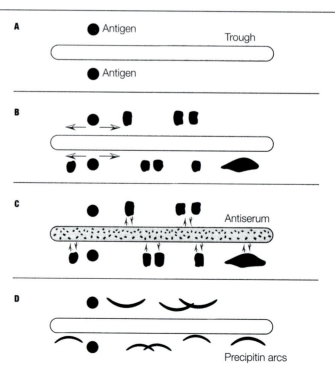

C.   PRACTICAL ASPECTS OF ELECTROPHORESIS

Instrumentation

The basic components required for electrophoresis are a power supply and an electrophoresis chamber (gel box). A power supply that provides a constant current is suitable for most conventional electrophoresis experiments. Power supplies that generate both constant voltage (up to 4 kV) and constant current (up to 200 mA) are commercially available. In order to implement the many modifications of electrophoresis, such versatile power supplies are essential in a research laboratory. If isoelectric focusing experiments are planned, a power supply that furnishes a constant voltage is necessary. DNA sequencing experiments require power supplies capable of generating 2–4 kV. Because of the high power requirements of these experiments, the glass plates sandwiching the gel must be covered with conductive aluminum plates to dissipate heat and prevent gel melting and glass plate breakage. CE also requires the use of high voltage, which generates heat as well. However, the heat is quickly dissipated through the thin walls of the capillary tubing. Pulsed field electrophoresis experiments have special requirements, including a high-voltage power supply, electrical switching devices for control of the field, equipment for temperature control, and a specially designed gel box with an array of electrodes. All types of electrophoresis chambers for both horizontal and vertical placements of gels are available from commercial suppliers.

Reagents

High-quality, electrophoresis grade chemicals must be used, since impurities may influence both the gel polymerization process and electrophoretic mobility. The reagents used for gel formation should be stored in a refrigerator.

▶ **CAUTION**

Acrylamide, N,N,N',N'-tetramethylethylenediamine, N,N'-methylene-bis-acrylamide, and ammonium persulfate are toxic and must be used with care. Acrylamide is a neurotoxin, a cancer suspect agent, and a potent skin irritant, so gloves and a mask must be worn while handling it in the unpolymerized form.

Buffers appropriate for electrophoresis gels include Tris-glycine, Tris-acetate, Tris-phosphate, and Tris-borate at concentrations of about 0.05 M.

Several advances in gel electrophoresis have recently been made that have streamlined and improved many of the established techniques. Some of those improvements will be outlined here.

1. Precast Gels   As previously mentioned, precast gels of polyacrylamide and agarose are now commercially available. A wide variety of gel sizes, types, configurations, and compositions may be purchased. Costs are reasonable, beginning at about \$6 for an 8 × 10 cm single percentage poly-acylamide gel. Precast gels not only offer convenience but also increase safety, save time, and provide more reproducible electrophoretic runs.

2. Bufferless Precast Gels   One of the major inconveniences of running gels is the necessity of having liquid buffer reservoirs to saturate gels during electrophoresis. There is always the chance of leaking reservoirs and spilling solutions perhaps containing acrylamide or ethidium bromide. Precast gels (agarose and acrylamide) are now available in dry, plastic cassettes. They do not require liquid buffers because the gels contain ion-exchange matrices, which sustain the electric field.

3. Reusable Precast Gels   Precast agarose gels are now available that may be recycled. After a run, the DNA samples on the gel are removed by reversing the direction of the electric field. The gels are then reloaded with new samples and reused.

Staining and Detecting Electrophoresis Bands

During the electrophoretic process, it is important to know when to stop applying the voltage or current. If the process is run for too long, the desired components may pass entirely through the medium and into the buffer; if too short a period is used, the components may not be completely resolved. It is common practice to add a "tracking dye," usually bromophenol blue and/or xylene cyanol, to the sample mixture. These dyes, which are small and anionic, move rapidly through the gel ahead of most proteins or nucleic acids. After electrophoresis, bands have a tendency to widen by diffusion. Because this broadening may decrease resolution, gels should be analyzed as soon as possible after the power supply has been turned off. In the case of proteins, gels can be treated with an agent that "fixes" the proteins in their final positions, a process that is often combined with staining.

Reagent dyes that are suitable for visualization of biomolecules after electrophoresis were discussed earlier. The most commonly used stain for proteins is **Coomassie Brilliant Blue.** This dye can be used as a 0.25% aqueous solution. It is followed by destaining (removing excess background dye) by repeated washing of the paper or gel with 7% acetic acid. Alternatively, gels may be stained by soaking in 0.25% dye in H_2O, methanol, acetic acid (5:5:1) followed by repeated washings with the same solvent.

The most time-consuming procedure in the visualization process is destaining, which often requires days of washing. Rapid destaining of gels may be brought about electrophoretically. The gel, after soaking in the stain, is subjected to electrophoresis again, using a buffer of higher concentration to remove excess stain.

The search for more rapid and sensitive methods of protein detection after electrophoresis led to the development of fluorescent staining techniques. Two commonly used fluorescent reagents are fluorescamine and anilinonaphthalene sulfonate. New dyes based on silver salts (silver diamine or silver–tungstosilicic acid complex) have been developed for protein staining. They are 10 to 100 times more sensitive than Coomassie Blue (Fig. 4.7).

There is often a need for a visualization procedure that is specific for a certain biomolecule, for example, an enzyme. If the enzyme remains in an active form while in the gel, any substrate that produces a colored product could be used to locate the enzyme on the gel. Although it is less desirable for detection, the electrophoresis support medium may be cut into small segments and each part extracted with buffer and analyzed for the presence of the desired component.

Nucleic acids are visualized in agarose and polyacrylamide gels using the fluorescent dye ethidium bromide. The gel is soaked in a solution of the dye and washed to remove excess dye. Illumination of the rinsed slab with UV light reveals red-orange stains where nucleic acids are located.

Although ethidium bromide stains both single- and double-stranded nucleic acids, the fluorescence is much greater for double-stranded molecules. The electrophoresis may be performed with the dye incorporated in the gel and buffer. This has the advantage that the gel can be illuminated with UV light during electrophoresis to view the extent of separation. Precast gels are made in plastic cassettes that are UV transparent. The mobility of double-stranded DNA may be reduced 10 to 15% in the presence of ethidium bromide. Destaining of the gel is not necessary because ethidium bromide–DNA complexes have a much greater fluorescent yield than free ethidium bromide, so relatively small amounts of DNA can be detected in the presence of free ethidium bromide. The detection limit for DNA is 10 ng.

CAUTION

Ethidium bromide must be used with great care as it is a potent mutagen. Gloves should be worn at all times while using dye solutions or handling gels.

Because single-stranded nucleic acids do not stain deeply with ethidium bromide, other techniques must be used for detection, especially when only small amounts of biological material are available for analysis. One of the most sensitive techniques is to use radiolabeled molecules. For nucleic acids, this usually means labeling the 5' or 3' end with ^{32}P, a strong β emitter (see Chapter 6). Bands of labeled nucleic acids on an electrophoresis gel can easily be located by **autoradiography.** For this technique, the electrophoresed slab gel is transferred to heavy chromatography paper. After covering the gel and paper with plastic wrap, they are placed on X-ray film and wrapped in a cassette to avoid external light exposure. This procedure must be done in a darkroom. The gel-film combination is stored at $-70°C$ for exposure. The low temperature maintains the gel in a rigid form and prevents diffusion of gel bands. The exposure time depends on the amount of radioactivity but can range from a few minutes to several days. Figure 4.15 shows the autoradiogram of a typical DNA sequencing gel. The autoradiogram also provides a permanent record of the gel for storage and future analysis. The actual gel is very fragile and difficult to store. Also, because of the short half-life of ^{32}P (14 days), an autoradiogram of the gel cannot be

Figure 4.15

Autoradiogram of a DNA sequencing gel. *From Zyskind and Bernstein,* Recombinant DNA Laboratory Manual *(1989), Academic Press (San Diego, CA), Figure 7.4.*

obtained in the distant future. Proteins labeled with ^{32}P or ^{125}I can be dealt with in a manner similar to nucleic acids.

Because of concerns about the safety of radioisotope use, researchers are developing fluorescent and chemiluminescent methods for detection of small amounts of biomolecules on gels. One attractive approach is to label biomolecules before analysis with the coenzyme biotin. Biotin forms a strong complex with enzyme-linked streptavidin. Some dynamic property of the enzyme is then measured to locate the biotin-labeled biomolecule on the gel. These new methods approach the sensitivity of methods involving radiolabeled molecules, and rapid advances are being made.

If an autoradiogram of a gel can be prepared, a permanent record of the experimental data is available. For biomolecules on a gel stained with Coomassie Blue (proteins) or ethidium bromide (nucleic acids), the best method for permanent storage is conventional photography.

Protein and Nucleic Acid Blotting

Only a minute amount of protein or nucleic acid is present in bands on electropherograms. In spite of this, there is often a need to extract the desired biomolecule from the gel for further investigation. This sometimes involves the tedious and cumbersome process of crushing slices of the gel in a buffer to release the trapped proteins or nucleic acids. Techniques are now available for removing nucleic acids and proteins from gels and characterizing them using probes to detect certain structural features or functions. After electrophoresis, the biomolecules are transferred or "blotted" out of the gel onto a nitrocellulose filter or nylon membrane. The desired biomolecule is now accessible on the filter for further analysis. The first blotting technique was reported by E. Southern in 1975. Using labeled complementary DNA probes, he searched for certain nucleotide sequences among DNA molecules blotted from the gel. This technique of detecting DNA-DNA hybridization is called Southern blotting. The general blotting technique has now been extended to the transfer and detection of specific RNA with labeled complementary DNA probes (Northern blotting) and the transfer and detection of proteins that react with specific antibodies (Western blotting). In practice, the electropherogram is alkali treated, neutralized, and placed in contact with the filter or nylon membrane. A buffer is used to facilitate the transfer. Figure 4.16 shows the setup for a blotting experiment. The location of the desired nucleic acid or protein is then detected by incubation of the membrane with a radiolabeled probe and autoradiography, by use of a biotinylated probe or by linkage to an enzyme-catalyzed reaction that generates a color (Experiment 7).

Blotting techniques have many applications, including mapping the genes responsible for inherited diseases by using restriction fragment length polymorphisms (RFLPs), screening collections of cloned DNA fragments (DNA libraries), "DNA fingerprinting" for analysis of biological material remaining at the scene of a crime, and identification of specific proteins.

Figure 4.16

Diagram of a blotting experiment. *From J. Watson, Molecular Biology of the Gene, 4th ed. (1987), Benjamin/Cummings Publishing Company (Redwood City, CA).*

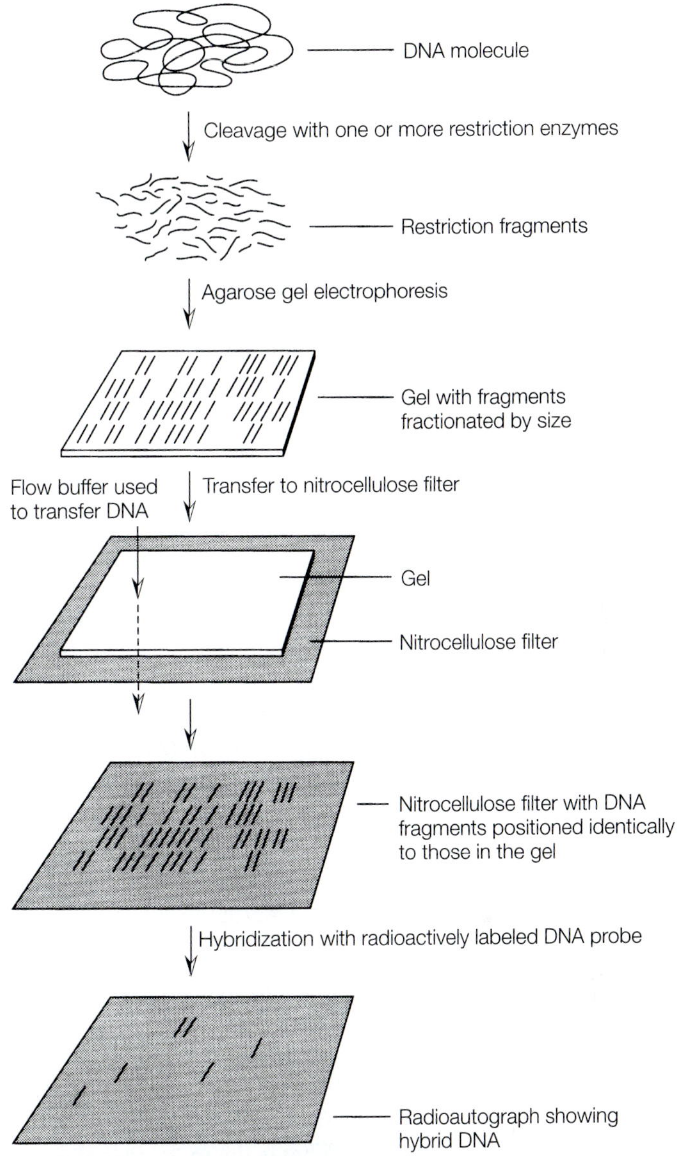

DNA molecule

Cleavage with one or more restriction enzymes

Restriction fragments

Agarose gel electrophoresis

Gel with fragments fractionated by size

Flow buffer used to transfer DNA

Transfer to nitrocellulose filter

Gel

Nitrocellulose filter

Nitrocellulose filter with DNA fragments positioned identically to those in the gel

Hybridization with radioactively labeled DNA probe

Radioautograph showing hybrid DNA

Analysis of Electrophoresis Results

By separating biochemicals on the basis of charge, size, and conformation, electrophoresis can provide valuable information, such as purity, identity, and molecular weight. Purity is indicated by the number of stained bands in the electropherogram. One band usually means that only one detectable component is present; that is, the sample is homogeneous or "electrophoretically pure." Two or more bands usually indicate that the sample contains two or more components, contaminants, or impurities and is therefore heterogeneous or impure. There are, of course, exceptions to this description. Other proteins or nucleic acids may be present in what appears to be a homogeneous sample, but they may be below the limit of detection of the staining method. Occasionally a homogeneous sample may result in two or more bands because of degradation during the electrophoresis process. Information on purity may be obtained by all of the electrophoresis methods discussed.

The identity of unknown biomolecules can be confirmed by electrophoresis on the same gel, the unknown alongside known standards. This is similar to the identification of unknowns by gas chromatography and HPLC as discussed in Chapter 3.

As previously discussed in this chapter, the molecular size of protein or nucleic acid samples may be determined by electrophoresis. This requires the preparation of standard curves of log molecular weight versus μ (mobility) using standard proteins or nucleic acids.

Study Problems

1. What physical characteristics of a biomolecule influence its rate of movement in an electrophoresis matrix?

2. Draw a slab gel to show the results of nondenaturing electrophoresis of the following mixture of proteins. The molecular weight is given for each.
   Lysozyme (13,930)   Egg white albumin (45,000)
   Chymotrypsin (21,600)   Serum albumin (65,400)

3. Each of the proteins listed below is treated with sodium dodecyl sulfate and separated by electrophoresis on a polyacrylamide slab gel. Draw pictures of the final results.
   (a) Myoglobin
   (b) Hemoglobin (two α subunits, molecular weight = 15,500; two β subunits, molecular weight = 16,000)

4. Explain the purpose of each of the chemical reagents that are used for PAGE.
   (a) acrylamide   (d) sodium dodecyl sulfate
   (b) N, N'-methylene-bis-acrylamide   (e) Coomassie Blue dye
   (c) TEMED   (f) bromophenol blue

5. What is the main advantage of slab gels over column gels for PAGE?

6. Is it possible to use polyacrylamide as a matrix for electrophoresis of nucleic acids? What are the limitations, if any?

7. Explain the purposes of protein and nucleic acid "blotting."

8. Can polyacrylamide gels be used for the analysis of plasmid DNA with greater than 3000 base pairs? Why or why not?

9. Describe the toxic characteristics of acrylamide and outline precautions necessary for its use.

10. The dye ethidium bromide is often used to detect the presence of nucleic acids on electrophoresis supports. Explain how it functions as an indicator.

Further Reading

W. Bauer and J. Vinograd, *J. Mol. Biol.* **33,** 141–171 (1968). "The Interaction of Closed Circular DNA with Intercalative Dyes."

R. Boyer, *Concepts in Biochemistry* (1999). Brooks/Cole (Pacific Grove, CA) pp. 104–105. Introduction to electrophoresis.

M. Brush, *The Scientist,* May 11, 16–23 (1998). "Protein Gel Staining Products."

M. Brush, *The Scientist,* July 20, 18–21 (1998). "A Profile of Precast Acrylamide Gels."

M. Brush, *The Scientist,* June 7, 15–17 (1999). "A Profile of New Precast Gels for Nucleic Acid Analysis."

C. Bustamante, S. Gurrieri, and S. Smith, *Tibtech* **11,** 23–30 (1993). "Towards a Molecular Description of Pulsed-Field Gel Electrophoresis."

R. Garrett and C. Grisham, *Biochemistry,* 2nd ed. (1999), Saunders (Orlando, FL), pp. 141–142; 154–156; 405–412. Basic principles of electrophoresis.

P. Grabar and C. Williams, *Biochim. Biophys. Acta* **10,** 193–194 (1953). Application of immunoelectrophoresis to protein separation.

B. Hames and D. Rickwood, Editors, *Gel Electrophoresis of Proteins: A Practical Approach,* 2nd ed. (1990), IRL Press (Oxford). An excellent book for starters.

P. Hayes, C. Wolf, and J. Hayes, *Br. Med. J.* **299,** 965–968 (1989). "Blotting Techniques for the Study of DNA, RNA and Proteins."

D. Holme and H. Peck, *Analytical Biochemistry,* 3rd ed. (1998), Addison Wesley Longman (Essex), pp. 132–146. Introduction to all techniques of electrophoresis.

M. Khaledi, *High Performance Capillary Electrophoresis: Theory, Techniques, and Applications* (1998), John Wiley & Sons (New York). Up-to-date coverage of CE.

W. Kuhr, *Anal. Chem.* **62,** 403R–414R (1990). "Capillary Electrophoresis."

A. Lehninger, D. Nelson, and M. Cox, *Principles of Biochemistry,* 3rd ed. (1999), Worth Publishers (New York), pp. 134–136. Introduction to electrophoresis.

T. Maniatis, E. Fritsch, and J. Sambrook, *Molecular Cloning, a Laboratory Manual,* 2nd ed. (1989), Cold Spring Harbor Press (Cold Spring Harbor, NY). An excellent, detailed manual for techniques and experimental protocols in molecular biology.

C. Mathews, K. van Holde, and K. Ahern, *Biochemistry,* 3rd ed. (2000), Benjamin/ Cummings (San Francisco), pp. 52–56, 252–254, 976–978.

P. O'Farrell, *J. Biol. Chem.* **250,** 4007–4021 (1975). "High Resolution Two-Dimensional Electrophoresis of Proteins."

B. Sinclair, *The Scientist,* April 17, 29–31 (2000). "An Eye for a Dye."

E. Southern, in *Methods in Enzymology,* R. Wu, Editor, Vol. 68 (1979), Academic Press (New York), pp. 152–176. "Gel Electrophoresis of Restriction Fragments."

L. Stryer, *Biochemistry,* 4th ed. (1995), Freeman (New York), pp. 46–48. An introduction to electrophoresis.

D. Voet and J. Voet, *Biochemistry,* 2nd ed. (1995), John Wiley & Sons (New York), pp. 94–100. An introduction to electrophoresis.

D. Voet, J. Voet, and C. Pratt, *Fundamentals of Biochemistry* (1999), John Wiley & Sons (New York), pp. 58–59, 104–105, 748–749. An introduction to electrophoresis.

G. Weill, *Trends Polym. Sci.* **2,** 132–136 (1994). "DNA Pulsed-Field Gel Electrophoresis."

D. Wilkinson, *The Scientist,* May 11, (1998), pp. 20–21. "Products for the Chemiluminescent Detection of DNA."

D. Wilkinson, *The Scientist,* May 15, pp. 21–24 (2000). "Capillary Action."

Electrophoresis on the Web

http://www.uni-giessen.de/~gh43electrophoresis.html
    The World of Electrophoresis and other links.

http://gslc.genetics.utah.edu
    Genetic Science Learning Center

http://www.uct.ac.za/microbiology/sdspage.html
    Discussion of SDS-PAGE.

http://biotech.biology.arizona.edu/labs/Electrophoresis_dyes_stude.html
    The University of Arizona Biotech Project, "Agarose Gel Electrophoresis with Dyes."

http://www.ceandcec.com
    CE and CEC Web site. Several links to journals, theory, applications, and books.

http://www.bio.davidson.edu/Biology/courses/Molbio/tips/trblDNAgel.html
    "Troubleshooting DNA Agarose Gel Electrophoresis."

http://grimwade.biochem.unimelb.edu.au/bfjones/gen7/m7a1.htm
    Agarose Gel Electrophoresis of RNA.

http://www.neptune.net/~whatley/capelec.htm
    Harry's CE Page. Discussion of the technique.

5

SPECTROSCOPIC ANALYSIS OF BIOMOLECULES

Very early experimental measurements on biomolecules involved studies of their interactions with electromagnetic radiation of all wavelengths including x-ray, ultraviolet-visible, and infrared. It was experimentally observed that when light impinges on solutions or crystals of molecules, at least two distinct processes occur: **light scattering** and **light absorption.** Both processes have led to the development of fundamental techniques for characterizing and analyzing biomolecules. Absorption in the ultraviolet-visible regions is an especially valuable procedure for molecular structure elucidation. With some molecules, the process of absorption is followed by emission of light of a different wavelength. This process, called **fluorescence,** depends on molecular structure and environmental factors and assists in the characterization and analysis of biologically significant molecules and dynamic processes occurring between molecules. Nuclear magnetic resonance spectroscopy and mass spectrometry techniques are also being applied to the study of biological molecules and processes. NMR is especially versatile because, in addition to proton spectra, monitoring the presence of ^{13}C, ^{15}N, ^{19}F, and ^{31}P nuclei in biomolecules is possible. Multidimensional NMR is being applied to the study of protein secondary and tertiary structure. Accurate measurements of protein molecular weight may now be carried out by mass spectrometry.

A.   ULTRAVIOLET-VISIBLE ABSORPTION SPECTROPHOTOMETRY

Principles

The electromagnetic spectrum, as shown in Figure 5.1, is composed of a continuum of waves with different properties. Several regions of the electromagnetic spectrum are of importance in biochemical studies including x-ray (x-ray crystallography, up to 7 nm), the ultraviolet (UV, 180–340 nm), the visible (VIS, 340–800 nm), the infrared (IR, 1000–100,000 nm), and radio waves (NMR, 10^6–10^{10} nm). In this section, we will concentrate on the UV and VIS regions. Light in these regions has sufficient energy to excite the valence electrons of molecules. Figure 5.2 shows that the propagation of light is due to an electrical field component, *E*, and a magnetic field component, *H*, that are perpendicular to each other. The **wavelength** of light, defined by Equation 5.1, is the distance between adjacent wave peaks as shown in Figure 5.2.

$$\lambda = \frac{c}{v}$$

Equation 5.1

where

> λ = wavelength
>
> c = speed of light
>
> v = frequency, the number of waves passing a certain point per unit time

Light also behaves as though it were composed of energetic particles. The amount of energy, *E*, associated with these particles (or photons) is given by Equation 5.2.

$$E = hv$$

Equation 5.2

where *h* is Planck's constant

Figure 5.1

The electromagnetic spectrum. The wavelengths of light associated with important spectroscopy techniques are shown in the figure. *From D. M. Freifelder,* Physical Biochemistry, *W. H. Freeman (San Francisco). Copyright © 1982. All rights reserved.*

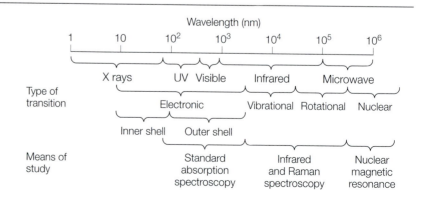

Figure 5.2

An electromagnetic wave, showing the E and H components.

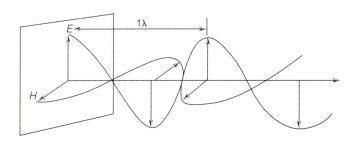

When a photon of specified energy interacts with a molecule, one of two processes may occur. The photon may be **scattered,** or it may transfer its energy to the molecule, producing an **excited state** of the molecule. The former process, called **Rayleigh scattering,** occurs when a photon collides with a molecule and is diffracted or scattered with unchanged frequency. Light scattering is the physical basis of several experimental methods used to characterize macromolecules. Before the development of electrophoresis, light scattering techniques were used to measure the molecular weights of macromolecules. The widely used techniques of X-ray diffraction (crystal and solution), electron microscopy, laser light scattering, and neutron scattering all rely in some way on the light scattering process.

The other process mentioned above, the transfer of energy from a photon to a molecule, is **absorption.** For a photon to be absorbed, its energy must match the energy difference between two energy levels of the molecule. Molecules possess a set of quantized energy levels, as shown in Figure 5.3. Although several states are possible, only two electronic states are shown, a **ground state,** G, and the **first excited state,** S_1. These two states differ in the distribution of valence electrons. When electrons are promoted from a ground state orbital in G to an orbital of higher energy in S_1, an **electronic transition** is said to occur. The energy associated with ultraviolet and visible light is sufficient to promote molecules from one electronic state to another, that is, to move electrons from one orbital to another.

Within each electronic energy level is a set of **vibrational levels.** These represent changes in the stretching and bending of covalent bonds. The importance of these energy levels will not be discussed here, but transitions between the vibrational levels are the basis of infrared spectroscopy.

The electronic transition for a molecule from G to S_1, represented by the vertical arrow in Figure 5.3, has a high probability of occurring if the energy of the photon corresponds to the energy necessary to promote an electron from energy level E_1 to energy level E_2:

>>
$$E_2 - E_1 = \Delta E = \frac{hc}{\lambda}$$

Equation 5.3

A transition may occur from any vibrational level in G to some other vibrational level in S_1, for example, $v = 3$; however, not all transitions have equal

Figure 5.3

Energy-level diagram showing the ground state, *G*, and the
first excited state, S_1.

- -

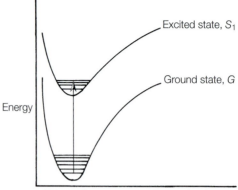

probability. The probability of absorption is described by quantum mechanics and will not be discussed here.

A UV-VIS spectrum is obtained by measuring the light absorbed by a sample as a function of wavelength. Since only discrete packets of energy (specific wavelengths) are absorbed by molecules in the sample, the spectrum theoretically should consist of sharp discrete lines. However, the many vibrational levels of each electronic energy level increase the number of possible transitions. This results in several spectral lines, which together make up the familiar spectrum of broad peaks as shown in Figure 5.4.

An absorption spectrum can aid in the identification of a molecule because the wavelength of absorption depends on the functional groups or arrangement of atoms in the sample. The spectrum of oxyhemoglobin in Figure 5.4 is due to the presence of the iron porphyrin moiety and is useful for the characterization of heme derivatives or hemoproteins. Note that the spectrum consists of several peaks at wavelengths where absorption reaches a maximum (415, 542, and 577 nm). These points, called λ_{max}, are of great significance in the identification of unknown molecules and will be discussed later in the chapter.

Quantitative measurements in spectrophotometry are evaluated using the Beer-Lambert law:

>> $A = Elc$ **Equation 5.4**

where

A = absorbance or $-\log (I/I_0)$

I_0 = intensity of light irradiating the sample

I = intensity of light transmitted through the sample

Figure 5.4

The visible absorbance
spectrum of hemoglobin.

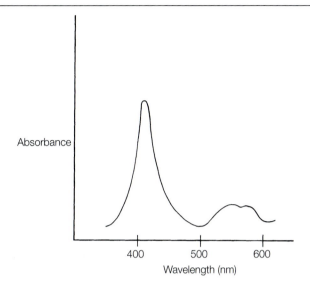

E = absorption coefficient or absorptivity

l = path length of light through the sample, or thickness of the cell

c = concentration of absorbing material in the sample

Since the absorbance, A, is derived from a ratio ($-\log I/I_0$), it is unit-less. The term E, which is a proportionality constant, defines the efficiency or extent of absorption. If this is defined for a particular chromophore at a specific wavelength, the term **absorption coefficient** or **absorptivity** is used. However, students should be aware that in the older biochemical literature, the term extinction coefficient is often used. The units of E depend on the units of l (usually cm) and c (usually molar) in Equation 5.4. For biomolecules, E is often used in the form **molar absorption coefficient, ϵ**, which is defined as the absorbance of a 1 M solution of pure absorbing material in a 1-cm cell under specified conditions of wavelength and solvent. The units of ϵ are M^{-1} cm^{-1}. To illustrate the use of Equation 5.4, consider the following calculation.

Example 1   The absorbance, A, of a 5×10^{-4} M solution of the amino acid tyrosine, at a wavelength of 280 nm, is 0.75. The path length of the cuvette is 1 cm. What is the molar absorption coefficient, ϵ?

$A = \epsilon l c = 0.75$

$l = 1$ cm

$c = 5 \times 10^{-4} M$

(continued)

$$\epsilon = \frac{0.75}{(1 \text{ cm})(5 \times 10^{-4} \text{ mole/liter})}$$

$$= 1500 \frac{\text{liter}}{\text{mole} \times \text{cm}} = 1500 \ M^{-1} \ \text{cm}^{-1}$$

Notice that the units of ϵ are defined by the concentration units of the tyrosine solution (M) and the dimension units of the cuvette (cm). Although E is most often expressed as a molar absorption coefficient, you may encounter other units such as $E_\lambda^{\%}$, which is the absorbance of a 1% (w/v) solution of pure absorbing material in a 1-cm cuvette at a specified wavelength, λ.

Instrumentation

The **spectrophotometer** is used to measure absorbance experimentally. This instrument produces light of a preselected wavelength, directs it through the sample (usually dissolved in a solvent and placed in a cuvette), and measures the intensity of light transmitted by the sample. The major components are shown in Figure 5.5. These consist of a light source, a monochromator (including various filters, slits, and mirrors), a sample chamber, a detector, and a meter or recorder. All of these components are usually under the control of a computer.

Light Source

For absorption measurements in the ultraviolet region, a high-pressure hydrogen or deuterium lamp is used. These lamps produce radiation in the 200 to 340 nm range. The light source for the visible region is the tungsten-halogen lamp, with a wavelength range of 340 to 800 nm. Instruments with both lamps have greater flexibility and can be used for the study of most biologically significant molecules.

Monochromator

Both lamps discussed above produce continuous emissions of all wavelengths within their range. Therefore, a spectrophotometer must have an op-

Figure 5.5

A diagram of a conventional, single-beam spectrophotometer.

- - - - - - - - - - - - - - - -

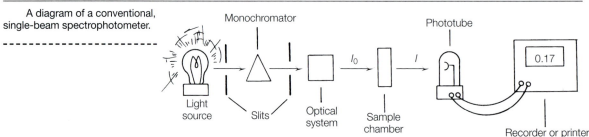

Monochromator · Phototube · I_0 · I · 0.17 · Light source · Slits · Optical system · Sample chamber · Recorder or printer

tical system to select monochromatic light (light of a specific wavelength). Modern instruments use a prism or, more often, a diffraction grating to produce the desired wavelengths. It should be noted that light emitted from the monochromator is not entirely of a single wavelength but is enhanced in that wavelength. That is, most of the light is of a single wavelength, but shorter and longer wavelengths are present.

Before the monochromatic light impinges on the sample, it passes through a series of slits, lenses, filters, and mirrors. This optical system concentrates the light, increases the spectral purity, and focuses it toward the sample. The operator of a spectrophotometer has little control over the optical manipulation of the light beam, except for adjustment of slit width. Light passing from the monochromator to the sample encounters a "gate" or slit. The slit width, which is controlled by a computer, determines both the intensity of light impinging on the sample and the spectral purity of that light. Decreasing the slit width increases the spectral purity of the light, but the amount of light directed toward the sample decreases. The efficiency or sensitivity of the detector then becomes a limiting factor.

In instruments equipped with photodiode array detectors (see Figure 5.6 and the following section on detectors), polychromatic light from a source passes through the sample and is focused on the entrance slit of a polychromator (a holographic grating that disperses the light into its wavelengths). The wavelength-resolved light beam is then focused onto the photodiode array detector. The relative positions of the sample and grating are reversed compared to conventional spectrometry; hence the new configuration is often called reversed optics. (Compare Figures 5.5 and 5.6.)

Figure 5.6

Optical diagram for a diode array spectrophotometer.

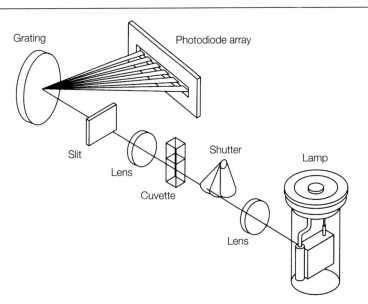

Sample Chamber

The processed monochromatic light is then directed into a sample chamber, which can accommodate a wide variety of sample holders. Most UV-VIS measurements on biomolecules are taken on solutions of the molecules. The sample is placed in a tube or cuvette made of glass, quartz, or other transparent material. Figure 5.7 shows the design of the most common sample holders and the transmission properties of several transparent materials used in cuvette construction.

Glass cuvettes are the least expensive but, because they absorb UV light, they can be used only above 340 nm. Quartz or fused silica cuvettes may be used throughout the UV and visible regions (~200–800 nm). Disposable cuvettes are now commercially available in polymethacrylate (280–

Figure 5.7

A An assortment of cuvettes. *Courtesy of VWR Scientific, Division of Univar.* **B** A standard 3-mL cuvette. *Courtesy of Beckman Instruments, Inc.* **C** The transmission properties of several materials used in cuvettes. A = silica (quartz); B = NIR silica; C = polymethacrylate; D = polystyrene; E = glass.

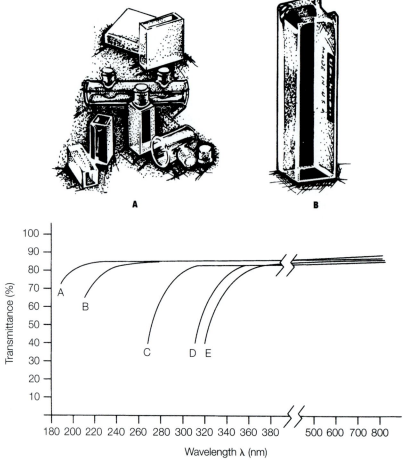

800 nm) and polystyrene (350–800 nm). The care and use of cuvettes were discussed in Chapter 1.

Sample chambers for spectrometers come in two varieties—those holding only one cuvette at a time (single-beam) and those holding two cuvettes, one for a reference, usually solvent, and one for sample (double-beam). In a double-beam instrument, the sample spectrum is continuously corrected by subtraction of the reference spectrum. In the past, single-beam instruments were usually less expensive but more cumbersome to use because reference and sample cuvettes required constant exchange. However, modern single-beam instruments with computer control and analysis can be programmed to correct automatically for the reference spectrum, which may be stored in a memory file. The use of both types of instruments is outlined in the applications section.

Detector

The intensity of the light that passes through the sample under study depends on the amount of light absorbed by the sample. Intensity is measured by a light-sensitive detector, usually a photomultiplier tube (PMT). The PMT detects a small amount of light energy, amplifies this by a cascade of electrons accelerated by dynodes, and converts it into an electrical signal that can be fed into a meter or recorder.

A new technology has been introduced during the past few years that greatly increases the speed of spectrophotometric measurements. New detectors called photodiode arrays are being used in modern spectrometers. Photodiodes are composed of silicon crystals that are sensitive to light in the wavelength range 170–1100 nm. Upon photon absorption by the diode, a current is generated in the photodiode that is proportional to the number of photons. Linear arrays of photodiodes are self-scanning and have response times on the order of 100 milliseconds; hence, an entire UV-VIS spectrum can be obtained with an extremely brief exposure of the sample to polychromatic light. New spectrometers designed by Hewlett-Packard and Perkin-Elmer use this technology and can produce a full spectrum from 190 to 820 nm in one-tenth of a second.

Printers and Recorders

Less expensive instruments give a direct readout of absorbance and/or transmittance in analog or digital form. These instruments are suitable for single-wavelength measurements; however, if a scan of absorbance vs. wavelength (Figure 5.4) is desired, some type of device to display the spectrum must be available.

Modern, research-grade spectrometers are available that offer the latest in technology. All of the components discussed above are integrated into a single package and are completely under the control of a computer. By simply pushing a button, one can obtain the UV-VIS spectrum of a sample displayed on a computer screen in less than 1 second. In addition, these modern instruments with computers can be programmed to carry out several functions, such as subtraction of solvent spectrum, spectral overlay, storage, difference spectra, derivative spectra, and calculation of concentrations and rate constants.

Applications

Now that you are familiar with the theory and instrumentation of absorption spectrophotometry, you will more easily understand the actual operation and typical applications of a spectrophotometer. Since virtually all UV-VIS measurements are made on samples dissolved in solvents, only those applications will be described here. Although many different types of operations can be carried out on a spectrophotometer, all applications fall in one of two categories: (1) measurement of absorbance at a fixed wavelength and (2) measurement of absorbance as a function of wavelength. Measurements at a fixed wavelength are most often used to obtain quantitative information such as the concentration of a solute in solution or the absorption coefficient of a chromophore. Absorbance measurements as a function of wavelength provide qualitative information that assists in solving the identity and structure of a pure substance by detecting characteristic groupings of atoms in a molecule.

For fixed-wavelength measurements with a single-beam instrument, a cuvette containing solvent only is placed in the sample beam and the instrument is adjusted to read "zero" absorbance. A matched cuvette containing sample plus solvent is then placed in the sample chamber and the absorbance is read directly from the display. The adjustment to zero absorbance with only solvent in the sample chamber allows the operator to obtain a direct reading of absorbance for the sample.

Fixed-wavelength measurements using a double-beam spectrophotometer are made by first zeroing the instrument with no cuvette in either the sample or reference holder. Alternatively, the spectrophotometer can be balanced by placing matched cuvettes containing water or solvent in both sample chambers. Then, a cuvette containing pure solvent is placed in the reference position and a matched cuvette containing solvent plus sample is set in the sample position. The absorbance reading given by the instrument is that of the sample; that is, the absorbance due to solvent is subtracted by the instrument.

An **absorbance spectrum** of a compound is obtained by scanning a range of wavelengths and plotting the absorbance at each wavelength. Most double-beam spectrophotometers automatically scan the desired wavelength range and record the absorbance as a function of wavelength. If solvent is placed in the reference chamber and solvent plus sample in the sample position, the instrument will continuously and automatically subtract the solvent absorbance from the total absorbance (solvent plus sample) at each wavelength; hence, the recorder output is really a difference spectrum (absorbance of sample plus solvent, minus absorbance of solvent).

Both types of measurements (fixed wavelength and absorbance spectrum) are common in biochemistry, and you should be able to interpret results from each. The following examples are typical of the kinds of problems readily solved by spectrophotometry.

Measurement of the Concentration of a Solute in Solution

According to the Beer-Lambert law, the absorbance of a material in solution is directly dependent on the concentration of that material. Two methods are commonly used to measure concentration. If the absorption coefficient is known for the absorbing species, the concentration can be calculated after experimental measurement of the absorbance of the solution.

Example 2   A solution of the nucleotide base uracil, in a 1-cm cuvette, has an absorbance at λ_{max}(260 nm) of 0.65. Pure solvent in a matched quartz cuvette has an absorbance of 0.07. What is the molar concentration of the uracil solution? Assume the molar absorption coefficient, ϵ, is 8.2×10^3 M^{-1} cm^{-1}.

$A = \epsilon l c$

$A = $ (absorbance of solvent + sample) $-$ (absorbance of solvent)

$A = 0.65 - 0.07 = 0.58$

$\epsilon = 8.2 \times 10^3\ M^{-1}\ cm^{-1}$

$l = 1$ cm

$$c = \frac{A}{\epsilon l} = \frac{0.58}{(8.2 \times 10^3\ M^{-1}\ cm^{-1})(1\ cm)}$$

$c = 7.1 \times 10^{-5} M$

If the absorption coefficient for an absorbing species is known, the concentration of that species in solution can be calculated as outlined. However, there are limitations to this application. Most spectrophotometers are useful for measuring absorbances up to 1, although more sophisticated instruments can measure absorbances as high as 2. Also, some substances do not obey the Beer-Lambert law; that is, absorbance may not increase in a linear fashion with concentration. Reasons for deviation from the Beer-Lambert law are many; however, the majority are instrumental, chemical, or physical. Spectrophotometers often display a nonlinear response at high absorption levels because of stray light. Physical reasons for nonlinearity include hydrogen bonding of the absorbing species with the solvent and intermolecular interactions at high concentrations. Chemical reasons may include reaction of the solvent with the absorbing species and the presence of impurities. Linearity is readily tested by preparing a series of concentrations of the absorbing species and measuring the absorbance of each. A plot of A vs. concentration should be linear if the Beer-Lambert law is valid. If the absorption coefficient for a species is unknown, its concentration in solution can be measured if the absorbance of a standard solution of the compound is known.

Example 3 The absorbance of a 1% (w/v) solution of the enzyme tyrosinase, in a 1-cm cell at 280 nm, is 24.9. What is the concentration of a tyrosinase solution that has an A_{280} of 0.25?

Since the absorption coefficient, $E\%$, is the same for both solutions, the concentration can be calculated by a direct ratio:

$$\frac{A_{std}}{C_{std}} = \frac{A_x}{C_x}$$

A_{std} = absorbance of the 1% standard solution = 24.9

C_{std} = concentration of the standard solution = 1% (1 g/dL)

A_x = absorbance of the unknown solution = 0.25

C_x = concentration of the unknown solution in %

$$\frac{24.9}{1\%} = \frac{0.25}{C_x}$$

$$C_x = 0.01\% = 0.01 \text{ g/dL} = 0.1 \text{ mg/mL}$$

Alternatively, the concentration of a species in solution can be determined by preparing a standard curve of absorbance vs. concentration.

Example 4 The Bradford protein assay is one of the most used spectrophotometric assays in biochemistry. (For a discussion of the Bradford assay, see Chapter 2.) Solutions of varying amounts of a standard protein are mixed with reagents that cause the development of a color. The amount of color produced depends on the amount of protein present. The absorbance at 595 nm of each reaction mixture is plotted against the known protein concentration. A protein sample of unknown concentration is treated with the Bradford reagents and the color is allowed to develop.

The following absorbance measurements are typical for the standard curve of a protein:

Protein (μg per assay)	A_{595}
15	0.07
25	0.15
50	0.28
100	0.55
150	0.90
0.1 mL unknown protein solution	0.10
0.2 mL unknown protein solution	0.22

A standard curve for the data is plotted in Figure 2.6 on page 45. Note the linearity, indicating that the Beer-Lambert law is obeyed over this concentration range of standard protein. Two different volumes of unknown protein were tested. This was to ensure that one volume would be in the concentration range of the standard curve. Since the accuracy of the assay is dependent on identical times for color development, the unknowns must be assayed at the same time as the standards. From graphical analysis, an A_{595} of 0.22 corresponds to 40 μg of protein per 0.20 mL. This indicates that the original protein solution concentration was approximately 200 μg/mL or 0.20 mg/mL. Using computer graphics, students can now quickly visualize experimental data and determine the need for further analysis. Most modern computer programs use the method of least squares to calculate automatically the slope, intercepts, and correlation coefficients.

Identification of Unknown Biomolecules by Spectrophotometry

The UV-VIS spectrum of a biomolecule reveals much about its molecular structure. Therefore, a spectral analysis is one of the first experimental measurements made on an unknown biomolecule. Natural molecules often contain chromophoric (color-producing) functional groups that have characteristic spectra. Figure 5.8 displays spectra of the well-characterized biomolecules DNA, FMN, $FMNH_2$, NAD, NADH, and nucleotides. Spectral analysis in the visible region is used in Experiment 8 to identify pigments isolated from plants.

The procedure for obtaining a UV-VIS spectrum begins with the preparation of a solution of the species under study. A standard solution should be prepared in an appropriate solvent. An aliquot of the solution is transferred to a cuvette and placed in the sample chamber of a spectrophotometer. A cuvette containing solvent is placed in the reference holder. The spectrum is scanned over the desired wavelength range and an absorption coefficient is calculated for each major λ_{max}.

Kinetics of Biochemical Reactions

Spectrophotometry is one of the best methods available for measuring the rates of biochemical reactions. Consider a general reaction as shown in Equation 5.5.

$$A + B \rightleftharpoons C + D$$

Equation 5.5

If reactants A or B absorb in the UV-VIS region of the spectrum at some wavelength λ_1 the rate of the reaction can be measured by monitoring the decrease of absorbance at λ_1 due to loss of A or B. Alternatively, if products C or D absorb at a specific wavelength λ_2, the kinetics of the reaction can be evaluated by monitoring the absorbance increase at λ_2. According to the Beer-Lambert law, the absorbance change of a reactant or product is proportional to the concentration change of that species occurring during the reaction. This method is widely used to assay enzyme-catalyzed processes. Since the

Figure 5.8

UV-VIS absorbance spectra of significant biomolecules.
A DNA;   **B** FMN and FMNH$_2$;
C NAD$^+$ and NADH;   **D** d-GMP;
E thymine.

- -

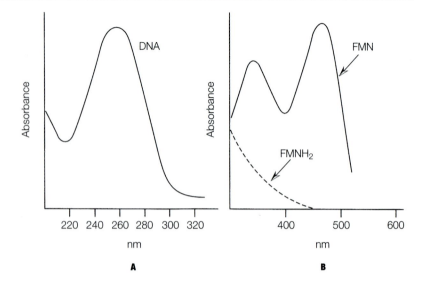

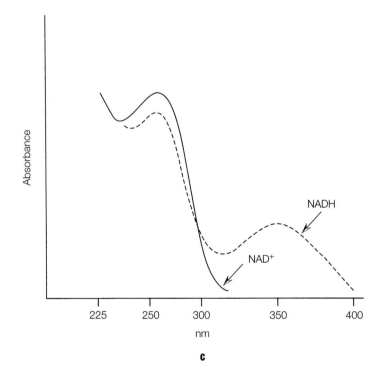

Figure 5.8

continued

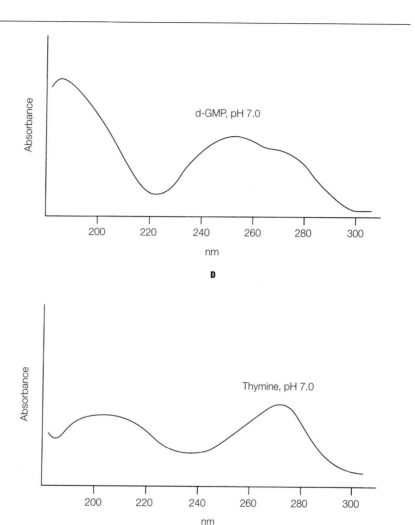

rates of chemical reactions vary with temperature, the sample cuvette containing the reaction mixture must be held in a thermostated chamber. Experiment 5 utilizes spectrophotometry to characterize the kinetics of the tyrosinase-catalyzed oxidation of 3,4-dihydroxyphenylalanine.

Characterization of Macromolecule-Ligand Interactions by Difference Spectroscopy

Many biological processes depend on a specific interaction between molecules. The interaction often involves a macromolecule (protein or nucleic acid) and a smaller molecule, a **ligand.** Specific examples include enzyme-substrate interactions and receptor protein–hormone interactions. One of the most effective and convenient methods for detecting and characterizing such interactions is **difference spectroscopy.** The interaction of small molecules with the transport protein hemoglobin is a classic example of the utility of difference spectroscopy. If a small molecule or ligand, such as inositol hexaphosphate, binds to hemoglobin, there is a change in the heme spectral properties. The spectral change is small and would be difficult to detect if the experimenter recorded the spectrum in the usual fashion. Normally, one would obtain a spectrum of a heme protein by placing a solution of the protein in a cuvette in the sample compartment of a spectrophotometer and the neat solvent in the reference compartment. Any absorption due to solvent is subtracted because the solvent is present in both light beams, so the spectrum is that due to the heme protein. Then the ligand to be tested would be added to the heme protein, and the spectrum would be obtained for this mixture vs. solvent. If the free or bound ligand molecule did not absorb light in the wavelength range studied, there would be no need to have ligand in the reference cell. The two spectra (heme protein in solvent vs. solvent and heme protein and ligand in solvent vs. solvent) can then be compared and differences noted.

A difference spectrum is faster than the preceding method because only one spectral recording is necessary. Two cuvettes are prepared in the following manner. The reference cuvette contains heme protein and solvent, whereas the sample cuvette contains heme protein, solvent, and ligand. There must be equal concentrations of the heme protein in the two cuvettes. (Why?) The two cuvettes are placed in a double-beam spectrophotometer and the spectrum is recorded. If the spectrum of the heme protein is not influenced by the ligand, the result would be a zero difference spectrum, that is, a straight line (Figure 5.9A). Both samples have identical spectral properties, indicating that there is probably little or no interaction between heme protein and ligand. However, such data should be treated with caution because it is possible that the heme group is not affected by ligand binding. A nonzero difference spectrum indicates that the ligand interacts with the heme protein and induces a change in the environment of the heme group (Figure 5.9B).

A difference spectrum can be analyzed and used in several ways. This is a useful technique for demonstrating qualitatively whether an interaction occurs between a macromolecule and a ligand. Quantitative analysis of difference spectra requires measurement of λ_{max} and ΔA at λ_{max}. This method can be used to quantify the strength of ligand-protein interaction. Note that every time you record a spectrum in a double-beam spectrometer, you are obtaining a difference spectrum between sample and reference. Although a heme protein was used in this example, this does not imply that only heme interactions can be characterized by difference spectroscopy. Any protein that contains a chromophoric group, whether it be an aromatic amino acid, cofactor, prosthetic group, or metal ion, can be studied by difference spectroscopy.

Limitations and Precautions in Spectrophotometry

The use of a spectrophotometer is relatively straightforward and can be mastered in a short period of time. There are, however, difficulties that must

Figure 5.9

Difference spectroscopy.
A Hemoglobin vs. hemoglobin.
B Hemoglobin vs. hemoglobin +
inositol hexaphosphate.

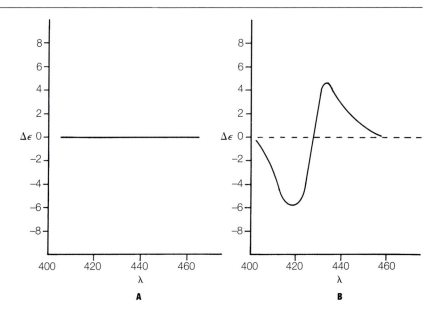

be considered. A common problem encountered with biochemical measurements is turbidity or cloudiness of biological samples. This can lead to great error in absorbance measurements because much of the light entering the cuvette is not absorbed but is scattered. This causes artificially high absorbance readings. Occasionally, absorbance readings on turbid solutions are desirable (as in measuring the rate of bacterial growth in a culture), but in most cases turbid solutions must be avoided or clarified by filtration or centrifugation.

A difficulty encountered in measuring the concentration of an unknown absorbing species in solution is deviation from the Beer-Lambert law. For reasons stated earlier in this chapter, some absorbing species do not demonstrate an increase in absorbance that is proportional to an increase in concentration. (In reality, most compounds follow the Beer-Lambert relationship over a relatively small concentration range.) When measuring solution concentration, adherence to the Beer-Lambert law must always be tested in the concentration range under study.

B.   FLUORESCENCE SPECTROPHOTOMETRY

Principles

In our discussion of absorption spectroscopy, we noted that the interaction of photons with molecules resulted in the promotion of valence electrons from ground state orbitals to higher energy level orbitals. The molecules were said to be in an excited state.

Molecules in the excited state do not remain there long, but spontaneously relax to the more stable ground state. With most molecules, the relaxation process is brought about by collisional energy transfer to solvent or other molecules in the solution. Some excited molecules, however, return to the ground state by emitting the excess energy as light. This process, called **fluorescence,** is illustrated in Figure 5.10. The solid vertical arrow in the figure indicates the photon absorption process in which the molecule is excited from *G* to some vibrational level in *S*. The excited molecule loses vibrational energy by collision with solvent and ground state molecules. This relaxation process, which is very rapid, leaves the molecule in the lowest vibrational level of *S*, as indicated by the wavy arrow. The molecule may release its energy in the form of light (fluorescence, dashed arrow) to return to some vibrational level of *G*.

Two important characteristics of the emitted light should be noted: (1) It is usually of longer wavelength (lower energy) than the excitation light. This is because part of the energy initially associated with the *S* state is lost as heat energy, and the energy lost by emission may be sufficient only to return the excited molecule to a higher vibrational level in *G*. (2) The emitted light is composed of many wavelengths, which results in a fluorescence spectrum as shown in Figure 5.11. This is due to the fact that fluorescence from any particular excited molecule may return the molecule to one of many vibrational levels in the ground state. Just as in the case of an absorption spectrum, a wavelength of maximum fluorescence is observed, and the spectrum is composed of a wavelength distribution centered at this emission maximum. Note that all emitted wavelengths are longer and of lower energy than the initial excitation wavelength.

In our discussion above, it was pointed out that a molecule in the excited state can return to lower energy levels by collisional transfer or by light emission. Since these two processes are competitive, the **fluorescence intensity** of a fluorescing system depends on the relative importance of each process. The fluorescence intensity is often defined in terms of **quantum yield,** represented by Q. This describes the efficiency or probability of the fluorescence process. By definition, Q is the ratio of the number of photons emitted to the number of photons absorbed (Equation 5.6).

Figure 5.10

Energy-level diagram outlining the fluorescence process.

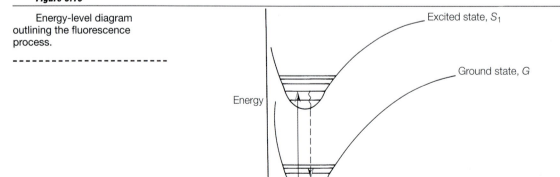

Distance between atoms in a molecule

Figure 5.11

Absorption (——) and fluorescence (-----) spectra of tryptophan. *From D. M. Freifelder, Physical Biochemistry, W. H. Freeman (San Francisco).*

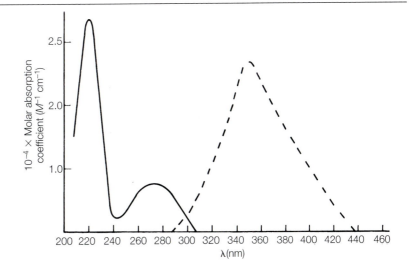

$$Q = \frac{\text{number of photons emitted}}{\text{number of photons absorbed}}$$

Equation 5.6

Measurement of quantum yield is often the goal in fluorescence spectroscopy experiments. Q is of interest because it may reveal important characteristics of the fluorescing system. Two types of factors affect the intensity of fluorescence, internal and external (environmental) influences. Internal factors, such as the number of vibrational levels available for transition and the rigidity of the molecules, are associated with properties of the fluorescent molecules themselves. Internal factors will not be discussed in detail here because they are of more interest in theoretical studies. The external factors that affect Q are of great interest to biochemists because information can be obtained about macromolecule conformation and molecular interactions between small molecules (ligands) and larger biomolecules (proteins, nucleic acids). Of special value is the study of experimental conditions that result in **quenching** or **enhancement** of the quantum yield. Quenching in biochemical systems can be caused by chemical reactions of the fluorescent species with added molecules, transfer of energy to other molecules by collision (actual contact between molecules), and transfer of energy over a distance (no contact, resonance energy transfer). The reverse of quenching, enhancement of fluorescent intensity, is also observed in some situations. Several fluorescent dye molecules are quenched in aqueous solution, but their fluorescence is greatly enhanced in a nonpolar or rigidly bound environment (the interior of a protein, for example). This is a convenient method for characterizing ligand binding. Both fluorescence quenching and fluorescence enhancement studies can yield important information about biomolecular structure and function. Several applications will be described in a later section of this chapter and in Experiment 13.

Instrumentation

The basic instrument for measuring fluorescence is the **spectrofluorometer.** It contains a light source, two monochromators, a sample holder, and a detector. A typical experimental arrangement for fluorescence measurement is shown in Figure 5.12. The setup is similar to that for absorption measurements, with two significant exceptions. First, there are two monochromators, one for selection of the excitation wavelength, another for wavelength analysis of the emitted light. Second, the detector is at an angle (usually 90°) to the excitation beam. This is to eliminate interference by the light that is transmitted through the sample. Upon excitation of the sample molecules, the fluorescence is emitted in all directions and is detected by a photocell at right angles to the excitation light beam.

The lamp source used in most instruments is a xenon arc lamp that emits radiation in the ultraviolet, visible, and near-infrared regions (200 to 1400 nm). The light is directed by an optical system to the excitation monochromator, which allows either preselection of a wavelength or scanning of a certain wavelength range. The exciting light then passes into the sample chamber, which contains a fluorescence cuvette with dissolved sample. Because of the geometry of the optical system, a typical fused absorption cuvette with two opaque sides cannot be used; instead, special fluorescence cuvettes with four translucent quartz or glass sides must be used. When the excitation light beam impinges on the sample cell, molecules in the solution are excited and some will emit light.

Light emitted at right angles to the incoming beam is analyzed by the emission monochromator. In most cases, the wavelength analysis of emitted light is carried out by measuring the intensity of fluorescence at a preselected

Figure 5.12

A diagram of a typical fluorometer.

- -

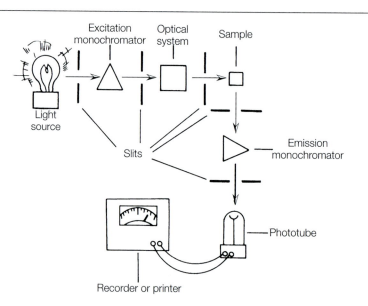

wavelength (usually the wavelength of emission maximum). The analyzer monochromator directs emitted light of only the preselected wavelength toward the detector. A photomultiplier tube serves as a detector to measure the intensity of the light. The output current from the photomultiplier is fed to some measuring device that indicates the extent of fluorescence. The final readout is not in terms of Q but in units of the photomultiplier tube current (microamperes) or in relative units of percent of full scale. Therefore, the scale must be standardized with a known. Some newer instruments provide, as output, the ratio of emitted light to incident light intensity.

Applications

Two types of measurements are most common in fluorescence experiments, measurements of **relative fluorescence intensities** and measurements of the **quantum yield.** Experiments introduced in this book will require only relative fluorescence intensity measurements, and they proceed as follows. The fluorometer is set to "zero" or "full scale" fluorescence intensity (microamps or %) with the desired biochemical system under standard conditions. Some perturbation is then made in the system (pH change, addition of a chemical agent in varying concentrations, change of ionic strength, etc.) and the fluorescence intensity is determined relative to the standard conditions. This is a straightforward type of experiment because it consists of replacing one solution with another in the fluorometer and reading the detector output for each. For these experiments, the excitation wavelength and the emission wavelength are preselected and set for each monochromator.

The measurement of quantum yield is a more complicated process. Before these measurements can be made, the instrument must be calibrated. A thermopile or chemical actinometer may be used to measure the absolute intensity of incident light on the sample. Alternatively, quantum yields may be measured relative to some accepted standard. Two commonly used fluorescence standards are quinine sulfate in 0.5 M H_2SO_4 ($Q = 0.70$) and fluorescein in 0.1 M NaOH ($Q = 0.93$). The quantum yield of the unknown, Q_x, is then calculated by Equation 5.7.

 $$\frac{Q_x}{Q_{std}} = \frac{F_x}{F_{std}}$$ *Equation 5.7*

where

Q_x = quantum yield of unknown

Q_{std} = quantum yield of standard

F_x = experimental fluorescence intensity of unknown

F_{std} = experimental fluorescence intensity of standard

Some biomolecules are **intrinsic fluors;** that is, they are fluorescent themselves. The amino acids with aromatic groups (phenylalanine, tyrosine, and tryptophan) are fluorescent; hence, proteins containing these amino acids have intrinsic fluorescence. The purine and pyrimidine bases in nucleic acids (adenine, guanine, cytosine, uracil) and some coenzymes (NAD, FAD) are also intrinsic fluors. Intrinsic fluorescence is most often used to study protein conformational changes and to probe the location of active sites and coenzymes in enzymes.

Valuable information can also be obtained by the use of **extrinsic fluors.** These are fluorescent molecules that are added to the biochemical system under study. Many fluorescent dyes have enhanced fluorescence when they are in a nonpolar solution or bound in a rigid hydrophobic environment. Some of these dyes bind to specific sites on proteins or nucleic acid molecules, and the resulting fluorescence intensity depends on the environmental conditions at the binding site. Extrinsic fluorescence is of value in characterizing the binding of natural ligands to biochemically significant macromolecules. This is because many of the extrinsic fluors bind in the same sites as natural ligands. Extrinsic fluorescence has been used to study the binding of fatty acids to serum albumin, to characterize the binding sites for cofactors and substrates in enzyme molecules, to characterize the heme binding site in various hemoproteins, and to study the intercalation of small molecules into the DNA double helix (Experiments 13, 14, 15).

Figure 5.13 shows the structures of extrinsic fluors that have been of value in studying biochemical systems. ANS, dansyl chloride, and fluorescein are used for protein studies, whereas ethidium, proflavine, and various acridines are useful for nucleic acid characterization. Ethidium bromide has the unique characteristic of enhanced fluorescence when bound to double-stranded DNA but not to single-stranded DNA. Aminomethyl coumarin (AMC) is of value as a fluorogenic leaving group in measuring peptidase activity.

Difficulties in Fluorescence Measurements

Fluorescence measurements have much greater sensitivity than absorption measurements. Therefore, the experimenter must take special precautions in making fluorescence measurements because any contaminant or impurity in the system can lead to inaccurate results. The following factors must be considered when preparing for a fluorescence experiment.

Preparation of Reagents and Solutions

Since fluorescence measurements are very sensitive, dilute solutions of biomolecules and other reagents are appropriate. Special precautions must be taken to maintain the integrity of these solutions. All solvents and reagents must be checked for the presence of fluorescent impurities, which can lead to large errors in measurement. "Blank" readings should be taken on all solvents and solutions, and any background fluorescence must be subtracted from the fluorescence of the complete system under study. Solutions should be stored in the dark, in clean glass-stoppered containers, in order to avoid photochemical breakdown of the reagents and contamination by corks

Figure 5.13

Fluorescent molecules useful in biochemical studies.

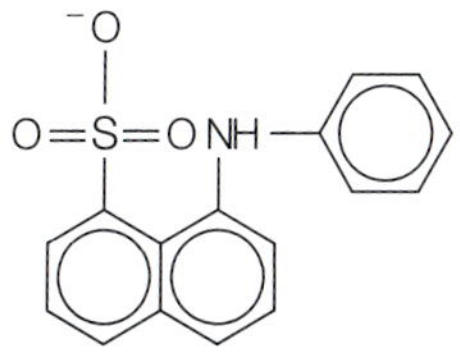

1-Anilino-8-naphthalene sulfonate (ANS)

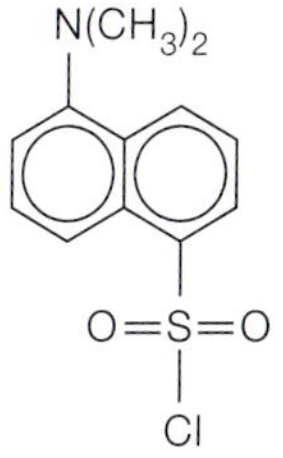

Dansyl chloride

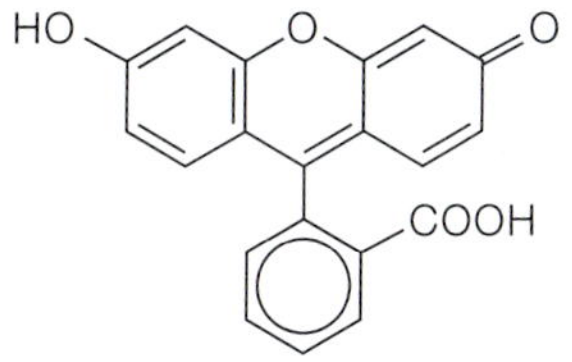

Fluorescein

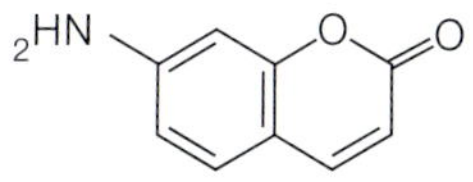

Aminomethyl coumarin (AMC)

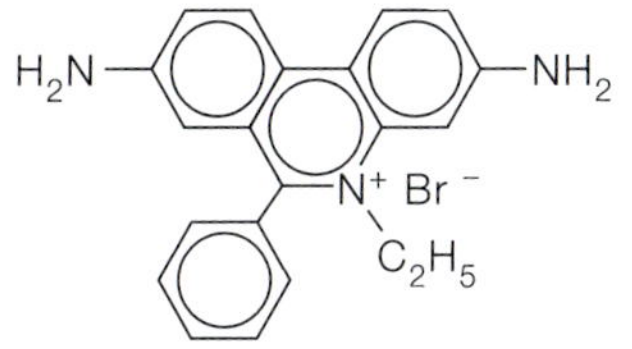

Ethidium bromide

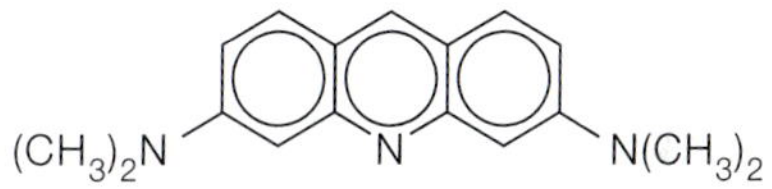

Acridine orange

and rubber stoppers. Some biomolecules, especially proteins, tend to adsorb to glass surfaces, which can lead to loss of fluorescent material or to contamination of fluorescence cuvettes. All glassware must be scrupulously cleaned. Turbid solutions must be clarified by centrifugation or filtration.

Control of Temperature

Fluorescence measurements, unlike absorption, are temperature dependent. All solutions, especially if relative fluorescent measurements are taken, must be thermostated at the same temperature.

C.   OTHER SPECTROSCOPIC METHODS

Nuclear Magnetic Resonance Spectroscopy

Nuclear magnetic resonance (NMR) spectroscopy is a method of absorption spectroscopy that has some characteristics similar to ultraviolet and visible spectroscopy but also some that are unique. In NMR, a molecular sample, usually dissolved in a liquid solvent, is placed in a magnetic field and the absorption of radio-frequency waves by certain nuclei (protons and others) is measured. NMR spectroscopy was originally developed as an analytical tool to determine molecular structure by monitoring the environment of individual protons. The relative positions and intensity of absorption signals provide detailed information from which chemical structures may be elucidated. Early proton NMR studies focused primarily on small organic molecules, some of which had biological origin and significance. Since most biomolecules have a very large number of protons, the observed spectra are extremely complex and difficult to interpret. With the advent of modern techniques—superconducting magnets, Fourier transform analysis, multidimensional spectra, and powerful computer control—NMR has become an important method for the study of biological macromolecules. The fundamental principles, instrumentation, and biochemical applications for NMR will be outlined here.

All nuclei possess a positive charge. For some nuclei, this charge confers the property of spin, which causes the nuclei to act like tiny magnets. The angular momentum of the spin is described by the quantum spin number I. If I is an integral number ($I = 0, 1, 2$, etc.), then there is no net spin and no NMR signal for that nucleus. However, if I is half-integral ($I = \frac{1}{2}, \frac{3}{2}, \frac{5}{2}$, etc.), the nuclei have spins, and when placed in a magnetic field, the spins orient themselves with (parallel) or opposed (antiparallel) to the external magnetic field. The nuclei aligned with the magnetic field have lower energy (are more stable) than those opposed. Energy in the radio-frequency range is sufficient to flip the nuclei from the parallel to the antiparallel alignment. The NMR instrument is designed to measure the energy difference between the nuclear spin states. Absorption of energy may be detected by scanning the radio-frequency range and measuring the absorption that causes spin state transition (resonance). Modern NMR instruments instead maintain a constant radio-frequency and electrically induce small changes in the strength of the magnetic field until resonance is attained. The point of resonance for a nucleus is dependent upon the electronic environment of that nucleus, so an NMR spectrum provides information that helps elucidate biochemical structures.

NMR has found a wide variety of applications in biochemistry. Proton NMR has a long and rich history in organic chemistry and biochemistry. The structures of many small but significant biomolecules were elucidated by proton NMR. Protons on different atoms and in different molecular environments absorb energy of different levels (measured by radio-frequency units or magnetic field units). Two experimentally measured characteristics, **chemical shifts** and **spin-spin coupling,** provide important structural information. The chemical shift (δ) of an absorbing nucleus, measured in parts per million (ppm), is the spectral position of resonance relative to a standard signal, usually tetramethylsilane. The NMR signal for a proton is "split" by interactions with neighboring protons. This characteristic, called spin-spin coupling, helps to determine positions and numbers of equivalent and nonequivalent protons.

NMR experiments are not limited to the study of protons. Resonance signals from other atomic nuclei including ^{13}C, ^{15}N, ^{19}F, and ^{31}P can be

| Table 5.1 | | |

Nuclei Important in Biochemical NMR

Isotope[1]	Natural Abundance (%)[2]	Relevant Biomolecules
^1H	100	Most biomolecules
^{13}C	1	Most biomolecules
^{15}N	0.4	Amino acids, proteins, nucleotides
^{19}F	100	Substitute for H
^{31}P	100	Nucleotides, nucleic acids, and other phosphorylated compounds

[1] *All have a spin of ½.*

[2] *The percentage of this isotope in naturally occurring molecules containing this element.*

detected and measured (see Table 5.1). ^{13}C NMR techniques have been especially valuable for the study of carbohydrate, amino acid, and fatty acid structures. Just as with protons, each distinct carbon atom in a molecule yields a signal that is split by neighboring interacting nuclei (see Figure 5.14). Because of the low natural abundance of the ^{13}C isotope, it is necessary either to enrich the ^{13}C content by chemical synthesis or to use powerful computers and magnets (500–600 MHz). The ^{31}P spectra of phosphorylated biomolecules display a peak for each type of phosphorus (Figure 5.15). NMR instruments now have the sensitivity necessary to measure the *in vivo* concentrations and reactions of biomolecules in cells. For example, the catabolism of glucose by glycolysis in erythrocytes has been monitored by ^{13}C NMR, and the involvement of ATP in phosphoryl-group transfer processes has been studied by ^{31}P NMR (see Figure 5.15).

New techniques for data analysis and improvements in instrumentation have now made it possible to carry out structural and conformational studies of biopolymers including proteins, polysaccharides, and nucleic acids. NMR, which may be done on noncrystalline materials in solution, provides a technique complementary to X-ray diffraction, which requires crystals for analysis. One-dimensional NMR, as described to this point, can offer structural data for smaller molecules. But proteins and other biopolymers with large numbers of protons will yield a very crowded spectrum with many overlapping lines. In multidimensional NMR (2-D, 3-D, 4-D), peaks are spread out through two or more axes to improve resolution. The techniques of correlation spectroscopy (COSY), nuclear Overhausser effect spectroscopy (NOESY), and transverse relaxation-optimized spectroscopy (TROSY) depend on the observation that nonequivalent protons interact with each other. By using multiple-pulse techniques, it is possible to perturb one nucleus and observe the effect on the spin states of other nuclei. The availability of powerful computers and Fourier transform (FT) calculations makes it possible to elucidate structures of proteins up to 40,000 daltons in molecular mass and there is future promise for studies on proteins over 100,000

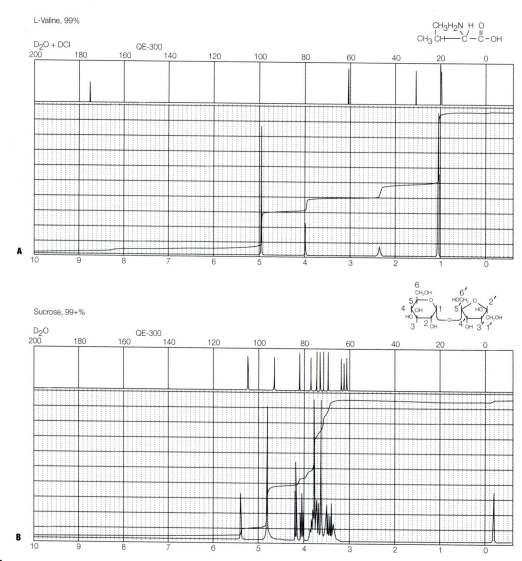

Figure 5.14

¹H and ¹³C FT-NMR spectra of biomolecules. The lower spectra are for ¹H showing a ppm scale of 0–10. TMS standard is at 0 ppm. The upper spectra are for ¹³C showing a ppm scale of 0–200. **A** L-Valine in D_2O + DCl. Can you assign each peak to the correct protons and carbon atoms in the valine structure? Hint: The carboxyl carbon of valine has a peak at about 175 ppm. **B** Sucrose in D_2O. Carbon numbers in the chemical structure correspond to the following peaks in order from 0 to 120 ppm: C-6; C-1′; C-6′; C-4; C-2; C-5; C-3; C-4′; C-3′; C-5′; C-1; C-2′. *Reprinted with permission of Aldrich Chemical Co., Inc.*

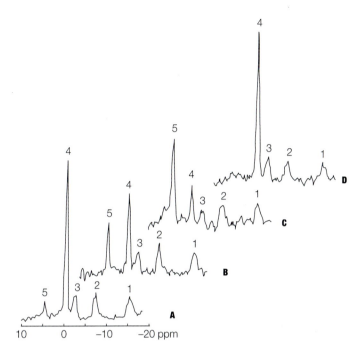

Figure 5.15

^{31}P NMR spectra of human forearm muscle showing the effect of exercise. **A** Before exercise; **B** and **C** during 19 minutes of exercise; **D** 5–6 minutes after C. Peak assignments: 1, β-phosphorus of ATP; 2, α-phosphorus of ATP; 3, γ-phosphorus of ATP; 4, phosphocreatine; 5, Pi. Phosphocreatine is used as a major source of energy during exercise. It is hydrolyzed to creatine and Pi. Note that the level of ATP remains relatively constant during exercise because it is produced and used at about the same rate. After G. Radda, *Science* **233,** 641 (1986). *Reprinted with permission from the American Association for the Advancement of Science.*

daltons. In addition to structural studies, NMR will be applied to studies of conformational changes in macromolecules and interactions between ligands and biopolymers.

Mass Spectrometry

Mass spectrometry (MS) is similar to NMR in that it has historically been of great value in organic chemistry, but only recently has it received attention in the analysis of biological molecules. The MS analysis of biopolymers was initially hindered because samples are usually measured in the gas phase and it was difficult to vaporize these large molecules. The development of new techniques for sample preparation, tandem MS instruments, and powerful computers now makes it possible to study large molecules, especially proteins.

In mass spectrometry, neutral molecules are ionized and their positively charged ion products are directed through an electric and/or magnetic field, where they are separated on the basis of their mass-to-charge ratio (m/z). Ionization of small organic molecules can be done by electron impact; however, ionization of nonvolatile, biological macromolecules requires the use of electrospray techniques or matrix-assisted laser desorption (MALDI). In electrospray ionization, a solution of the macromolecules is sprayed from a metal needle held at a potential of about $+5000$ volts. This results in a spray of tiny droplets containing positively charged ions. The solvent evaporates and the ions are directed into the spectrometer for analysis. In matrix-assisted laser desorption, the macromolecule is placed in a matrix of small organic molecules and irradiated by a laser flash. The resulting positively charged ions are directed into the mass spectrometer for mass analysis.

The most important ions resulting from the removal of an electron by ionization procedures are the positively charged **molecular ions** ($M^{+\cdot}$) or **protonated molecular ions** $(M + nH)^{n+}$. Measurement of the molecular mass of these species provides the molecular weight of the original molecule. The accuracy of this method for molecular weight determination is approximately 0.01%. Molecular mass measurements using gel filtration are typically no better than 5–10%. Some molecular ions are unstable and disintegrate to produce fragment ions. These fragmentation processes are useful in structural elucidation of smaller molecules.

Mass spectrometry may also be used to sequence polypeptides of 25 residues or fewer. Here ionization is accomplished by fast atom bombardment (FAB) and mass analysis carried out by two coupled spectrometers (tandem mass spectrometers). In FAB, the macromolecule sample is dissolved in a viscous, nonvolatile solvent such as glycerol and directed into a mass spectrometer with a stream of neutral atoms such as Ar or Xe. FAB generates predominately the protonated molecular species of the macromolecule $(M + nH)^{n+}$.

Study Problems

1. For each pair of wavelengths listed below, specify which one is higher in energy.
   (a) 1 nm (X-ray) or 10,000 nm (IR)
   (b) 280 nm (UV) or 360 nm (VIS)
   (c) 200,000 nm (microwave) or 800 nm (VIS)

2. In a laboratory experiment, you are asked to determine the molar concentration of a solution of an unknown compound, X. The solution diluted in half by water (1 mL of X and 1 mL of H_2O) has an absorbance at 425 nm of 0.8 and a molar extinction coefficient of 1.5×10^3 M^{-1} cm^{-1}. What is the molar concentration of the original solution of X?

3. Match the spectral region listed below with the appropriate molecular transition that occurs. The first problem is worked as an example.

Spectral region	Transition
__a__ 1. X-rays	a. inner-shell electrons
_____ 2. UV	b. molecular rotations
_____ 3. VIS	c. valence electrons
_____ 4. IR	d. molecular vibrations
_____ 5. microwave	

➤ 4. Several spectroscopic techniques were studied in this chapter. Which experimental techniques involve an actual measurement of radiation absorbed?
(a) UV-VIS spectroscopy
(b) NMR spectroscopy
(c) MS
(d) Fluorescence spectroscopy

5. Explain the differences between a spectrophotometer that uses a photo-tube for a detector and one that uses a photodiode array detector.

➤ 6. Why can you not use a glass cuvette for absorbance measurements in the UV spectral range?

➤ 7. What is the single structural characteristic that all of the fluorescent molecules in Figure 5.13 possess?

➤ 8. Why must a cuvette with four translucent sides be used for fluorescence measurements?

9. Study the ^{1}H spectrum of valine in Figure 5.14A and match each peak to the corresponding proton in the chemical structure. Explain any spin-spin coupling.

10. Study the ^{13}C spectrum of valine in Figure 5.14A. Identify the carbon atom that produces each peak in the spectrum.

Further Reading

R. Garrett and C. Grisham, *Biochemistry,* 2nd ed. (1999), Saunders (Philadelphia), pp. 99–100. Introduction to UV-VIS and NMR.

B. Giles et al., *J. Chem. Educ.* **76,** 1564–1566 (1999). "An In Vivo ^{13}C NMR Analysis of the Anaerobic Yeast Metabolism of 1-^{13}C-Glucose."

R. Hester and R. Girling, Editors, *Spectroscopy of Biological Molecules* (1991), Royal Society of Chemistry Special Publication (Cambridge). Spectroscopy focused on biomolecules.

J. Hollas, *Modern Spectroscopy,* 3rd ed. (1996), John Wiley & Sons (London). Up-to-date theory and applications.

D. Holme and H. Peck, *Analytical Biochemistry* (1998), Addison Wesley Longman (Essex, UK), pp. 36–90, 125–129. An introduction to UV-VIS, NMR, and MS.

A. Lehninger, D. Nelson, and M. Cox, *Principles of Biochemistry,* 3rd ed. (1999), Worth Publishers (New York), pp. 121, 146–149, 180–181. Introduction to spectroscopy.

C. Mathews, K. van Holde, and K. Ahern, *Biochemistry,* 3rd ed. (2000), Benjamin/Cummings (San Francisco), pp. 202–208. Introduction to spectroscopic methods.

F. Mirabella, *Modern Techniques in Applied Molecular Spectroscopy* (1998), John Wiley & Sons (New York). Theory and applications in all areas of spectroscopy.

S. Mostad and A. Glasfeld, *J. Chem. Educ.* **70,** 504–506 (1993). "Using High Field NMR to Determine Dehydrogenase Stereospecificity with Respect to NADH."

A. Owen, *The Diode-Array Advantage in UV/VIS Spectroscopy* (1988), Hewlett-Packard Co. Theory and practice of diode array spectroscopy.

K. Peterman, K. Lentz, and J. Duncan, *J. Chem. Educ.* **75,** 1283–1284 (1998). "A ^{19}F NMR Study of Enzyme Activity."

G. Radda, *Science* **233,** 640–645 (1986). "The Use of NMR Spectroscopy for the Understanding of Disease."

D. Skoog, F. Holler, and T. Nieman, *Principles of Instrumental Analysis,* 5th ed. (1998), Saunders (Philadelphia). An introductory coverage of analytical spectroscopy.

E. Solomon and K. Hodgson, *Spectroscopic Methods in Bioinorganic Chemistry* (1998), Oxford University Press (New York). An excellent, specialized book.

L. Stryer, *Biochemistry,* 4th ed. (1995), Freeman (New York), pp. 52–53, 66–68, 457–458. Application of NMR and MS to biochemistry.

D. Voet and J. Voet, *Biochemistry,* 2nd ed. (1995), John Wiley & Sons (New York), pp. 112, 117–118. Introduction to NMR and MS.

K. Wüthrich, *J. Biol. Chem.* **265,** 22059–22062 (1990). "Protein Structure Determination in Solution by NMR Spectroscopy."

J. Yarger, R. Nieman, and A. Bieber, *J. Chem. Educ.* **74,** 243–246 (1997). "NMR Titration Used to Observe Specific Proton Dissociation in Polyprotic Tripeptides."

Spectroscopy on the Web

http://www.scimedia.com/chem-ed/analytic/ac-meths.htm
   Introduction to all forms of spectroscopy including UV-VIS, NMR, and MS.

http://www.scimedia.com/chem-ed/spec/uv-vis/uv-vis.htm
   Theory and practice of spectroscopy.

http://chemistry.gsu.edu/post_docs/koen/wuv.html
   Read Theory and Spectra Interpretation.

http://chem.external.hp.com/cag/products/uvistech.html
   Introduction and applications of diode array technology

http://www.cis.rit.edu/htbooks/nmr/
   Study the Basics of NMR spectroscopy: Introduction, Math Background, Fourier Transforms, Practical Considerations.

6

RADIOISOTOPES IN BIOCHEMICAL RESEARCH

The use of radioactive isotopes in experimental biochemistry has provided us with a wealth of information about biological processes. In the earlier days of biochemistry, radioisotopes were primarily used to elucidate metabolic pathways. In modern biochemistry, there is probably no experimental technique that offers such a diverse range of applications as that provided by radioactive isotopes. The measurement of radioactivity is now a tool used in all areas of experimental biochemistry, including enzyme assays, biochemical pathways of synthesis and degradation, analysis of biomolecules, measurements of antibodies, binding and transport studies, and many others. Radioisotope use has two advantages over other analytical techniques: (1) sensitive instrumentation allows the detection of minute quantities of radioactive material (some radioactive substances can be measured in the picomole, 10^{-12}, range), and (2) radioisotope techniques offer the ability to differentiate physically between substances that are chemically indistinguishable.

This chapter will explore the origin, properties, detection, biochemical applications and safety concerns of radioisotopes.

A.   ORIGIN AND PROPERTIES OF RADIOACTIVITY

Introduction

Radioactivity results from the spontaneous nuclear disintegration of unstable isotopes. The hydrogen nucleus, consisting of a proton, is represented as ^1_1H. Two additional forms of the hydrogen nucleus contain one and two neutrons; they are represented by ^2_1H and ^3_1H. These **isotopes** of hydrogen are commonly called deuterium and tritium, respectively. All isotopes of hydrogen have an identical number of protons (constant charge, $+1$) but they differ in the number of neutrons. Only tritium is radioactive.

The stability of a nucleus depends on the ratio of neutrons to protons. Some nuclei are unstable and undergo spontaneous nuclear disintegration accompanied by emission of particles. Unstable isotopes of this type are called **radioisotopes.** Three main types of radiation are emitted during nuclear decay: α particles, β particles, and γ rays. The α particle, a helium nucleus, is emitted only by elements of mass number greater than 140. These elements are seldom used in biochemical research.

Most of the commonly used radioisotopes in biochemistry are β emitters. The β particles exist in two forms (β^+, positrons, and β^-, electrons) and are emitted from a given radioisotope with a continuous range of energies. However, an average energy (called the *mean* energy) of β particles from an element can be determined. The mean energy is a characteristic of that isotope and can be used to identify one β emitter in the presence of a second (see Figure 6.1). Equations 6.1 and 6.2 illustrate the disintegration of two β emitters. Here $\bar{v}_e$, represents the antineutrino and v_e the neutrino.

$$\text{>>}\qquad {}^{32}_{15}\text{P} \longrightarrow {}^{32}_{16}\text{S} + \beta^- + \bar{v}_e \qquad\qquad \textit{Equation 6.1}$$

$$\text{>>}\qquad {}^{65}_{30}\text{Zn} \longrightarrow {}^{65}_{29}\text{Cu} + \beta^+ + v_e \qquad\qquad \textit{Equation 6.2}$$

A few radioisotopes of biochemical significance are γ emitters. Emission of a γ ray (a photon of electromagnetic radiation) is often a secondary process occurring after the initial decay by β emission. The disintegration of the isotope ^{131}I is an example of this multistep process.

$$\text{>>}\qquad {}^{131}_{53}\text{I} \longrightarrow \beta^- + {}^{131}_{54}\text{Xe} \longrightarrow {}^{131}_{54}\text{Xe} + \gamma \qquad\qquad \textit{Equation 6.3}$$

Each radioisotope emits γ rays of a distinct energy, which can be measured for identification of the isotope.

The spontaneous disintegration of a nucleus is a first-order kinetic process. That is, the rate of radioactive decay of N atoms ($-dN/dt$, the change of N with time, t) is proportional to the number of radioactive atoms present (Equation 6.4).

$$\text{>>}\qquad -\frac{dN}{dt} = \lambda N \qquad\qquad \textit{Equation 6.4}$$

The proportionality constant, λ, is called the **disintegration** or **decay constant.** Equation 6.4 can be transformed to a more useful equation by integration within the limits of $t = 0$ and t. The result is shown in Equation 6.5.

Figure 6.1

The energy spectra for the β emitters ^3H and ^{14}C. The dashed lines indicate the upper and lower limits for discrimination counting.

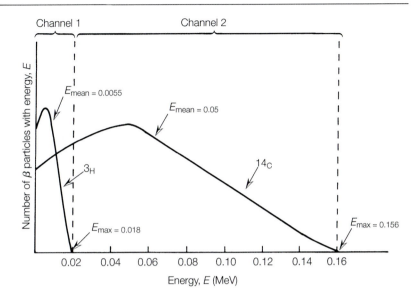

$$N = N_0 e^{-\lambda t}$$

Equation 6.5

where

N_0 = the number of radioactive atoms present at $t = 0$

N = the number of radioactive atoms present at time = t

Equation 6.5 can be expressed in the natural logarithmic form as in Equation 6.6.

$$\ln \frac{N}{N_0} = -\lambda t$$

Equation 6.6

Combining Equations 6.5 and 6.6 leads to a relationship that defines the **half-life.** Half-life, $t_{1/2}$, is a term used to describe the time necessary for one-half of the radioactive atoms initially present in a sample to decay. At the end of the first half-life, N in Equation 6.6 becomes $\frac{1}{2}N_0$, and the result is shown in Equations 6.7 and 6.8.

$$\ln \frac{\frac{1}{2}N_0}{N_0} = -\lambda t_{1/2}$$

Equation 6.7

$$\ln \frac{1}{2} = -\lambda t_{1/2}$$

$$t_{1/2} = \frac{0.693}{\lambda}$$

Equation 6.8

Equations 6.6 and 6.8 allow calculation of the ratio N/N_0 at any time during an experiment. This calculation is especially critical when radioisotopes with short half-lives are used.

Isotopes in Biochemistry

The properties of several radioisotopes that are important in biochemical research are listed in Table 6.1. Note that many of the isotopes are β emitters; however, a few are γ emitters.

  The half-life, defined in the previous section and listed for each isotope in Table 6.1, is an important property when designing experiments using radioisotopes. Using an isotope with a short half-life (for example, ^{24}Na with $t_{1/2} = 15$ hr) is difficult because the radioactivity lost during the course of the experiment is significant. Quantitative measurements made before and after the experiment must be corrected for this loss of activity. Radioactive phosphorus, ^{32}P, an isotope of significant value in biochemical research, has a relatively short half-life (14 days), so if quantitative measurements are made they must be corrected as described in Equations 6.7 and 6.8. More information about the choice of a radioisotope in an experiment, the detec-

Table 6.1

Properties of Biochemically Important Radioisotopes

Isotope	Particle Emitted	Energy of Particle (MeV)	Half-Life, $t_{1/2}$
^{3}H	β^-	0.018	12.3 yr
^{14}C	β^-	0.155	5570 yr
^{22}Na	β^+	0.55	2.6 yr
	γ	1.28	
^{24}Na	β^-	1.39	15 hr
	γ	1.7, 2.75	
^{32}P	β^-	1.71	14.2 days
^{35}S	β^-	0.167	87.1 days
^{36}Cl	β^-	0.714	3×10^5 days
^{40}K	β^-	1.33	1.3×10^9 yr
^{45}Ca	β^-	0.25	165 days
^{59}Fe	β^-	0.46	45 days
	γ	1.1	
^{60}Co	β^-	0.318	5.3 yr
	γ	1.3	
^{65}Zn	β^+	0.33	245 days
	γ	1.14	
^{90}Sr	β^-	0.54	29 yr
^{125}I	γ	0.035	60 days
	γ	0.027	
^{131}I	β^-	0.61, 0.33	8.1 days
	γ	0.64, 0.36, 0.28	
^{137}Cs	β^-	0.51	30 yr
	γ	0.66	
^{226}Ra	α	4.78	1600 yr
	γ	0.19	

tion and measurement of that radioisotope, and safety rules for handling radioisotopes will be given later in the chapter.

Mention should also be made of short-lived isotopes that are important in biotechnology and medical biochemistry. The isotopes ^{11}C, ^{13}N, ^{15}O, and ^{18}F, which are positron emitters, are crucial for use in positron emission tomography (PET).

Units of Radioactivity

The basic unit of radioactivity is the **curie,** Ci. One curie is the amount of radioactive material that emits particles at a rate of 3.7×10^{10} disintegrations per second (dps), or 2.2×10^{12} min^{-1} (dpm). Amounts that large are seldom used in experimentation, so subdivisions are convenient. The **millicurie** (mCi, 2.2×10^9 min^{-1}) and **microcurie** (μCi, 2.2×10^6 min^{-1}) are standard units for radioactive measurements (see Table 6.2). The radioactivity unit of the meter-kilogram-seconds (MKS) system is the **becquerel** (Bq). A becquerel, named in honor of Antoine Becquerel, who studied uranium radiation, represents one disintegration per second. The two systems of measurement are related by the definition 1 curie $= 3.70 \times 10^{10}$ becquerels. Since the becquerel is such a small unit, radioactive units are sometimes reported in MBq (mega, 10^6) or TBq (tera, 10^{12}). Both unit systems are in common use today, and radioisotopes received through commercial sources are labeled in curies and bequerels.

The number of disintegrations emitted by a radioactive sample depends on the purity of the sample (number of radioactive atoms present) and the decay constant, λ. Therefore, radioactive decay is also expressed in terms of **specific activity,** the disintegration rate per unit mass of radioactive atoms. Typical units for specific activity are mCi/mmole and μCi/μmole.

Although radioactivity is defined in terms of nuclear disintegrations per unit of time, rarely does one measure this absolute number in the laboratory. Instruments that detect and count emitted particles respond to only a small fraction of the particles. Data from a radiation counter are in counts per minute (cpm), which can be converted to actual disintegrations per minute if

Table 6.2

Units of Radioactivity

Units Name	Multiplication Factor (relative to curie)	Activity (dps)
Curie (Ci)	1.0	3.70×10^{10}
Millicurie (mCi)	10^{-3}	3.70×10^7
Microcurie (μCi)	10^{-6}	3.70×10^4
Nanocurie (nCi)	10^{-9}	3.70×10
Becquerel (Bq)	2.7×10^{-11}	1.0
Mega becquerel (MBq)	2.7×10^{-5}	1.0×10^6

the counting efficiency of the instrument is known. The percent efficiency of an instrument is determined by counting a standard compound of known radioactivity and using the ratio of detected activity (observed cpm) to actual activity (disintegrations per minute). Equation 6.9 illustrates the calculation using a standard of 1 μCi.

$$\% \text{ efficiency} = \frac{\text{observed cpm of standard}}{\text{dpm of 1 } \mu\text{Ci of standard}} \times 100$$

$$= \frac{\text{observed cpm}}{2.2 \times 10^6 \text{ dpm}} \times 100 \qquad \textit{Equation 6.9}$$

B. DETECTION AND MEASUREMENT OF RADIOACTIVITY

Since most of the radioisotopes used in biochemical research are β emitters, only methods that detect and measure β particles will be emphasized. Two counting techniques are in current use, scintillation counting and Geiger-Müller counting.

Liquid Scintillation Counting

Samples for liquid scintillation counting consist of three components: (1) the radioactive material; (2) a solvent, usually aromatic, in which the radioactive substance is dissolved or suspended; and (3) one or more organic fluorescent substances. Components 2 and 3 make up the **liquid scintillation system** or cocktail. β particles emitted from the radioactive sample interact with the scintillation system, producing small flashes of light or scintillations. The light flashes are detected by a photomultiplier tube (PMT). Electronic pulses from the PMT are amplified and registered by a counting device called a scaler. A schematic diagram of a typical scintillation counter is shown in Figure 6.2.

The scintillation process, in detail, begins with the collision of emitted β particles with solvent molecules, S (Equation 6.10).

$$\beta + S \longrightarrow S^* + \beta \text{ (less energetic)} \qquad \textit{Equation 6.10}$$

Contact between the energetic β particles and S in the ground state results in transfer of energy and conversion of an S molecule into an excited state, S^*. Aromatic solvents are most often used because their electrons are easily promoted to an excited state orbital (see discussion of fluorescence, Chapter 5). The β particle after one collision still has sufficient energy to excite several more solvent molecules. The excited solvent molecules normally return to the ground state by emission of a photon, $S^* \longrightarrow S + h\nu$. Photons from the typical aromatic solvent are of short wavelength and are not efficiently detected by photocells. A convenient way to resolve this technical

Figure 6.2

Diagram of a typical scintillation counter, showing coincidence circuitry.

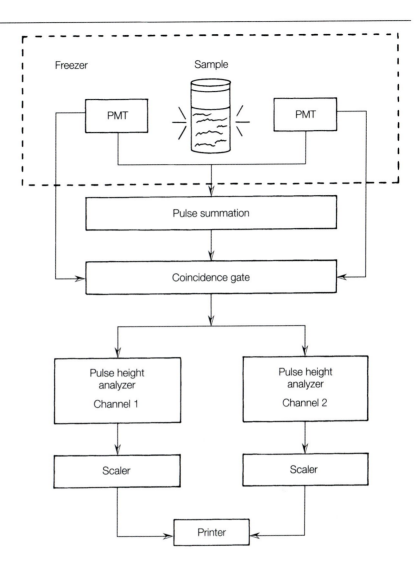

problem is to add one or more fluorescent substances (fluors) to the scintillation mixture. Excited solvent molecules interact with a **primary fluor,** F_1, as shown in Equation 6.11.

>> $$S^* + F_1 \longrightarrow S + F_1^*$$ *Equation 6.11*

Energy is transferred from S^* to F_1, resulting in ground state S molecules and excited F_1 molecules, F_1^*. F_1^* molecules are fluorescent and emit light of a longer wavelength than S^* (Equation 6.12).

>> $F_1^* \longrightarrow F_1 + h\nu_1$ *Equation 6.12*

If the light emitted during the decay of F_1^* is still of a wavelength too short for efficient measurement by a PMT, a **secondary fluor,** F_2, that accepts energy from F_1^* may be added to the scintillation system. Equations 6.13 and 6.14 outline the continued energy transfer process and fluorescence of F_2^*.

>> $F_1^* + F_2 \longrightarrow F_1 + F_2^*$ *Equation 6.13*

>> $F_2^* \longrightarrow F_2 + h\nu_2$ *Equation 6.14*

The light, $h\nu_2$, from F_2^* is of longer wavelength than $h\nu_1$ from F_1^* and is more efficiently detected by a PMT. Two widely used primary and secondary fluors are 2,5-diphenyloxazole (PPO) with an emission maximum of 380 nm and 1,4-bis-2-(5-phenyloxazolyl)benzene (POPOP) with an emission maximum of 420 nm.

The most basic elements in a liquid scintillation counter are the PMT, a pulse amplifier, and a counter, called a **scaler.** This simple assembly may be used for counting; however, there are many problems and disadvantages with this setup. Many of the difficulties can be alleviated by more sophisticated instrumental features. Some of the problems and practical solutions are outlined below.

Thermal Noise in Photomultiplier Tubes

The energies of the β particles from most β emitters are very low. This, of course, leads to low-energy photons emitted from the fluors and relatively low-energy electrical pulses in the PMT. In addition, photomultiplier tubes produce thermal background noise with 25 to 30% of the energy associated with the fluorescence-emitted photons. This difficulty cannot be completely eliminated, but its effects can be lessened by placing the samples and the PMT in a freezer at -5 to $-8°C$ in order to decrease thermal noise.

A second way to help resolve the thermal noise problem is to use two photomultiplier tubes for detection of scintillations. Each flash of light that is detected by the photomultiplier tubes is fed into a **coincidence circuit.** A coincidence circuit counts only the flashes that arrive simultaneously at the two photodetectors. Electrical pulses that are the result of simultaneous random emission (thermal noise) in the individual tubes are very unlikely. A schematic diagram of a typical scintillation counter with coincidence circuitry is shown in Figure 6.2.

Counting More Than One Isotope in a Sample

The basic liquid scintillation counter with coincidence circuitry can be used only to count samples containing one type of isotope. Many experiments in biochemistry require the counting of just one isotope; however, more valuable experiments can be performed if two radioisotopes can be simultaneously counted in a single sample (double-labeling experiments). The basic scintillation counter just described has no means of discriminating between electrical pulses of different energies.

The size of the current generated in a photocell is nearly proportional to the energy of the β particle initiating the pulse. Recall that β particles from different isotopes have characteristic energy spectra with an average energy (see Figure 6.1). Modern scintillation counters are equipped with **pulse height analyzers** that measure the size of the electrical pulse and count only the pulses within preselected energy limits set by **discriminators.** The circuitry required for pulse height analysis and energy discrimination of β particles is shown in Figure 6.2. Discriminators are electronic "windows" that can be adjusted to count β particles within certain energy or voltage ranges called **channels.** The channels are set to a **lower limit** and an **upper limit,** and all voltages within those limits are counted. Figure 6.1 illustrates the function of discriminators for the counting of ^{3}H and ^{14}C in a single sample. Discriminator channel 1 is adjusted to accept typical β particles emitted from ^{3}H and channel 2 is adjusted to receive β particles of the energy characteristic of ^{14}C.

Quenching

Any chemical agent or experimental condition that reduces the efficiency of the scintillation and detection process leads to a reduced level of counting, or **quenching.** Quenching may be caused by a decrease in the amount of fluorescence from the primary and secondary scintillators (fluors) or a decrease in light activating the PMT. There are four common origins of quenching.

Color quenching is a problem if chemical substances that absorb photons from the secondary fluors are present in the scintillation mixture. Since the secondary fluors emit light in the visible region between 410 and 420 nm, colored substances may absorb the emitted light before it is detected by the photocells. Radioactive samples may be treated to remove colored impurities before mixing with the scintillation solvent.

Chemical quenching occurs when chemical substances in the scintillation solution interact with excited solvent and fluor molecules and decrease the efficiency of the scintillation process. To avoid this type of quenching the sample can be purified or the fluors can be increased in concentration. Modern scintillation counters have computer programs to correct for color and chemical quenching.

Point quenching occurs if the radioactive sample is not completely dissolved in the solvent. The emitted β particles may be absorbed before they have a chance to interact with solvent molecules. The addition of solubilizing agents such as Cab-O-Sil or Thixin decreases point quenching by converting the liquid scintillator to a gel.

Dilution quenching results when a large volume of liquid radioactive sample is added to the scintillation solution. In most cases, this type of quenching cannot be eliminated, but it can be corrected by one of the techniques discussed below.

Since quenching can occur during all experimental counting of radioisotopes, it is important to be able to determine the extent of the reduced counting efficiency. Two methods for quench correction are in common use.

In the **internal standard ratio** method, the sample, X, under study is first counted; then a known amount of a standard radioactive solution is added to the sample, and it is counted again. The absolute activity of the original sample A_x, is represented by Equation 6.15.

>> $$A_x = A_s \frac{C_x}{C_s}$$ **Equation 6.15**

where

A_x = activity of the sample, X
C_x = radioactive counts from sample
A_s = activity of the standard
C_s = radioactive counts from standard

The absolute value of C_s is determined from Equation 6.16.

>> $$C_T = C_x + C_s$$ **Equation 6.16**

where

C_T = total radioactive counts from sample plus standard

Equation 6.15 can then be modified to Equation 6.17.

>> $$A_x = \frac{A_s C_x}{C_T - C_x}$$ **Equation 6.17**

The internal standard ratio method for quench correction is tedious and time-consuming and it destroys the sample, so it is not an ideal method. Scintillation counters are equipped with a standard radiation source inside the instrument but outside the scintillation solution. The radiation source, usually a gamma emitter, is mechanically moved into a position next to the vial containing the sample, and the combined system of standard and sample is counted. Gamma rays from the standard excite solvent molecules in the sample, and the scintillation process occurs as previously described. However, the instrument is adjusted to register only scintillations due to γ particle collisions with solvent molecules. This method for quench correction, called the **external standard method,** is fast and precise.

Scintillation Cocktails and Sample Preparation

Many of the quenching problems discussed earlier can be lessened if an experiment is carefully planned. Many sample preparation methods and scintillation liquids are available for use. Only a few of the more common techniques will be mentioned here. Most liquid or solid radioactive samples can be counted by mixing with a **scintillation cocktail,** a mixture of solvent and fluor(s). The radioactive sample and scintillation solution are placed in a glass, polyethylene, polypropylene, polyester, or polycarbonate vial for counting. Traditional counting vials include a 20-mL-standard vial, 6-mL miniature vial, 1-mL-Eppendorf tube, or 200-μL microfuge tube. If glass is used, it must contain a low potassium content because naturally occurring ^{40}K is a β emitter. Two major types of solvent systems are available: (1) those that are immiscible with water, in which only organic samples may be used, and (2) those that dissolve aqueous samples. During the early years of liquid scintillation counting the most common solvents used were toluene and xylene for organic samples and dioxane for aqueous samples. These solvents provided high counting efficiency. However, on the downside, they were flammable (flash points between 4 and 25°C), highly toxic, and presented major disposal costs and problems. In addition, they penetrated plastic counting vials, releasing vapors and liquids into the laboratory. A new environmental awareness has prompted the development of biosafe liquid scintillation cocktails. The use of highly alkylated aromatic solvents has produced cocktails that are fully biodegradable (converted to CO_2 and H_2O), have higher flash points (above 120°C) and offer high counting efficiency. The newer solvents may be disposed of after use by incineration, which greatly reduces costs compared to road transport to a disposal site. Occupational risks are also lessened because the solutions are defined by the Environmental Protection Agency as nontoxic.

Solid radioactive samples or those that are insoluble in either type of solvent may be quantified by collection onto small pieces of a solid support (filter paper or cellulose membrane) and added directly to the cocktail for counting. The efficiency of counting these samples depends on the support but is usually less than that of counting a homogeneous sample.

Liquid scintillation counting as described above is especially useful for counting weak β emitters such as ^3H, ^{14}C, and ^{35}S. Counting high-energy β emission, such as that of ^{32}P, does not require a fluor because the β particle can be detected directly by the PMT. Samples containing ^{32}P may be counted directly with relatively high efficiency in water solution. This mode of direct measurement is called **Cerenkov counting.**

Geiger-Müller Counting of Radioactivity

Another method of radiation detection that has value in biochemistry is the use of gas ionization chambers. The most common device that uses this technique is the **Geiger-Müller tube** (G-M tube). When β particles pass through a gas, they collide with atoms and may cause ejection of an electron from a gas atom. This results in the formation of an ion pair made up

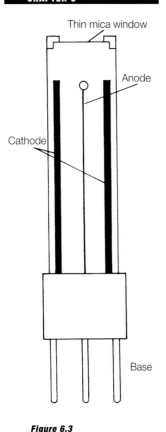

Figure 6.3

A Geiger-Müller tube.

of the negatively charged electron and the positively charged atom. If this ionization occurs between two charged electrodes (an anode and a cathode), the electron will be attracted to the anode and the positive ion to the cathode. This results in a small current in the electrode system. If only a low voltage difference exists between the anode and cathode, the ion pairs will move slowly and will, most likely, recombine to form neutral atoms. Clearly, this will result in no pulse in the electric circuit because the individual ions do not reach the respective electrodes. At higher voltages, the charged particles are greatly accelerated toward the electrodes and collide many times with un-ionized gas atoms. This leads to extensive ionization and a cascade or avalanche of ions. If the voltage is high enough (1000 volts for most G-M tubes), all ions are collected at the electrodes. A Geiger-Müller counting system uses this voltage region for ion acceleration and detection. A typical G-M tube is diagramed in Figure 6.3. It consists of a mica window for entry of β particles from the radiation source, an anode down the center of the tube, and a cathode surface inside the walls. A high voltage is applied between the electrodes. Current generated from electron movement toward the anode is amplified, measured, and converted to counts per minute. The cylinder contains an inert gas that readily ionizes (argon, helium, or neon) plus a quenching gas (Q gas, usually butane) to reduce continuous ionization of the inert gas. Beta particles of high energy emitted from atoms such as ^{24}Na, ^{32}P, and ^{40}K have little difficulty entering the cylinder by penetrating the mica window. Particles from weak β emitters (especially ^{14}C and ^{3}H) cannot efficiently pass through the window to induce ionization inside the chamber. Modified G-M tubes with thin Mylar windows, called **flow window tubes,** may be used to count weak β emitters.

G-M counting has several disadvantages compared to liquid scintillation counting. The counting efficiency of a G-M system is not as high. The response time for G-M tubes is longer than for photomultiplier tubes; therefore, samples of high radioactivity are not efficiently counted by the G-M tube. G-M meters are seldom used when careful, accurate measurements are required. They are most useful in the biochemical laboratory as survey meters to monitor and detect radioactive contamination on lab benches, glassware, equipment, and laboratory personnel.

Scintillation Counting of γ Rays

^{24}Na and ^{131}I are both β and γ emitters and are often used in biochemical research. Gamma rays are more energetic than β particles, and denser materials are necessary for absorption. A gamma counter consists of a sample well, a sodium iodide crystal as fluor, and a photomultiplier tube (see Figure 6.4). The high-energy γ rays are not absorbed by the scintillation solution or vial, but they interact with a crystal fluor, producing scintillations. The scintillations are detected by photomultiplier tubes and electronically counted.

Figure 6.4

Diagram of a γ counter.

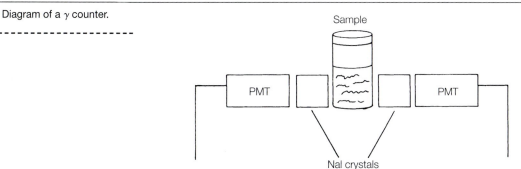

Background Radiation

All radiation counting devices register counts even if there is no specific or direct source of radiation near the detectors. This is due to **background radiation.** It has many sources, including natural radioactivity, cosmic rays, radioisotopes in the construction materials of the counter, radioactive chemicals stored near the counter, and contamination of sample vials or counting equipment. Background count in a scintillation counter will depend on the scintillation solution used, but is usually 30 to 50 cpm. Background counts that are the result of natural sources cannot be eliminated; however, those due to contamination can be greatly reduced if the laboratory and counting instrument are kept clean and free of radioactive contamination. Since it is difficult to regulate background radiation, it is usually necessary to monitor its level. The typical procedure is to count a radioactive sample and then count a background sample containing the same scintillation system but no radioisotope. The actual or corrected activity is obtained by subtracting the background counts from the results for the radioactive sample.

Applications of Radioisotopes

The applications of radioisotopes in biochemical measurements are too numerous to outline here. Excellent reference books, some of which are listed at the end of this chapter, provide experimental procedures for radioisotope utilization. The technique of **autoradiography,** however, should be mentioned. This procedure allows the detection and localization of a radioactive substance (molecule or atom) in a tissue, cell, or cell organelle. Briefly, the technique involves placing a radioactive material directly onto or close to a photographic emulsion. Radioisotopes in the material emit radiation that impinges on the photographic plate, activating silver halide crystals in the emulsion. Upon development of the plate, a pattern or image is displayed that yields information about the location and amount of radioactive material in the sample. A densitometer may be used to scan the autoradiograph and quantify the amount of radioactivity in each region. One of the most common uses of autoradiography in biochemistry and molecular biology is in the detection and quantification of ^{32}P-labeled nucleic acids on polyacrylamide or agarose gels after electrophoresis (Figure 6.5). Autoradiography has also been useful in concentration and localization studies of biomolecules in cells and cell organelles.

Statistical Analysis of Radioactivity Measurements

Radioactive decay is a random process. It is impossible to predict when a radioactive event will occur. Therefore, counting measurements provide only average rates of decay. However, counting measurements may be treated by Gaussian distribution analysis to determine an average counting rate, standard deviation, percent confidence level, and other statistical parameters. An introduction to statistical analysis of radioactivity counting data and other experimental measurements is given in Chapter 1.

C.   RADIOISOTOPES AND SAFETY

Safety should be of major concern in performing any chemistry experiment, but when radioisotopes are involved, special precautions must be taken. Many chemistry and biochemistry departments have specially equipped laboratories for radioisotope work. Alternatively, your instructor should set aside a specific area of your laboratory for using radioactive materials. In either situation, specific guidelines should be followed.

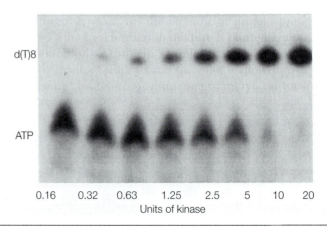

Figure 6.5

Autoradiograph showing the transfer of ^{32}P from γ-^{32}P–labeled ATP to the 5' end of a polynucleotide, $(dT)_8$, catalyzed by polynucleotide kinase. Reagents were incubated for 30 min and electrophoresed on a 20% polyacrylamide, *7 M* urea gel and autoradiographed. Note that as more units of kinase are added, more ^{32}P-labeled polynucleotide is made. The dark regions indicate the presence of ^{32}P. *Source: Copyright © 1992* New England Biolabs Catalog. *Reprinted with permission.*

Preparation for the Experiment

> **CAUTION**
>
> If you are pregnant or may become pregnant during this course, special precautions in the use of radioisotopes must be considered. Since each institution has its own rules and regulations, consult your instructor. Some institutions may require the use of a personnel badge in order to monitor the exposure for each individual. Your instructor will provide this if necessary.

1.   The first responsibility of the student is to become knowledgeable about the properties and hazards of the radioactive substances to be used. You must know which radioisotopes are to be handled and the form, gas, liquid or solid, of the material. It is also important to know whether the isotope is a β and/or γ emitter and whether it is "weak" or "strong." ^{3}H and ^{14}C are considered weak β emitters and ^{32}P is a strong β emitter.

2.   Confine your work with radioisotopes to a small area in the laboratory. A convenient plan is to use a stainless steel tray lined with absorbent blotter paper coated on the bottom side with polyethylene. The paper must be replaced every day. If the radioactive materials are volatile, the work should be done in a fume hood. If spills occur, a small work area such as a tray is much easier to clean than a large lab bench. If ^{32}P or other strong β emitter is used, it is necessary to work at all times with shielding between yourself and the radioactive samples. The most cost-effective and convenient shielding material is Plexiglas. The thicknesses of shielding required for various materials are given in Table 6.3.

3.   Label all glassware and equipment that will be used with special adhesive tape labeled "Radioactive." Plan the experiment so that a minimal number of transfers of radioactive materials is required. This will reduce the amount of contaminated glassware.

Table 6.3	
Maximum Range of β Particles from ^{32}P Through Various Materials	
Material	Maximum Range
Air	6 m
Water	0.85 cm
Plexiglas	0.64 cm
Glass	0.38 cm
Lead	0.045 cm

4. Two types of containers should be available for disposal purposes. One should be labeled "Liquid Radioactive Waste" and used for all waste solutions; the other, "Solid Radioactive Waste," for blotter paper, broken glassware, etc. **Liquid wastes must not be poured down any drain, nor solid wastes deposited in normal trash cans.**

5. Take all precautions against contaminating yourself or fellow workers. Always wear safety goggles, gloves, and a lab coat. No smoking, eating, or drinking is allowed in the laboratory. It is especially hazardous to ingest radioactive materials.

6. The laboratory should be equipped with a portable G-M survey meter in order to check spills or possible self-contamination.

7. A specific area with a sink should be set aside for washing weakly contaminated glassware and equipment.

8. All radioactive materials should be stored in well-labeled, glass containers. The label must include your name, the type of isotope, the radioactive compound, the total amount of radioactivity, the specific activity, and the date of measurement.

Performing the Experiment

1. **Never pipet any solutions by mouth; always use a dedicated mechanical pipet filler. Wear gloves, lab coat, and goggles.**

2. Contaminated glassware should be kept separated from uncontaminated. Contaminated beakers and flasks are placed in the special sink or other container for washing. Clean and wash all equipment with soap and water immediately after the experiment has been completed. If water-insoluble materials are being used, the first washing should be done with an organic solvent such as acetone. Soak contaminated pipets in a container filled with water. All broken glassware is disposed of in the "Solid Radioactive Waste" container.

3. Minor spills must be cleaned first with absorbent blotter paper and then *thoroughly* rinsed with water. Always wear gloves during cleanup. Dispose of the blotter paper in the "Solid Radioactive Waste" container. The spill area should then be checked with a portable G-M counter.

4. All skin cuts, accidental ingestion of radioactive materials, and major spills must be reported immediately to the instructor.

5. After final clean-up of all contaminated glassware and the work area, remove your gloves and thoroughly wash your hands with soap and warm water. Use a portable G-M counter to check for contamination on your hands, clothes, and work area in the laboratory. Report any areas of unusually high count (above 200 cpm) to your instructor.

Study Problems

▶ *1.* Calculate the half-life of an isotope from the following experimental measurements. At time = 0, the activity is 300 disintegrations per minute. After 1 hour, the activity is 200 disintegrations per minute.

▶ *2.* Calculate λ, the decay constant, for ^{32}P.

▶ *3.* How can you determine if the carbon isotope $^{12}_{6}C$ is a stable isotope?

▶ *4.* How long will it take a sample of ^{32}P that contains 65,000 disintegrations per minute to decay to 1500 disintegrations per minute?

▶ *5.* Which is a higher energy β emitter, ^{32}P or ^{35}S?

   6. Define the following terms related to radioactivity units.
   (a) Becquerel
   (b) Millicurie
   (c) Specific activity

▶ *7.* What type of shielding is essential if you are working with ^{32}P-labeled ATP?

▶ *8.* List five radioisotopes that may be measured using a liquid scintillation counter.

   9. Assume that a radioisotope sample, counted in one measurement, yielded 1450 counts per minute over a period of 15 minutes. Use information in Chapter 1 to determine the accuracy for the 95% confidence level.

▶ *10.* What is the quantitative relationship between a becquerel and a curie?

Further Reading

D. Billington, G. Jayson and P. Maltby, *Radioisotopes* (1992), Bios Scientific Publishers (Oxford). Excellent reference on principles, methods, techniques, and applications.

W. Bonner, in *Methods in Enzymology,* Vol. 152, S. Berger and A. Kimmel, Editors (1987), Academic Press (Orlando, FL), pp. 55–61. "Autoradiograms: ^{35}S and ^{32}P."

G. Chase and J. Rabinowitz, *Principles of Radioisotope Methodology,* 3rd ed. (1967), Burgess (Minneapolis), pp. 75–108. Although somewhat outdated, this is still a useful book.

R. Garrett and C. Grisham, *Biochemistry,* 2nd ed. (1999), Saunders College Publishing (Orlando, FL), pp. 580–581. Isotopes in biochemistry.

D. Holme and H. Peck, *Analytical Biochemistry,* 3rd ed. (1998), Addison Wesley Longman (Essex, UK), pp. 196–209. Nature of radioactivity and use of isotopes.

C. Mathews, K. van Holde, and K. Ahern, *Biochemistry,* 3rd ed. (2000), Benjamin/Cummings (San Francisco), pp. 440–444. An introduction to radioisotopes and the liquid scintillation counter.

C. Pike and J. Berry, *Biochem. Educ.* **17,** 96–98 (1989). "Radiochemical Assay of Ribulose Bisphosphate Carboxylase."

J. Robyt and B. White, *Biochemical Techniques: Theory and Practice* (1987), Brooks/Cole (Monterey, CA), pp. 73–128. Theory, measurement, and use of radioisotopes.

R. Slater, Editor, *Radioisotopes in Biology—A Practical Approach* (1990), IRL Press (Oxford). Theory and techniques for the use of isotopes.

D. Voet and J. Voet, *Biochemistry,* 2nd ed. (1995), John Wiley & Sons (New York), pp. 424–427. Isotopes in biochemistry.

R. Zoon, in *Methods in Enzymology,* Vol. 152, S. Berger and A. Kimmel, Editors (1987), Academic Press (Orlando, FL), pp. 25–29. "Safety with ^{32}P- and ^{35}S-Labeled Compounds."

Radioisotopes on the Web

http://www.wallac.fi/catalog/beta.htm
> Instrumentation for radioisotope counting.

http://www.ruf.rice.edu/~bioslabs/methods.html
> Select option, Detection and use of Radioisotopes, and review appropriate topics in radioisotope counting.

http://nucleus.wpi.edu/Reactor/labs.html
> Review pertinent Laboratory Handouts.
> Beta particle absorption
> Compound radioisotopic decay
> Radioisotopic decay
> Introduction to Geiger detectors
> Introduction to counting statistics

CENTRIFUGATION IN BIOCHEMICAL RESEARCH

A centrifuge of some kind is found in every biochemistry laboratory. Centrifuges have many uses, but they are used primarily for the preparation of biological samples and the analytical measurement of the hydrodynamic properties (shape, size, density, etc.) of purified macromolecules or cellular organelles. This is done by subjecting a biological sample to an intense force by spinning the sample at high speed. These experimental conditions cause sedimentation of particles, cell organelles, or macromolecules at a rate that depends on their masses, sizes, and densities.

In this chapter we will explore the underlying principles of centrifugation and discuss the application of this technique to the isolation and characterization of biological molecules and cellular components.

A. BASIC PRINCIPLES OF CENTRIFUGATION

A particle, whether it is a precipitate, a macromolecule, or a cell organelle, is subjected to a centrifugal force when it is rotated at a high rate of speed. The **centrifugal force,** F, is defined by Equation 7.1.

>> $$F = m\omega^2 r$$ **Equation 7.1**

where

F = intensity of the centrifugal force

m = effective mass of the sedimenting particle

ω = angular velocity of rotation in rad/sec

r = distance of the migrating particles from the central axis of rotation

The force on a sedimenting particle increases with the velocity of the rotation and the distance of the particle from the axis of rotation. A more common measurement of F, in terms of the earth's gravitation force, g, is **relative centrifugal force**, RCF, defined by Equation 7.2.

>> $$RCF = (1.119 \times 10^{-5})(rpm)^2(r)$$ *Equation 7.2*

This equation relates RCF to revolutions per minute of the sample. Equation 7.2 dictates that the RCF on a sample will vary with r, the distance of the sedimenting particles from the axis of rotation (see Figure 7.1). It is convenient to determine RCF by use of the nomogram in Figure 7.2. It should be clear from Figures 7.1 and 7.2 that, since RCF varies with r, it is important to define r for an experimental run. Often an average RCF is determined using a value for r midway between the top and bottom of the sample container. The RCF value is reported as "a number times gravity, g." To illustrate this, consider a sample with an average r of 7 cm being rotated at 20,000 rpm. From Figure 7.2, RCF is 32,000 × g.

Although this introduction outlines the basic principles of centrifugation, it does not take into account other factors that influence the rate of particle sedimentation. Centrifuged particles migrate at a rate that depends on the mass, shape, and density of the particle and the density of the medium. The centrifugal force felt by the particle is defined by Equation 7.1. The term m is the **effective mass** of the particle, that is, the actual mass,

Figure 7.1

A diagram illustrating the variation of RCF with r, the distance of the sedimenting particles from the axis of rotation. *Courtesy of Beckman Instruments, Inc.*

- -

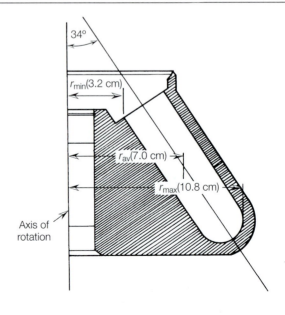

Figure 7.2

A nomogram for estimating RCF. To use, place a straight edge connecting the instrumental revolutions (right column) with the distance from the center of rotation (left column). The RCF is read where the straight edge crosses the middle column. *Courtesy of Beckman Instruments, Inc.*

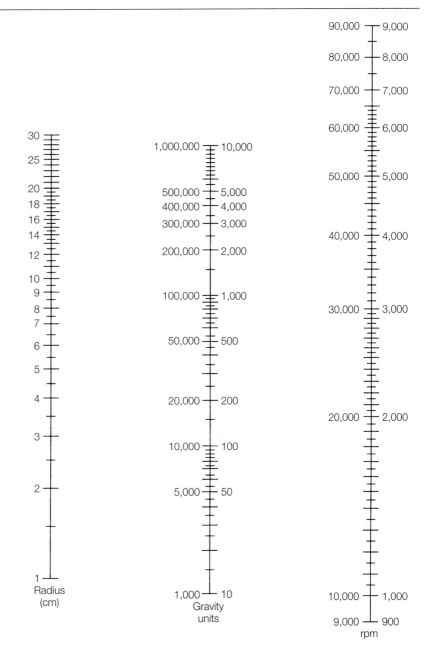

m_0, minus a correction factor for the weight of water displaced (buoyancy factor) (Equation 7.3).

>> $$m = m_0 - m_0\bar{v}\rho$$ *Equation 7.3*

where

m = effective mass of the sedimenting particle

m_0 = actual mass of the particle

$\bar{v}$ = partial specific volume, the volume change occurring when a particle is placed in a large excess of solvent

ρ = density of the solvent or medium

Equations 7.1 and 7.3 may be combined to describe the centrifugal force on a particle (Equation 7.4).

>> $$F = m_0(1 - \bar{v}\rho)\omega^2 r$$ *Equation 7.4*

As particles sediment under the influence of the centrifugal field, their movement is countered by a resistance force, the **frictional force.** The frictional force is defined by Equation 7.5.

>> Frictional force $= fv$ *Equation 7.5*

where

f = frictional coefficient

v = velocity of the sedimenting particle (sedimentation velocity)

The **frictional coefficient,** f, depends on the size and shape of the particle, as well as the viscosity of the solvent. The frictional force increases with the velocity of the particle until a constant velocity is reached. At this point, the two forces are balanced (Equation 7.6).

>> $$m_0(1 - \bar{v}\rho)\omega^2 r = fv = f\left(\frac{dr}{dt}\right)$$ *Equation 7.6*

The rate of sedimentation, sometimes called **sedimentation velocity,** v, is defined by Equation 7.7.

>> $$v = \frac{dr}{dt} = \frac{m_0(1 - \bar{v}\rho)\omega^2 r}{f}$$ *Equation 7.7*

It is cumbersome and sometimes impractical to express sedimentation velocity in terms of ρ, $\bar{v}$, and f, since these factors are difficult to measure. A new term, **sedimentation coefficient,** s (the ratio of sedimentation velocity to centrifugal force) is introduced by rearranging Equation 7.7 to Equation 7.8.

>>

$$s = \frac{v}{\omega^2 r} = \frac{m_0(1 - \bar{v}\rho)}{f}$$

Equation 7.8

The term s is most often defined under standard conditions, 20°C and water as the medium, and denoted by $s_{20,\,w}$. The s value is a physical characteristic used to classify biological macromolecules and cell organelles. Sedimentation coefficients are in the range 1×10^{-13} to $10,000 \times 10^{-13}$ second. For numerical convenience, sedimentation coefficients are expressed in Svedberg units, S, where $1\ S = 1 \times 10^{-13}$ second. Human hemoglobin has an s value of 4.5×10^{-13} second or 4.5 S. The value of S for several biomolecules, bacterial cells, and cell organelles is shown in Figure 7.3. Note in the figure that there appears to be a direct relationship between the S value and the molecular weight or particle size. This, however, is not always true, as in the case of nonspherical molecules.

The goal of many centrifugation experiments is the measurement of s. This value is important because it can be used to calculate the size (molecular weight, kilo base pairs, etc.) of a molecule or cell organelle. The units of s are not obvious from Equation 7.8. Dimensional analysis shows the following: v in cm/sec, ω in radians/sec, r in cm, m_0 in grams, $\bar{v}$ in cm³/g, ρ in g/cm³, and f in g/sec. Therefore, the unit for s is second.

B.   INSTRUMENTATION FOR CENTRIFUGATION

The basic centrifuge consists of two components, an electric motor with drive shaft to spin the sample and a **rotor** to hold tubes or other containers of the sample. A wide variety of centrifuges is available, ranging from a low-speed centrifuge used for routine pelleting of relatively heavy particles to sophisticated instruments that include accessories for making analytical measurements during centrifugation. Here we will describe three types, the low-speed or clinical centrifuge, the high-speed centrifuge, and the ultracentrifuge. Major characteristics and applications of each type are compared in Table 7.1.

Low-speed Centrifuges

Most laboratories have a standard low-speed centrifuge used for routine sedimentation of relatively heavy particles. The common centrifuge has a maximum speed in the range of 4000 to 5000 rpm, with RCF values up to $3000 \times g$. These instruments usually operate at room temperature with no means of temperature control of the samples. Two types of rotors, **fixed angle** and **swinging bucket,** may be used in the instrument. Centrifuge tubes or bottles that contain 12 or 50 mL of sample are commonly used. Low-speed

Figure 7.3

Range of S values for bio-molecules, cell organelles, and cells. kb = kilo base pairs (1000 base pairs). Molecular weight or kb is shown in parentheses.

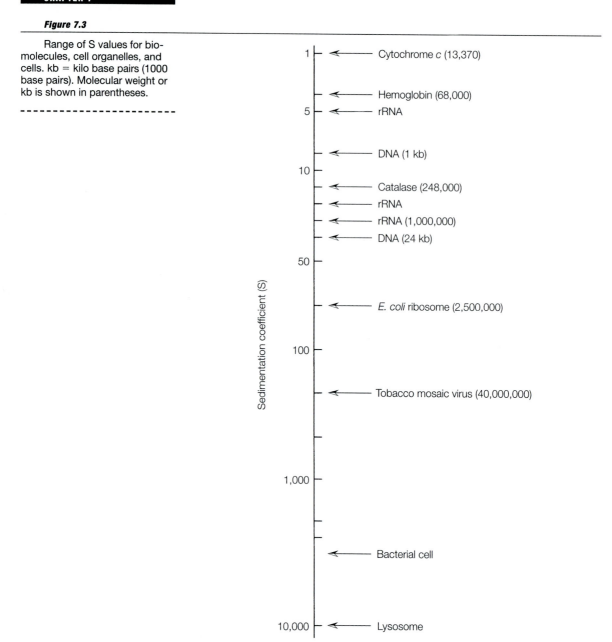

Sedimentation coefficient (S)

1 ← Cytochrome *c* (13,370)

← Hemoglobin (68,000)

5 ← rRNA

← DNA (1 kb)

10

← Catalase (248,000)

← rRNA

← rRNA (1,000,000)

← DNA (24 kb)

50

← *E. coli* ribosome (2,500,000)

100

← Tobacco mosaic virus (40,000,000)

1,000

← Bacterial cell

10,000 ← Lysosome

centrifuges are especially useful for the rapid sedimentation of coarse precip-itates or red blood cells. The sample is centrifuged until the particles are tightly packed into a **pellet** at the bottom of the tube. The upper, liquid por-tion, the **supernatant,** is then separated by decantation.

Table 7.1			
Types of Centrifuges and Applications			
	Type of Centrifuge		
Characteristic	Low-speed	High-speed	Ultracentrifuge
Range of speed (rpm)	1–6000	1000–25,000	20–80,000
Maximum RCF (*g*)	6000	50,000	600,000
Refrigeration	Some	Yes	Yes
Applications			
Pelleting of cells	Yes	Yes	Yes
Pelleting of nuclei	Yes	Yes	Yes
Pelleting of organelles	No	Yes	Yes
Pelleting of ribosomes	No	No	Yes
Pelleting of macromolecules	No	No	Yes

High-Speed Centrifuges

For more sophisticated biochemical applications, higher speeds and temperature control of the rotor chamber are essential. A typical high-speed centrifuge is shown in Figure 7.4. The operator of this instrument can carefully control speed and temperature, which is especially important for carrying out reproducible centrifugations of temperature-sensitive biological samples. Rotor chambers in most instruments are maintained at or near 4°C. Three types of rotors are available for high-speed centrifugation, the fixed-angle, the swinging-bucket, and the vertical rotor (Figure 7.5A–C). Fixed-angle rotors are especially useful for differential pelleting of particles (Figure 7.6A). In swinging-bucket rotors (Figure 7.5B), the sample tubes move to a position perpendicular to the axis of rotation during centrifugation, as shown in Figure 7.7. These are used most often for density gradient centrifugation (see below). In the vertical rotor (Figure 7.5C), the sample tubes remain in an upright position (Figure 7.8). These rotors are used often for gradient centrifugation. Prior to the early 1990s, rotors were constructed from metals such as aluminum and titanium. Although metal rotors have great strength, they do have several disadvantages: they are very heavy to handle, they are not corrosion resistant, and they become fatigued with use. Rotors are now available that are fabricated from carbon-fiber composite materials. They have several advantages over heavy metal rotors. These new rotors are 60% lighter than comparable aluminum and titanium rotors. Because of the lighter weight, acceleration and deceleration times are reduced, thus centrifuge run times are shorter. This also results in lower service and maintenance costs. Instruments are equipped with a brake to slow the rotor rapidly after centrifugation.

Widely used in the category of medium-speed centrifuges is the "microfuge" (Figure 7.9). These instruments, which are designed for the benchtop, are used for rapid pelleting of small samples. Fixed-angle rotors are available to hold up to eighteen 1.5- or 0.5-mL tubes. The maximum speed

Figure 7.4

A typical high-speed, refrigerated centrifuge. *Courtesy of Kendro Laboratory Products.*

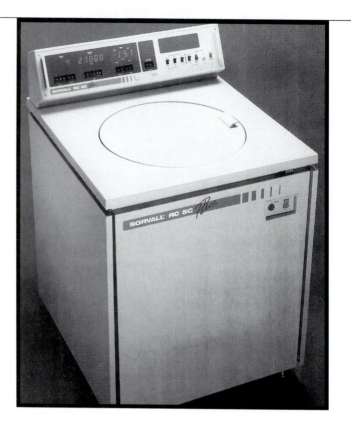

A   B   C

Figure 7.5

Rotors for a high-speed centrifuge. **A** Fixed angle; **B** swinging bucket; **C** vertical. *Courtesy of Beckman Instruments, Inc.*

Figure 7.6

Operation of a fixed angle rotor.   **A** Loading of sample. **B** Sample at start of centrifugation.   **C** Bands form as molecules sediment.   **D** Rotor at rest showing separation of two components.

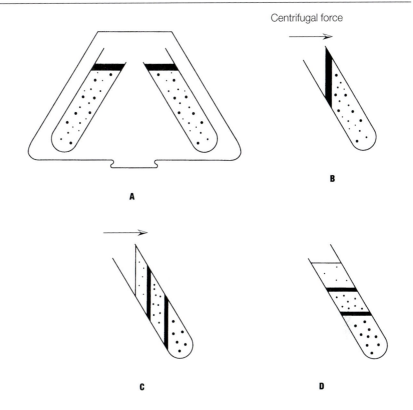

Centrifugal force

A

B

C

D

Figure 7.7

Operation of a swinging-bucket rotor.   **A** Loading of sample.   **B** Sample at start of centrifugation.   **C** Sample during centrifugation separates into two components.   **D** Rotor at rest.

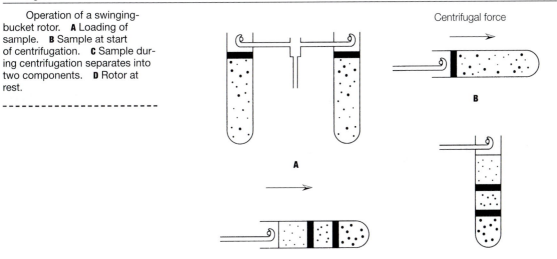

Centrifugal force

A

B

C

D

Figure 7.8

Operation of a vertical rotor. **A** Loading of sample. **B** Beginning of centrifugation. **C, D** During centrifugation. **E** Deceleration of sample. **F** Rotor at rest.

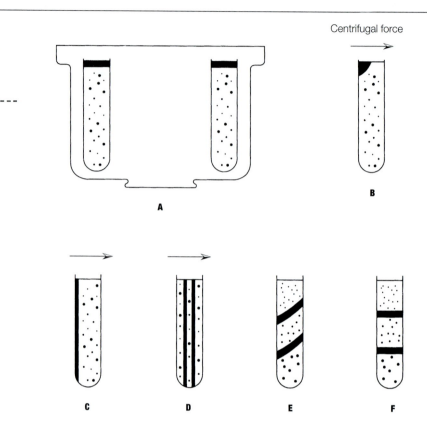

Centrifugal force

of most commercial microfuges is between 12,000 and 15,000 rpm, which delivers a force of 11,000–12,000 × *g*. Some instruments can accelerate to full speed in 6 seconds and decelerate within 18 seconds. Most instruments have a variable speed control and a momentary pulse button for minispins.

The preparation of biological samples almost always requires the use of a high-speed centrifuge. Specific examples will be described later, but high-speed centrifuges may be used to sediment (1) cell debris after cell homogenization, (2) ammonium sulfate precipitates of proteins, (3) microorganisms, and (4) cellular organelles such as chloroplasts, mitochondria, and nuclei.

Ultracentrifuges

The most sophisticated of the centrifuges are the **ultracentrifuges.** Because of the high speeds attainable (see Table 7. 1), intense heat is generated in the rotor, so the spin chamber must be refrigerated and placed under a high vacuum to reduce friction. The sample in a cell or tube is placed in a rotor, which is then driven by an electric motor. Although it is rare, metal rotors when placed under high stress sometimes break into fragments. The rotor chamber on all ultracentrifuges is covered with protective steel armor plate.

Figure 7.9

A variable-speed, refrigerated microcentrifuge. *Photo of
Eppendorf centrifuge courtesy
of Brinkmann Instruments, Inc.*

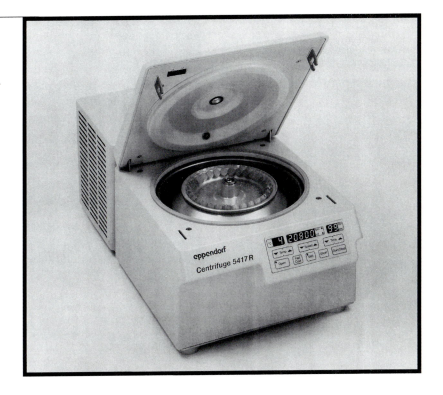

The drive shaft of the ultracentrifuge is constructed of a flexible material to
accommodate any "wobble" of the rotor due to imbalance of the samples.
It is still important to counterbalance samples as carefully as possible.

The previously discussed centrifuges—the low, medium, and high speed—
are of value only for preparative work, that is, for the isolation and separation of precipitates and biological samples. Ultracentrifuges can be used
both for preparative work and for analytical measurements. Thus, two types
of ultracentrifuges are available, **preparative models,** primarily used for separation and purification of samples for further analysis, and **analytical
models,** which are designed for performing physical measurements on the
sample during sedimentation. One of the most versatile models is the Beckman TLX, a microprocessor-controlled tabletop ultracentrifuge (Figure
7.10). With a typical fixed-angle rotor, which holds six, 0.2- to 2.2-mL samples, the instrument can generate 100,000 rpm and an RCF of 540,000 × *g.*

Analytical ultracentrifuges have the same basic design as preparative
models except that they are equipped with optical systems to monitor directly the sedimentation of the sample during centrifugation. The first commercial instrument of this type was the Beckman Model E, introduced in
1947. Because of a renewed interest in the use of ultracentrifugation, Beckman has begun marketing the Optima XL-A, which spins samples up to
60,000 rpm and generates up to 600,000 × *g.*

Figure 7.10

Outside **A** and inside **B** views of the Beckman Optima TLX ultracentrifuge. *Courtesy of Beckman Instruments, Inc.*

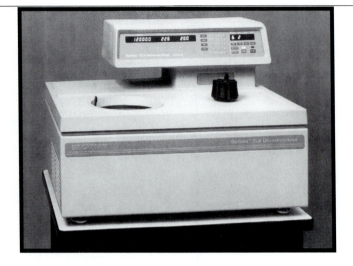

A

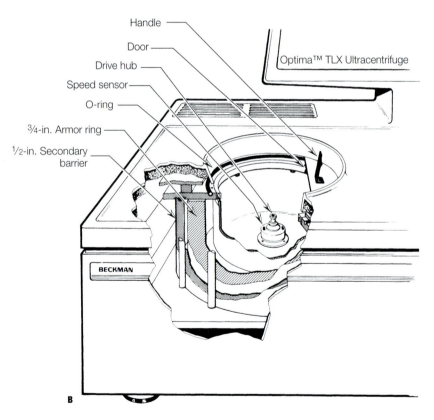

Handle

Door

Drive hub

Speed sensor

O-ring

¾-in. Armor ring

½-in. Secondary barrier

Optima™ TLX Ultracentrifuge

BECKMAN

B

For analysis, a sample of nucleic acid or protein (0.1 to 1.0 mL) is sealed in a special analytical cell and rotated. Light is directed through the sample parallel to the axis of rotation, and measurements of absorbance by sample molecules are made. (The Beckman instrument can scan the sample over the wavelength range 190 to 800 nm.) If sample molecules have no significant absorption bands in the wavelength range, then optical systems that measure changes in the refractive index may be used. Optical systems aided by computers are capable of relating absorbance changes or index of refraction changes to the rate of movement of particles in the sample. The optical system actually detects and measures the front edge or moving boundary of the sedimenting molecules. These measurements can lead to an analysis of concentration distributions within the centrifuge cell. Applications of these measurements will be discussed in the next section.

C. APPLICATIONS OF CENTRIFUGATION

Preparative Techniques

Centrifuges in undergraduate biochemistry laboratories are used most often for preparative-scale separation procedures. This technique is quite straightforward, consisting of placing the sample in a tube or similar container, inserting the tube in the rotor, and spinning the sample for a fixed period. The sample is removed and the two phases, pellet and supernatant (which should be readily apparent in the tube), may be separated by careful decantation. Further characterization or analysis is usually carried out on the individual phases. This technique, called **velocity sedimentation centrifugation,** separates particles ranging in size from coarse precipitates to cellular organelles. Relatively heavy precipitates are sedimented in low-speed centrifuges, whereas lighter organelles such as ribosomes require the high centrifugal forces of an ultracentrifuge.

Much of our current understanding of cell structure and function depends on separation of subcellular components by centrifugation. The specific method of separation, called **differential centrifugation,** consists of successive centrifugations at increasing rotor speeds. Figure 7.11 illustrates the differential centrifugation of a cell homogenate, leading to the separation and isolation of the common cell organelles. For most biochemical applications, the rotor chamber must be kept at low temperatures to maintain the native structure and function of each cellular organelle and its component biomolecules. A high-speed centrifuge equipped with a fixed-angle rotor is most appropriate for the first two centrifugations at $600 \times g$ and $20,000 \times g$. After each centrifuge run, the supernatant is poured into another centrifuge tube, which is then rotated at the next higher speed. The final centrifugation at $100,000 \times g$ to sediment microsomes and ribosomes must be done in an ultracentrifuge. The $100,000 \times g$ supernatant, the **cytosol,** is the soluble portion of the cell and consists of soluble proteins and smaller molecules. Differential centrifugation is used to isolate beef heart mitochondria for enzymatic characterization in Experiment 10.

Figure 7.11

Differential centrifugation of a cell homogenate. See text for description.

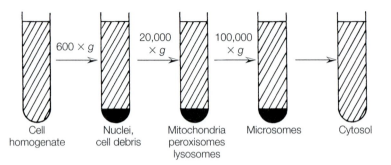

Cell homogenate → 600 × g → Nuclei, cell debris → 20,000 × g → Mitochondria peroxisomes lysosomes → 100,000 × g → Microsomes → Cytosol

Analytical Measurements

A variety of analytical measurements can be made on biological samples during and after a centrifuge run. Most often, the measurements are taken to determine molecular weight, density, and purity of biological samples. All analytical techniques require the use of an ultracentrifuge and can be classified as **differential** or **density gradient.**

Differential Centrifugation

Differential methods involve sedimentation of particles in a medium of homogeneous density. Although the technique is similar to preparative differential centrifugation as previously discussed, the goal of an experiment is to measure the sedimentation coefficient of a particle. The underlying principles of this technique are illustrated in Figure 7.12A. During centrifugation, a moving boundary is generated between pure solvent and sedimenting particles. An analytical ultracentrifuge is capable of detecting and measuring the rate of movement of the boundary. Hence, the sedimentation velocity, v, can be experimentally determined. By using Equation 7.8 the sedimentation coefficient, s, can be calculated. The value of s for a sedimenting particle is related to the molecular weight of that particle by the Svedberg equation. The Svedberg equation is derived from Equation 7.8 by recognizing that the frictional force, f, may be defined by Equation 7.9.

>> $$f = \frac{RT}{ND}$$ *Equation 7.9*

where

R = the gas constant, 8.3×10^7 g cm^2/sec/deg/mole

T = the absolute temperature

D = the diffusion coefficient of the solute in units of cm^2/sec

N = Avogadro's number

Figure 7.12

A comparison of differential and density gradient measurements. **A** Differential centrifugation in a medium of unchanging density. **B** Zonal centrifugation in a prepared density gradient. **C** Isopycnic centrifugation; the density gradient forms during centrifugation. *Illustration courtesy of Beckman Instruments, Inc.*

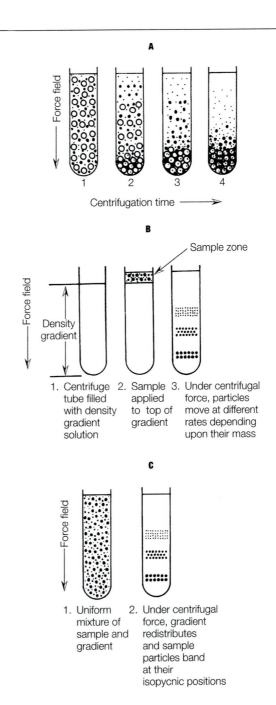

A

Centrifugation time ⟶

B

Sample zone

Density gradient

1. Centrifuge tube filled with density gradient solution
2. Sample applied to top of gradient
3. Under centrifugal force, particles move at different rates depending upon their mass

C

1. Uniform mixture of sample and gradient
2. Under centrifugal force, gradient redistributes and sample particles band at their isopycnic positions

Thus, Equation 7.8 may be transformed into Equation 7. 10.

>> $$s = \frac{m_0(1 - \bar{v}\rho)}{RT/ND}$$ *Equation 7.10*

$$RTs = m_0 ND(1 - \bar{v}\rho)$$

Since molecular weight, MW, is equal to $m_0 N$, Equation 7.10 is converted to Equation 7.11, the Svedberg equation.

>> $$MW = \frac{RTs}{D(1 - \bar{v}\rho)}$$ *Equation 7.11*

This equation provides an accurate calculation of molecular weight and is applicable to macromolecules such as proteins and nucleic acids. However, its usefulness is limited because diffusion coefficients are difficult to measure and are not readily available in the literature.

An alternative method sometimes used to determine molecular weights of macromolecules is **sedimentation equilibrium.** In the previous example, using the Svedberg equation, the sample is rotated at a rate sufficient to sediment the particles. Here, the sample is rotated at a lower rate, and the particles sediment until they reach an equilibrium position at the point where the centrifugal force is equal to the frictional component opposing their movement (see Equation 7.6). The molecular weight is then calculated using Equation 7.12.

>> $$MW = \frac{RT}{(1 - \bar{v}\rho)\omega^2}\left(\frac{1}{rc}\right)\left(\frac{dc}{dr}\right)$$ *Equation 7.12*

where

$$\frac{dc}{dr} = \text{change in particle concentration as a function of distance from the rotation center, as measured in the ultracentrifuge}$$

and MW, R, T, $\bar{v}$, ρ, ω, and r have the definitions previously given. This technique is time consuming, as the system may require 1 to 2 days to reach equilibrium. Also, the use of Equation 7.12 is complicated by the difficulty of making concentration measurements in the ultracentrifuge.

Differential ultracentrifugation methods may also be applied to analysis of the purity of macromolecular samples. If one sharp moving boundary is observed in a rotating centrifuge cell, it indicates that the sample has one component and therefore is pure. In an impure sample, each component would be expected to form a separate moving boundary upon sedimentation.

Density Gradient Centrifugation

In differential procedures, the sample is uniformly distributed in a cell before centrifugation, and the initial concentration of the sample is the same throughout the length of the centrifuge cell. Although useful analytical measurements can be made with this technique, it has disadvantages when applied to impure samples or samples with more than one component. Large particles that sediment faster pass through a medium consisting of solvent and particles of smaller size. Therefore, clear-cut separations of macromolecules are seldom obtained. This can be avoided if the sample is centrifuged in a fluid medium that gradually increases in density from top to bottom. This technique, called **density gradient centrifugation,** permits the separation of multicomponent mixtures of macromolecules and the measurement of sedimentation coefficients. Two methods are used, **zonal centrifugation,** in which the sample is centrifuged in a preformed gradient, and **isopycnic centrifugation,** in which a self-generating gradient forms during centrifugation.

Zonal Centrifugation

Figure 7.12B outlines the procedure for zonal centrifugation of a mixture of macromolecules. A density gradient is prepared in a tube prior to centrifugation. This is accomplished with the use of an automatic gradient mixer. Solutions of low-molecular-weight solutes such as sucrose or glycerol are allowed to flow into the centrifuge cell. The sample under study is layered on top of the gradient and placed in a swinging-bucket rotor. Sedimentation in an ultracentrifuge results in movement of the sample particles at a rate dependent on their individual *s* values. As shown in Figure 7.12B, the various types of particles sediment as *zones* and remain separated from the other components. The centrifuge run is terminated before any particles reach the bottom of the gradient. The various zones in the centrifuge tubes are then isolated by collecting fractions from the bottom of the tube and analyzing them for the presence of macromolecules. The zones of separated macromolecules are relatively stable in the gradient because it slows diffusion and convection. The gradient conditions can be varied by using different ranges of sucrose concentration. Sucrose concentrations up to 60% can be used, with a density limit of 1.28 g/cm³. The zonal method can be applied to the separation and isolation of macromolecules (preparative ultracentrifuge) and to the determination of *s* (analytical ultracentrifuge).

Isopycnic Centrifugation

In the isopycnic technique, the density gradient is formed during the centrifugation. Figure 7.12C outlines the operation of isopycnic centrifugation. The sample under study is dissolved in a solution of a dense salt such as cesium chloride or cesium sulfate. The cesium salts may be used to establish gradients to an upper density limit of 1.8 g/cm³. The solution of biological sample and cesium salt is uniformly distributed in a centrifuge tube and rotated in an ultracentrifuge. Under the influence of the centrifugal force, the cesium salt redistributes to form a continuously increasing density gradient from the top to the bottom. The macromolecules of the biological sample seek an area in the tube where the density is equal to their respective densities. That is, the macromolecules move to a region where the sum of the forces (centrifugal and frictional) is zero (Equation 7.6). The macromolecules either sediment or float to this region of equal density. Stable zones or bands of the individual components are formed in the gradient (see Figure 7.12C). These bands can be isolated as previously described. Cesium salt gradients are especially valuable for separation of nucleic acids.

Density gradients are widely used in separating and purifying biological samples. In addition to this preparative application, measurements of s can be made. Gradient techniques have been used to isolate and purify the subcellular components, microsomes, ribosomes, lysosomes, mitochondria, peroxisomes, chloroplasts, and others. After isolation, they have been biochemically characterized as to their protein, lipid, nucleic acid, and enzyme contents.

Nucleic acids, in particular, have been extensively studied by density gradient techniques. Both RNA and DNA are routinely classified according to their s values. The different structural forms of DNA discussed in Chapter 4 can be determined by density gradient centrifugation. In Experiment 14, plasmid DNA is purified by centrifugation in a cesium salt gradient. The DNA band in the gradient is detected with ethidium bromide, an intercalating dye.

Space does not allow an exhaustive review of centrifuge applications. Interested students should consult the references at the end of the chapter for recent developments.

Care of Centrifuges and Rotors

Centrifuge equipment represents a sizable investment for a laboratory, so proper maintenance is essential. In addition, poorly maintained equipment is especially dangerous. Since many instruments are now available, specific instructions will not be given here, but general guidelines are outlined.

1. Carefully read the operating manual or receive proper instructions before you use any centrifuge.

2. Select the proper operating conditions on the instrument. If refrigeration is necessary, set the temperature to the appropriate level and allow 1 to 2 hours for temperature equilibration.

3. Check the rotor chamber for cleanliness and for damage. Clean with soap and warm water and rinse with distilled water.

4. Select the proper rotor. Many sizes and types are available. Follow guidelines already stated in this chapter or consult your instructor.

5. Be sure the rotor is clean and undamaged. Observe any nicks, scratches, or other damage that may cause imbalance. If dirty, the rotor should be cleaned with warm water and a mild, nonbasic detergent. A soft brush

can be used inside the cavities. Rinse well with distilled water and dry. Scratches should not be made on the surface coating, as corrosion may result.

6. Filled centrifuge tubes or bottles should be weighed carefully and balanced before centrifugation.

7. Rotor manufacturers provide a maximum allowable speed limit for each rotor. Do not exceed that limit.

8. Keep an accurate record of centrifuge and rotor use. Just as your automobile needs service after a certain number of miles, the centrifuge should be serviced after certain intervals of use. Centrifuge maintenance is usually determined by hours of use and total revolutions of the rotor. It is also essential to maintain a record of the use for each rotor. Metal rotors weaken with use, and the maximum allowable speed limit decreases. Rotor manufacturers usually provide guidelines for decreasing the allowable speed for a rotor.

9. If an unusual noise or vibration develops during centrifugation, immediately turn the centrifuge off.

10. Carefully clean the rotor chamber and rotor after centrifugation.

Study Problems

1. Which of the following factors will have an effect on the sedimentation rate of a particle during centrifugation?
   (a) Mass of the sedimenting particle
   (b) Angular velocity of rotation
   (c) Atmospheric pressure
   (d) Density of the solvent

2. You wish to centrifuge a biological sample so that it experiences an RCF of 100,000 $\times$ g. At what rpm must you set the centrifuge assuming an average r value of 4?

3. Cytochrome c has an s value of 1×10^{-13} second and hemoglobin an s value of about 4.5×10^{-13} second. Which protein has the larger molecular weight?

4. An enzyme has a sedimentation coefficient of 3.5 S. When a substrate molecule is bound into the active site of the enzyme, the sedimentation coefficient decreases to 3.0 S. Explain this change.

5. A protein with molecular weight of 100,000 shows a single boundary when centrifuged in aqueous buffer. If the protein is centrifuged in a medium of the same buffer plus 6 M urea, two boundaries are observed, one corresponding to a molecular weight of 10,000, the other 30,000. The area of the slower peak is two-thirds that of the faster. Describe the subunit structure of the protein.

6. Describe how you would design a centrifuge experiment to isolate sediments containing cell nuclei.

▶   7. Explain the following observation. The density of DNA in 7 M CsCl containing 0.15 M $MgCl_2$ is less than the density of the same DNA in 7 M CsCl.

▶   8. Assume that you have centrifuged in a density gradient a sample of DNA that contained both closed, circular DNA and supercoiled DNA. Would you expect to see two bands in the sedimentation pattern? Explain.

▶   9. Assume that a centrifuge is operating at 43,000 rpm. What is the relative centrifugal force at a distance from the central axis of 6 cm?

▶   10. Could you use a low-speed centrifuge to sediment mitochondria? Explain.

Further Reading

C. Cantor and P. Schimmel, *Biophysical Chemistry,* Part II (1980), W. H. Freeman (San Francisco), pp. 591–641. Excellent but advanced treatment of theory and applications of ultracentrifugation.

R. Garrett and C. Grisham, *Biochemistry,* 2nd ed. (1999), Saunders (Orlando, FL), pp. 582–584. Techniques of centrifugation.

D. Holme and H. Peck, *Analytical Biochemistry,* 3rd ed. (1998), Addison Wesley Longman (Essex, UK), pp. 153–163. Principles of centrifugation.

C. Mathews, K. van Holde, and K. Ahern, *Biochemistry,* 3rd ed. (2000), Benjamin/Cummings (San Francisco), pp. 149–150, 983–984. Centrifugation in biochemistry.

D. Rickwood, Editor, *Centrifugation—A Practical Approach,* 3rd ed. (1993), IRL Press (Oxford). Theory and practice of centrifugation with many applications.

J. Robyt and B. White, *Biochemical Techniques: Theory and Practice* (1987), Brooks/Cole (Monterey, CA), pp. 256–260. Centrifugation in the preparation of biological materials.

L. Stryer, *Biochemistry,* 4th ed. (1995), Freeman (New York), pp. 51–52. Ultracentrifugation for separation and characterization of biomolecules.

D. Voet and J. Voet, *Biochemistry* (1995), John Wiley & Sons (New York), pp. 97–101. Use of centrifugation in the purification of proteins.

Centrifugation on the Web

http://www.ruf.rice.edu/~bioslabs/methods.html
     Review topics Centrifugation and Differential Centrifugation.

two

EXPERIMENTS

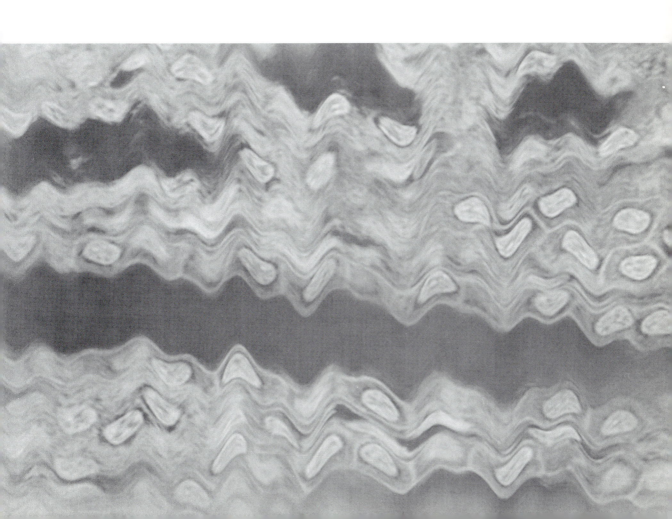

1

USING THE COMPUTER IN BIOCHEMICAL RESEARCH

- **Recommended Reading**

Chapter 1, Sections B and F.

- **Synopsis**

The computer has become an essential tool in biochemical research. A computer may be used for the routine jobs of word processing and data collection and analysis. In addition, if a computer is connected to the Internet, then it may be used for biochemical literature searching, accessing information about nucleic acid and protein sequences, predicting protein structure, and seeking research methodology. In this experiment, students will be introduced to all of these skills in bioinformatics.

I. INTRODUCTION AND THEORY

The modern computer has revolutionized the way we live. Not surprisingly, the computer has also changed the way we do biochemical research. Your first encounter with a computer in this laboratory will probably be while using an instrument that has a computer to control its operation, to collect data, and to analyze data. All major pieces of scientific equipment including UV-VIS spectrometers, high-performance liquid chromatographs, gas chromatographs, nuclear magnetic resonance spectrometers, and DNA sequencers are now controlled by computers. But your use of the computer will not end in the lab. You will use a computer to prepare each laboratory report including graphical analysis of experimental data. If the computer is connected to the Internet, you will greatly broaden its use to some of the following: (1) searching the biochemical literature for pertinent books and journal articles and (2) accessing biological databases that provide nucleic acid and protein sequences and protein structures.

Personal Computing in Biochemistry

It is now possible for most students to purchase a basic computer system at low cost. If a personal computer is not in the budget, most colleges and universities provide students access to campus-wide computer systems as part of tuition and fees. By this point in your studies, you are familiar with the use of a computer, but a few introductory comments are made just to help you get started with computing in the biochemistry laboratory. In terms of equipment, you will need a computer, monitor, printer, and some basic software. Some recommendations for specific hardware and software will be given here, but one must be aware that new products and important upgrades are continually being developed.

For word processing (writing lab reports), basic software programs including Microsoft Word and Word Perfect are most widely used. Software specialized for scientific writing is available but probably not necessary at this level. For many experiments that you complete, you will need to present data in a spreadsheet or graphical form. Current software programs for graphing or spreadsheet with graphing capability include Lotus, Excel, Sigmaplot, Quattropro, Kaleidagraph, and CricketGraph. Some graphs that you prepare from experimental data will be nonlinear. The most common example is a Michaelis-Menten graph from enzyme kinetics studies (substrate concentration vs. reaction rate). Since most computers and programs have different methods for dealing with nonlinearity, it is probably best not to connect the data points with a line. Rather, use a curve-fitting routine to get the appropriate line. Alternatively, one could analyze the data using a straight-line method such as the Lineweaver-Burk plot (see Experiment 5).

Throughout this course, your instructor will very likely have specialized software available on computers and will offer help in their application. It is important for your education and career that you become widely knowledgeable and skilled in the use of the computer and software. It will be a tool that you will continue to use at the home and office. Many new terms will be introduced in this experiment. All words in bold print are defined in a glossary at the end of the experiment.

The Computer and the Internet

If you are using the computer as described above, you are saving time and preparing good-looking lab reports. However, if your computer is not connected to the **Internet,** then you are not tapping into the vast wealth of biochemical tools and information available. The Internet can be defined, in simple terms, as a worldwide matrix that allows all computers and networks to communicate with each other. If the computer you are using is college owned, then it is probably already linked to the Internet, and the college pays the costs for that service. For your own home computer, you may need to subscribe to an Internet service and obtain a **modem** to transmit computer signals through a telephone line. Once you are connected to the Internet, many programs are available as **freeware,** software provided without charge by its creator. In this experiment, we will concentrate on accessing and using programs that are in the public domain (no charge).

After you are connected to the Internet, what are the basic facilities available for use? First, you will be able to communicate by **e-mail (electronic mail).** Messages containing text, files, and graphics may be sent to anyone who has a computer with an Internet link and an e-mail address. Addresses have three basic components, the user name, an @ sign, and the user's location or **domain.** Common domains that you will encounter usually have one of the following suffixes: edu (educational institution in the United States), ac (academic institution in the United Kingdom), gov (government), com (commercial organization), and org (other organization). You will need an e-mail program to collect, send, and organize messages. The most popular ones are Eudora and Pegasus. (Practice your e-mail skills by sending a message, perhaps a question, to your laboratory instructor and to the author of this book: boyer@hope.edu). Communication among scientists is now done primarily by e-mail. Connected to the Internet, you will also be able to join in list server discussion groups created to share ideas in a common area of interest or in news groups such as USENET. One of the most widely used facilities on the Internet is the ability to place and retrieve network data by **file transfer protocol (ftp).** More detail on ftp's will be given in later sections.

The World Wide Web

The newest and most rapidly growing component of the Internet is the **World Wide Web** (WWW, also called "the web"). This facility, which was launched in 1992, permits the transfer of data as pages in multimedia form consisting of text, graphs, audio, and video. The pages are linked together by **hypertext** pointers so that data stored on computers in different locations may be retrieved via the network by your computer. Web documents are written in a special coded language called **HyperText Markup Language (HTML).** To access all of the resources on the Web, you will need a **browser,** an interface program that reads hypertext and displays Web pages on your computer. The most commonly used Web browsers are Internet Explorer and Netscape Navigator. The use of these programs will not be described in detail here as they are constantly changing and students at this level are already familiar with their use. However, a brief summary will be presented.

To access the Web, the Web browser is activated. Displayed on the screen will be the **home page** or starting point for entry into the Web. On this page will be a dialogue box into which you can type text. The dialogue box may ask for "Address," "Netsite," "Location," or **"URL" (Uniform Resource Locator).** To request a specific Web page from another computer site, type in the

Web page address, which is usually in the form http://www.–. The home page, with instructions on the use of the Web site, will then be displayed on the screen. One important feature you will note is that some words on the page are highlighted. If you click the mouse on one of these words (called **hyperlinks**) your computer will connect to another, related, Web page that provides information on the hyperlink. This feature greatly enhances the use of the Web because related Web sites are connected or linked together and may be quickly accessed by a click of the mouse.

Web addresses that are useful for biochemical research are presented in Tables E1.1 and E1.2. Many of the current Web sites you will need are listed here; however, what about new Web sites that have been established since publication of this book? Millions of new Web sites are created every year. To access these new sites, you need the help of a **search engine,** a searchable directory that organizes Web pages by subject classification. Major search engines include AltaVista, Excite, HotBot, Lycos, Netscape Search, and Yahoo! As you "surf the Web," you may find sites you wish to save and review at a later date. You may use the **"bookmark"** (Netscape) or **"favorite"** (Explorer) function to save it for the future.

Applications of the Web

It is not necessary to have a complete understanding of the Internet in order to tap into its vast resources. The fundamental concepts provided here will allow you to take advantage of two essential activities: (1) biochemical literature searching and (2) using Web directories and biological databases.

The Biochemical Literature

Experimental biochemists do not spend all their working time in the laboratory. An important component of a biochemistry research project is a search of the biochemical literature. The library should be considered a tool for experimental biochemistry in the same way as any scientific instrument.

Table E1.1

Web Database Directories

Name	URL
Pedro's Biomolecular Research Tools	http://www.public.iastate.edu/~pedro/research_tools.html
Biology Workbench	http://biology.ncsa.uiuc.edu
CMS Molecular Biology Resources	http://www.sdsc.edu/ResTools/cmshp.html
BioTech	http://biotech.icmb.utexas.edu
Protocol Online	http://www.protocol-online.net
Chem Connection	http://chemconnect.com/news/journals.html
American Chemical Society	http://pubs.acs.org/

Table E1.2

Biochemical Databases and Tools

Name	Description	URL
Protein Data Bank (PDB)	Protein structures determined by X-ray and NMR	http://www.rcsb.org/pdb/
European Bioinformatics Institute (EBI)	DNA sequences	http://www.ebi.ac.uk/
National Center for Biotechnology Information (NCBI)	Variety of databases and resources	http://www.nlm.nih.gov/
Swiss-Protein	Protein sequences and analysis	http://www.expasy.ch/tools/
Biocatalysis/Biodegradation Database of the University of Minnesota	Microbial metabolism of many chemicals	http://www.labmed.umn.edu/umbbd/index.html
REBASE-The Restriction Enzyme Database	Restriction enzyme directory and action	http://rebase.neb.com/
Georgia Institute of Technology	Tutorials on PDB and RasMol	http://www.chemistry.gatech.edu/faculty/williams/bCourse_information/4582/labs/rasmol_pdb.html
The Institute for Genomic Research	Collection of genomic databases	http://www.tigr.org/
RasMol (RasMac)	Molecular graphics for proteins	http://www.umass.edu/microbio/rasmol/
Predict Protein	Protein sequence and structure prediction	http://www.embl-heidelberg.de/predictprotein/
Gene Quiz	Protein function analysis based on sequence	http://www.sander.ebi.dc.uk/gqsrv/submit

The use of the biochemical literature by the student in biochemistry laboratory is not as extensive as that of a full-time researcher, but you must be aware of what is available in the library and how to use it.

The library is used in all stages of research. Before an investigator can begin experimentation, a research idea must be generated. This idea develops only after extensive reading and study of the literature. A research project usually begins in the form of a question to be answered or problem to be solved. For ease of solution, a major project is subdivided into questions that may be answered by experimentation. Before laboratory work can begin, the researcher must have a knowledge of the past and current literature dealing with the research area. This can be reduced to two questions: What is the current state of knowledge in the area? and What are the significant unknowns? These questions can be answered only by developing a familiarity with the biochemical literature. The researcher will find that this knowledge of the literature is also invaluable for the design of experiments. The development of experiments requires knowledge of techniques and laboratory procedures. Excellent methods books and journals are available that provide experimental details. Finally, while performing experiments, the researcher often needs physical and chemical constants and miscellaneous information. Various handbooks and encyclopedias are excellent for this purpose. The beginning student in biochemistry laboratory will not be expected to proceed through all of these stages in the design of an experiment. However, a familiarity with the literature will increase your understanding of the experiment and may aid in the development of more effective methods. When you do begin a research program, you will be able to use the library to the fullest advantage.

The biochemical literature is massive and expanding rapidly. It is almost a full-time job just to maintain a current awareness of a specialized research area. There are few disciplinary boundaries in the study of biochemistry. The biochemical literature overlaps into the biological sciences, the physical sciences, and the basic medical sciences. The intent of the following discussion is to bring some order to the many textbooks, reference books, research journals, computer information retrieval services, and handbooks that are available.

Textbooks

The student's first exposure to biochemistry is probably a lecture course accompanied by the reading of a general textbook of biochemistry. By providing an in-depth survey of biochemistry, textbooks allow students to build a strong foundation of important principles and concepts. By the time most books are in print, the information is 1 to 2 years old, but textbooks still should be considered the starting point for mastery of the fundamentals of biochemistry.

Reference Books and Review Publications

For more specialized and detailed biochemical information that is not offered by textbooks, reference books must be used. Reference works range from general surveys to specialized series. The best works are multivolume sets that continue publication of volumes on a periodic basis. Each volume usually covers a specialized area with articles written by recognized authorities in the field. It should be noted that reference articles of interest to biochemists are often found in publications that are not strictly biochemical. The best known and most widely used review publication is *Annual Review of Biochemistry*. Each volume in this series, which was introduced in 1932, contains several detailed and extensive articles written by experts in the field. For shorter reviews emphasizing current topics, *Trends in the Biochemical Sciences* (TIBS) is widely read.

Research Journals

The core of the biochemical literature consists of research journals. It is essential for a practicing biochemist to maintain a knowledge of biochemical advances in his or her field of research and related areas. Scores of research journals are published with the intent of keeping scientists up to date. With the expansion of scientific information has come the need for efficient storage and use of research journals. Many publishers are now providing journals in forms such as microcards, microfilm, microfiche, and more recently CD-ROM disks and on line. Some research journals have achieved an especially excellent reputation, and articles therein are considered to be of the highest quality. A recent ranking of the biochemical journals, based on the number of citations received, produced the following order for the top six: *Journal of Biological Chemistry, Biochimica et Biophysica Acta, Biochemistry, Proceedings of the National Academy of Sciences of the United States of America, Biochemical Journal,* and *Biochemical and Biophysical Research Communications.* The core journals used by an individual depend on the area of specialty and are best determined from experience.

Methodology References

The active researcher has a continuing need for new methods and techniques. Several publications specialize in providing details of research methods, and many research methods are now available on the Web. Some of the useful biochemical methodology publications are:

Analytical Biochemistry, a monthly journal.

Analytical Chemistry, a monthly journal.

Biochemical Preparations, an annual volume.

Current Protocols in Molecular Biology, P. Ausabel et al., Editors. A manual of techniques in two volumes that are updated quarterly.

Laboratory Techniques in Biochemistry and Molecular Biology, T. S. Work and R. G. Burdon, Editors (formerly T. S. Work and E. Work). Each volume in the series is concentrated in an area of biochemistry and written by recognized authorities.

Methods of Enzymatic Analysis, H. Bergmeyer, Editor. Contains methods for enzyme purification and assay, in several volumes.

Methods in Enzymology, various editors. The most valuable methods series available. Each volume contains numerous articles describing biochemical techniques. The series is well indexed and easy to use. Over 200 volumes.

A Practical Guide to Molecular Cloning, 2nd ed., B. Perbal. Useful for setting up research projects in molecular cloning.

Handbooks of Chemical and Biochemical Data

Students in introductory biochemistry laboratory may use methodology books more than any other type, although much of the data is now on the Web. While doing biochemical experiments, you may need physical, chemical, and biochemical information such as definition of terms, R_f values, molecular weights, and physical constants. This information is easily found in the many handbooks and collections of biochemical data. Some useful handbooks with a brief description of contents are listed on the next page.

Dictionary of Biochemistry and Molecular Biology, 2nd ed., J. Stenesh (1989), Wiley Interscience (New York). Contains definitions of 16,000 terms.

Glossary of Biochemistry and Molecular Biology, D. Glick (1990), Raven Press (New York). Emphasis on words and phrases that are unique to biochemistry and molecular biology.

Merck Index, S. Budavari, Editor, Merck & Co. (Rahway, NJ). An encyclopedia of chemicals, drugs, and biological substances.

Practical Handbook of Biochemistry and Molecular Biology, G. Fasman, Editor (1989), CRC Press (Boca Raton, FL). Excellent source of chemical and physical data for nucleic acids, proteins, lipids, and carbohydrates.

Worthington Enzyme Manual, 2nd ed., C. Worthington, Editor (1988), Worthington Biochemical Corp. (Freehold, NJ). Contains detailed assay information and references for over 75 enzymes.

Computer-Based Searches and Other Aids to the Literature

As you study and work in biochemistry, you will often need to complete a thorough literature search on some specialized area or topic. It is not practical to survey the hundreds of books, journals, and reports that may contain information related to the topic. Two publications that provide brief summaries of published articles, reviews, and patents are *Chemical Abstracts* and *Biological Abstracts.* If you are not familiar with the use of these abstracts, ask your instructor or reference librarian for assistance.

Research articles of interest to biochemists may appear in many types of research journals. Research libraries do not have the funds necessary to subscribe to every journal, nor do scientists have the time to survey every current journal copy for articles of interest. Two publications that help scientists to keep up with published articles are *Chemical Titles* (published every 2 weeks by the American Chemical Society) and the weekly *Current Contents* available in hard copy and computer disks (published by the Institute of Science Information). The Life Science edition of *Current Contents* is the most useful for biochemists. The computer revolution has reached into the chemical and biochemical literature, and most college and university libraries now subscribe to computer bibliographic search services. One such service is STN International, the scientific and technical information network. This on-line system allows direct access to some of the world's largest scientific databases. The STN databases of most value to life scientists include BIOSIS Previews/RN (produced by Bio Sciences Information Service; covers original research reports, reviews, and U.S. patents in biology and biomedicine), CA (produced by Chemical Abstracts service; covers research reports in all areas of chemistry), MEDLINE, and MEDLARS (produced by the U.S. National Library of Medicine and *Index Medicus,* respectively; cover all areas of biomedicine). These networks provide on-line service and their databases can be accessed from personal computers in the office, laboratory, or library. Some of the computer bibliography services are freeware on the Internet, but others have user fees. For example, MEDLINE (PubMed) produced by the National Library of Medicine, available at http://www.ncbi.nlm.nih.gov/, may be used free of charge.

Web Directories, Tools, and Databases

Biochemical research generates huge amounts of data of interest to all scientists. For example, thousands of genes and proteins have been sequenced during the past several years and thousands more will be sequenced in the future. This number is being greatly expanded by the Human Genome Project, which has as its goal the sequencing of the entire human genome. In addition, determining the structures of proteins by X-ray diffraction and by NMR has become routine. Sequence and structural data are now being stored in computer networks for retrieval by biochemists throughout the world. Here, we will discuss the many biological databases and provide examples of their use. Our approach will be to focus on the use of databases readily available, free of charge, on the Web. However, it is important to recognize that many commercial hardware and software systems for analyzing biological database are available, but they are often very expensive and complicated to use.

A wide variety of databases are currently available including bibliographic, nucleic acid sequence, protein sequence and structure, metabolic pathways, transcription factors, enzymes, and many others. One of the best ways to find the resources suited to your needs is to use a directory that collects lists of information, tools, and other services. Several very good ones are available (Table E1.1). Some of these sites are hyperlinked to the database sites. This experiment will introduce you to some of the more general and useful sites. Specifically, they will include protein primary, secondary, and tertiary structure, sequence homology, sequence alignment, and structure prediction. The Web addresses for these resources are listed in Table E1.2. Because of the huge amount of data available, it is often necessary to use programs to help you analyze the data. Table E1.3 lists several software programs that are available and usually hyperlinked to the database sites. Those that we will introduce in this experiment are FASTA (protein amino acid sequences), BLAST (comparing protein sequence data), RasMol or Ras-Mac (coordinates for protein structure manipulation), Chime (protein structure coordinates), SWISS-MODEL (protein modeling), VAST (protein structure similarities), and Molecules R Us (protein structure coordinates).

Overview of the Experiment

In this experiment, students will be introduced to several uses of the computer and the Internet. Students are instructed in the use of bibliographic searches, sequence databases, and structural analytical tools available, free of charge, on the Web. For student practice, several worked examples are provided in the Study Problems. Instructors may modify the experiment by

Table E1.3

Useful Programs for Exploring Structures/Sequences

Program	Function
BLAST	Searches for similar protein and nucleic acid sequences
Chime	Protein structures on moving 3D coordinates
Entrez (NCBI)	Sequence retrieval system for cross-referencing databases
FASTA	Searches for similar protein sequences
GenBank (NCBI)	Database of gene sequences
Molecules R Us	Provides coordinates for protein 3D structure and manipulation
RasMol (Ras Mac)	Provides coordinates for protein 3D structure and manipulation
SRS (EMBL)	Sequence retrieval system for cross-referencing databases

adding new topics for bibliographic search or proteins/nucleic acids for sequence and structure analysis. The experiment may be modified to fill the necessary time slot.

II.   MATERIALS AND SUPPLIES

– Computer: Apple Macintosh or PC with printer; connected to the Internet.

– Software: Web browser such as Netscape Navigator or Internet Explorer; e-mail program such as Eudora.

III.   EXPERIMENTAL PROCEDURE

1. Searching the Biochemical Literature on MEDLINE

To illustrate the use of this search service, point your Web browser to the appropriate URL (http://www.nlm.nih.gov/). This will connect you to the National Center for Biotechnology Information. Click the mouse on the hyperlink "PubMed." Select MEDLINE in the upper dialogue box. Many features on display are available, but the most basic is the search capability. For bibliographic searching you may enter in the dialogue box under MEDLINE a search term, author name, or journal name. For example, you may want to type in "bovine alpha-lactalbumin," a protein you will isolate and characterize in Experiment 4. Clicking on "Search" will then provide over 500 citations (or articles). The lists are composed of author(s), title, and reference in reverse chronological order. By clicking on the author's name (in hypertext), you can retrieve the abstract of the article. Another useful and time-saving feature is the hypertext "(see Related Articles)." Clicking on this will provide a list of papers related to the specific citation. The 500 papers or so that you obtained in your original search are too many to screen; you may change the search parameters to reduce the number. Clicking on the "?" in the upper right-hand corner of the screen provides help for focusing the search process. Several more exercises for your practice are given in the

Study Problems. You may also practice the search method using terms, concepts, or scientists related to your biochemistry lecture class. Note that the NCBI home page offers other hypertexts, including entry to Entrez, BLAST, etc. We will use these in the next section; however, access will be through a different database Web site.

2. Using Web Tools and Biological Databases

The application of the primary databases and structural analytical tools will be introduced using a protein from a future experiment. In Experiment 4, you will extract, purify, and characterize α-lactalbumin from bovine milk. To prepare for this activity, here you will learn about the structure of a related protein, α-lactalbumin from humans. We will search databases to find and view its primary and secondary structure and also determine if there are other proteins with a similar amino acid sequence and structure. After completion of these exercises, you will be able to apply these computer tools to proteins of your own choice.

Point your Web browser to the Protein Data Bank (PDB) and the Research Collaboratory for Structural Bioinformatics (http://www.rcsb.org/pdb/). Become acquainted with the PDB by viewing the home page and perhaps clicking on some hyperlinks. Scroll until you find the term "Searchlite" under Search on the right side of the screen. Clicking on Searchlite will display a dialogue box for keywords. Type in "human alpha-lactalbumin" and click on Search. Your query will find at least seven structures that are listed. Click on the white square to the left and "EXPLORE" to the right of Structure 1A4V. This will display "Structure Explorer" with "Summary Information" about the structure of the protein. Clicking on the "?" will provide help if necessary. Review the functions possible on the left side of the screen. Click on "View Structure" to observe "Interactive 3D Display" and "Still Images." First, study the still images of human α-lactalbumin in ribbon or cylinder form. You may click on 250×250 or 500×500 to enlarge. Note the presence of α-helices and β-sheets in the structure. After studying the still images, click on "Chime" under Interactive 3D Display. Now, you will observe the ribbon structure rotating on an axis. Use "Chime Help" at the bottom of the screen to learn Mouse Controls of the rotating structure. Now return to the Summary Information list to try other functions. Click on "Sequence Details" to observe the amino acid sequence and definition of secondary structures. You may do an ftp download of this file by clicking on "Download in FASTA format." FASTA format is a listing of amino acid sequences using the standard single-letter abbreviation for each amino acid. Clicking on "Geometry" will display tables of bond angles and lengths. Similar sequence studies may be done by clicking on the function "Structural Neighbors." Several tools are available to search for similar structures. Try the VAST tool. Clicking on "VAST" will provide two options, Sequence Neighbors and Structure Neighbors. Clicking on "Sequence Neighbors: single chain" will display a list of many proteins with sequences similar to that of human α-lactalbumin. Note that most are α-lactalbumins from other species, but if you scroll far enough, you will see the enzyme lysozyme listed. Returning to the former screen and clicking on Structure Neighbors will display about eight structures similar to human α-lactalbumin. Note again the presence of lysozyme in the list. Clicking on "Other Sources" will display other data files with references to α-lactalbumin. It is interesting to note that the proteins α-lactalbumin and lysozyme have similar primary, secondary, and tertiary structures but they have quite different biochemical activities. The two proteins, which have about 40% sequence identity, may have been derived from a common ancestral gene.

Another useful structure tool is RasMol (or RasMac). This will allow you to view the detailed structure of a protein and rotate it on coordinates so you can see it from all perspectives. A hyperlink to RasMol is present under the "View Structure" function just above "Chime." You may need to study RasMol instructions provided under Help, or you may use a Ras-Mol tutorial listed in Table E1.2. Another useful protein viewer is the Swiss-Protein Pdv Viewer (Table E1.2). BLAST is an advanced sequence similarity tool available at NCBI. To access this, go to the NCBI home page (www.ncbi.nlm.nih.gov) and click on "BLAST." Then click on "Basic BLAST search" to obtain a dialogue box into which you may type the amino acid sequence of human α-lactalbumin. This process may be streamlined by downloading the amino acid sequence in FASTA format into a file and transferring the file into the BLAST dialogue box. BLAST will provide a list of proteins with sequences similar to the one entered.

Another approach to a study of protein (or nucleic acid) structure and sequence is through Entrez. This can be entered via the NCBI home page. Then click on "Proteins" to obtain a dialogue box where you can type "human alpha-lactalbumin" and then click on Search. You can retrieve about 25 documents for review. Note that you may also enter BLAST through Entrez.

IV. ANALYSIS OF DATA

Practice with the software introduced in the Experimental Procedure until you are familiar with it. Use proteins and nucleic acids discussed in your biochemistry class as exercises for bibliographic searches and structure/sequence studies. Several more examples are provided in the Study Problems. Continue to use the Web resources described in this experiment as you proceed through this laboratory course.

Study Problems

1. In Experiment 5, you will study kinetics and inhibition with the enzyme mushroom tyrosinase. Use PubMed bibliographic searches to learn the following aspects of the enzyme:

    (a) What other sources of the enzyme are there besides mushrooms?
    (b) What metal ion is present in the native enzyme?

(c) Find two references that study inhibition of the mushroom enzyme. What inhibitor molecules have been investigated?

(d) Find another substrate for the enzyme besides dihydroxyphenylalanine (Dopa).

2. Find two recent research articles published by Thomas R. Cech, who won the Nobel Prize for the discovery of ribozymes. Write brief summaries of the articles.

3. The technique immobilized metal ion affinity chromatography (IMAC) is widely used to purify proteins (Experiment 4). Find two proteins that have recently been purified by this technique and briefly describe the methods for isolation.

4. The Western blot procedure (Experiment 7) is now used to test human serum for the presence of antibodies to the AIDS virus. Find two publications that describe procedures for this assay.

5. In Experiment 9, you will study proton pumping in chloroplasts by measuring pH changes. Find an alternative experimental method to investigate proton pumping in chloroplasts.

6. Outline the pathway for microbial degradation of the detergent used in denaturing electrophoresis, sodium dodecyl sulfate (SDS). Hint: See the Web site on Biocatalysis/Biodegradation.

7. Use the REBASE site to determine the specificity of the restriction enzyme *Hin*dII.

8. Use the techniques outlined in the experimental procedure to explore two enzymes you will study in later experiments. Study the two enzymes malate dehydrogenase (Experiment 10) and tyrosinase (Experiment 5). View structures and look at amino acid sequences as you did for human α-lactalbumin.

9. Study the nucleotide sequence for the gene coding for human α-lactalbumin. Hint: Begin at the NCBI home page and enter Entrez. Click on "Nucleotides" and do a search on human α-lactalbumin. Review the GenBank report for the position of introns and exons. Obtain a FASTA report, transfer (download) the files, and complete a BLAST search for related sequences.

10. Use the BLAST tool to compare the amino acid sequences for human α-lactalbumin and lysozyme. Repeat the process using BLAST to compare the nucleotide sequences for the genes coding for human α-lactalbumin and lysozyme.

Further Reading

A. Baxevanis and B. Ouellette, Editors, *Bioinformatics: A Practical Guide to the Analysis of Genes and Proteins* (1998), John Wiley & Sons (New York). A new introduction to computing.

R. Doolittle, Editor, *Methods in Enzymology* (1996), "Computer Methods for Macromolecule Sequence Analysis," Vol. 266, Academic Press (San Diego). Several articles on current databases.

S. Hoersch, C. Leroy, N. Brown, M. Andrade, and C. Sander, *Trends Biochem. Sci.* **25,** 33–35 (2000). "The Gene Quiz Web Server."

D. Leon, S. Uridil, and J. Miranda, *J. Chem. Educ.* **75,** 731–734 (1998). "Structural Analysis and Modeling of Proteins on the Web."

L. Peruski and A. Peruski, *The Internet and the New Biology: Tools for Genomic and Molecular Research* (1997), ASM Press (Washington, DC). An excellent introductory book.

D. Ridley, *Online Searching: A Scientist's Perspective—A Guide for the Chemical and Life Sciences* (1996), John Wiley & Sons (New York). A review of database availability and use.

H. Salter, *Biochem. Educ.* **26,** 3–10 (1998). "Teaching Bioinformatics."

R. Sayle and E. Milner-White, *Trends Biochem. Sci.* **20,** 374–376 (1995). "RASMOL: Biomolecular Graphics for All."

C. Smith, *The Scientist,* August 31, pp. 17–19 (1998). "Molecular Modeling."

C. Smith, *The Scientist,* April 26, pp. 21–23 (1999). Bioinformatics software for the future.

Trends Guide to the Internet (1997), Elsevier Science Publishing (Cambridge, UK). Excellent introduction to the Internet.

T. Zielinski and M. Swift, Editors, *Using Compuers in Chemistry and Chemical Education* (1997), American Chemical Society (Washington, DC). A good instructional manual.

Glossary for the Internet

biological databases–computer sites that organize, store, and disseminate files that contain information consisting of literature references, nucleic acid sequences, protein sequences, and protein structures.

bookmark–a function to save a Web site address for later use (Netscape Navigator).

browser–an interface program that reads hypertext and displays Web pages on your computer.

domain–the computer user's location or local network.

e-mail–electronic mail; a means of exchanging messages via computer.

favorites–the Internet Explorer form of a bookmark.

freeware–software that is provided free of charge by its developer.

ftp–file transfer protocol; a mechanism for transferring data across a network.

home page–the beginning page for access to a Web site.

HTML–HyperText Markup Language; a special, coded language that is used to write Web pages.

hyperlink–link or connection between related Web pages.

hypertext–a language that connects similar documents on the Web.

Internet–the worldwide matrix that allows all computers and networks to communicate with each other.

Java–a programming language that allows the incorporation of multimedia into Web pages.

modem–electronic hardware that transmits computer signals through a telephone line.

multimedia–several forms of media including text, graphics, video, and audio.

search engine–a searchable directory that organizes Web pages by subject classification.

server–a computer that acts as the storage site for retrievable data.

URL–Uniform Resource Locator; a standard address form that identifies the location of a document on the Internet.

Web site–a collection of documents (Web pages) on a server.

WWW–World Wide Web ("the Web"); a component of the Internet that uses a hypertext-based language to provide resources.

2

STRUCTURAL ANALYSIS
OF A DIPEPTIDE

• **Recommended Reading**

Chapter 2, Section A; *Chapter 3,* Sections A, B, G; *Chapter 4,* Section B.

• **Synopsis**

The primary structure of a peptide or protein is defined by the sequence of amino acids. In this experiment the procedures that are in common use to determine protein primary structure are applied to an unknown dipeptide. Amino acid composition of the peptide will be determined by acid hydrolysis followed by HPLC, CE, or paper chromatography. The identity of the NH_2-terminal amino acid will be achieved by the dansyl method followed by thin-layer chromatography.

I. INTRODUCTION AND THEORY

Structural elucidation of natural macromolecules is an important step in understanding the relationships between the chemical properties of a biomolecule and its biological function. The techniques used in organic structure determination (NMR, IR, UV, and MS) are quite useful when applied to biomolecules, but the unique nature of natural molecules also requires the application of specialized chemical techniques. Proteins, polysaccharides, and nucleic acids are polymeric materials, each composed of hundreds or sometimes thousands of monomeric units (amino acids, monosaccharides, and nucleotides, respectively). But there is only a limited number of these types of units from which the biomolecules are synthesized. For example, only 20 different amino acids are found in proteins but these different amino acids may appear several times in the same protein molecule. Therefore, the structure of a peptide or protein can be recognized only after the amino acid composition *and* sequence have been determined.

Amino Acid Composition

All proteins found in nature are constructed by amide bond linkages between L-α-amino acids. The amino acids isolated from peptides and proteins all have common structural characteristics. They have at least one carboxyl group and at least one amino group (Figure E2.1). The distinctive physical, chemical, and biological properties associated with an amino acid are the result of the R group, which is unique for each amino acid. A list of the 20 common amino acids and their three-letter and one-letter abbreviations is given in Table E2.1. If you are not yet familiar with the structures of the common amino acids, refer to your biochemistry textbook. The structure and biological function of a protein depend on its amino acid content. It is a matter of basic importance to understand practical methods used for the separation and identification of the 20 amino acids found in proteins.

The amide bonds in peptides and proteins can be hydrolyzed in strong acid or base. Treatment of a peptide or protein under either of these conditions yields a mixture of the constituent amino acids. Neither acid- nor base-catalyzed hydrolysis of a protein leads to ideal results because both tend to destroy some constituent amino acids. Acid-catalyzed hydrolysis destroys tryptophan and cysteine, causes some loss of serine and threonine, and converts asparagine and glutamine to aspartic acid and glutamic acid, respectively. Base-catalyzed hydrolysis leads to destruction of serine, threonine, cysteine, and cystine and also results in racemization of the free amino acids. Because acid-catalyzed hydrolysis is less destructive, it is often the method of choice. The hydrolysis procedure consists of dissolving the protein sample in aqueous acid, usually 6 M HCl, and heating the solution in a sealed, evacuated vial at 100°C for 12 to 24 hours.

Now that the free amino acids present in a peptide or protein have been released by hydrolysis, they must be separated and identified. The most versatile, economical, and convenient techniques for separation are based on chromatographic methods. Earlier workers relied on paper chromatography (PC) and thin-layer chromatography (TLC); however, more sensitive techniques are now available. Automated ion-exchange chromatography (amino acid analyzers), gas chromatography, capillary electrophoresis (CE), and high-performance liquid chromatography (HPLC) are now powerful tools for the qualitative and quantitative analysis of amino acids and derivatives. Because there is still a demand for rapid, qualitative, routine analysis of

---

Figure E2.1

The general structure of the zwitterionic form of an L-amino acid. R represents the side chain.

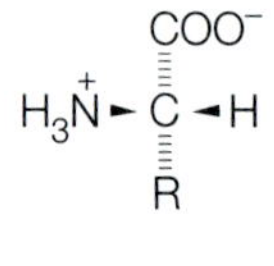

$$\begin{array}{c} COO^- \\ | \\ H_3\overset{+}{N} - C - H \\ | \\ R \end{array}$$

Table E2.1

Abbreviations of the 20 Common Amino Acids Found in Proteins

Name	Abbreviation	
	One-letter	Three-letter
Glycine	G	Gly
Alanine	A	Ala
Valine	V	Val
Leucine	L	Leu
Isoleucine	I	Ile
Methionine	M	Met
Phenylalanine	F	Phe
Proline	P	Pro
Serine	S	Ser
Threonine	T	Thr
Cysteine	C	Cys
Asparagine	N	Asn
Glutamine	Q	Gln
Tyrosine	Y	Tyr
Tryptophan	W	Trp
Aspartate	D	Asp
Glutamate	E	Glu
Histidine	H	His
Lysine	K	Lys
Arginine	R	Arg

amino acids, thin-layer and paper chromatographic methods are still being developed and improved. Amino acids and derivatives may be analyzed directly by PC and TLC without further derivatization. Several support materials are available, but most analyses are carried out on silica gel or cellulose. The free amino acids can be detected on the developed chromatographic plates by reaction with ninhydrin. A pink-purple color is obtained for all amino acids except proline. A yellow color develops with proline. Since many derivatives of amino acids, including phenylthiohydantoin and dansyl derivatives, contain aromatic moieties, most can be detected on TLC plates by fluorescence under an ultraviolet lamp.

Gas chromatography is used to analyze volatile derivatives of amino acids. Phenylthiohydantoins (products of Edman degradation) may be analyzed directly by GC but are better resolved if converted to their trimethylsilyl derivatives with N,O-bis(trimethylsilyl) acetamide. Free amino acids are generally converted to their N-trifluoroacetyl-n-butyl esters or trimethylsilyl derivatives before GC analysis. For best results, all gas chromatography of amino acid derivatives should be done with a glass column and injection port, as contact with metals causes extensive decomposition of the derivatives.

High-performance liquid chromatographic techniques have been applied with success to the analysis of phenylthiohydantoin, 2,4-dinitrophenyl, and dansyl amino acid derivatives.

If only very small samples of amino acids are available for analysis, fluorescence is used for detection. One of the most sensitive methods of microanalysis is based on the reaction of amino acids with *o*-phthalaldehyde and β-mercaptoethanol (Equation E2.1). The isoindole derivative is fluorescent and amounts as small as 10^{-12} mole may be measured.

| *o*-Phthalaldehyde | Amino acid | β-Mercaptoethanol | **Equation E2.1** |

Isoindole derivative
of amino acid

Another important derivatizing reagent for amino acids is 9-fluorenylmethyl chloroformate (FMOC). This reagent has the advantages that (1) the reaction with amino acids is very fast, occurring in less than 1 min; (2) the products are stable for long periods of time; (3) the derivatives can be separated by reversed-phase column chromatography and capillary electrophoresis procedures; and (4) the products are fluorescent for easy detection. FMOC derivatives of amino acids are usually prepared in borate buffer under slightly basic conditions (Equation E2.2).

FMOC   Amino acid   N-FMOC
derivative   **Equation E2.2**

End Group Analysis

Sequential analysis of amino acids in purified peptides and proteins is best initiated by analysis of the terminal amino acids. A peptide has one amino acid with a free α-amino group (NH$_2$-terminus) and one amino acid with a free α-carboxyl group (COOH-terminus). Many chemical methods have been developed to selectively tag and identify these terminal amino acids.

NH$_2$-terminal amino acid analysis is achieved by the use of (1) 2,4-dinitrofluorobenzene (Sanger reagent), (2) 1-dimethylaminonaphthalene-5-sulfonyl chloride (dansyl chloride), or (3) phenylisothiocyanate (Edman reagent). Figure E2.2 shows the structures of these reagents. Although the chemistry is different for each of these reagents, the same general concept is used. These compounds react with the NH$_2$-terminal amino acid of a peptide or protein to produce covalent derivatives that are stable to acid-catalyzed hydrolysis. The process is illustrated in Figure E2.3. The mixture of modified amino acid and free amino acids is separated, and the NH$_2$-terminal amino acid is identified by thin-layer, paper, gas, or high-performance liquid chromatography or capillary electrophoresis. The first NH$_2$-terminal method to be widely used was based on the Sanger reagent, which produces yellow-colored 2,4-dinitrophenyl (DNP) derivatives of NH$_2$-terminal amino acids. The Sanger method has several disadvantages, including poor yield of the DNP-peptide derivative, low sensitivity of analysis, and instability of some DNP–amino acids during acid hydrolysis. The dansyl chloride method has largely replaced the Sanger method because very sensitive fluorescence techniques may be used for detection and analysis of the dansyl amino acid derivatives and the derivatives are more stable during acid hydrolysis.

Figure E2.2

Reagents useful in NH$_2$-terminal analysis.

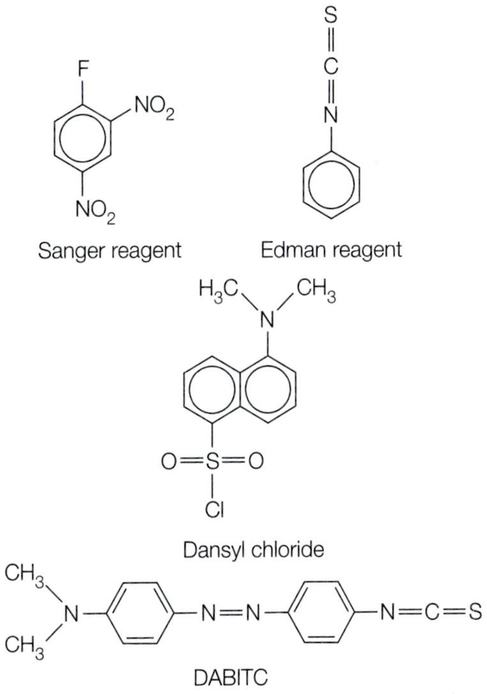

Sanger reagent   Edman reagent

Dansyl chloride

DABITC

Analysis of NH$_2$-terminal amino acids.

N-terminal reagent

Derivatized peptide

Free amino acids
or
intact peptide (Edman)

Even more versatile than the dansyl method is the Edman method (Figure E2.4). The NH$_2$-terminal amino acid is removed as its phenylthiohydantoin (PTH) derivative under anhydrous acid conditions, while all other amide bonds in the peptide remain intact. The derivatized amino acid is then extracted from the reaction mixture and identified by paper, thin-layer, gas, or high-performance liquid chromatography. The intact peptide (minus the original NH$_2$-terminal amino acid) may be isolated and recycled by reaction with phenylisothiocyanate. Since this method is nondestructive to the remaining peptide (aqueous acid hydrolysis is not required) and results in good yield, it can be used for stepwise sequential analysis of peptides. The method is now automated.

Under all NH$_2$-terminal labeling conditions, internal amino acid residues in a peptide may be modified, but in ways different from the NH$_2$-terminal amino acid. For example, if the NH$_2$-terminal amino acid is alanine, only the free α-amino group is reactive to the selective reagent. However, if lysine, serine, or other amino acids with nucleophilic side chains are present at the NH$_2$-terminus, both nucleophilic groups (α-amino and ϵ-amino for lysine; α-amino and β-hydroxyl for serine) have the potential for reaction with the labeling reagent. If lysine or serine residues are located internally, only the side chain nucleophilic group (ϵ-amino for lysine, β-hydroxyl for serine) has the potential for reaction. Chromatographic techniques may be used to separate and identify each of these three types of amino acid modification (α-amino, side chain, and both).

The development of new chemical reagents and instrumentation has now made it possible to achieve end group analysis and sequencing on extremely small samples of protein (microsequencing). By using a chromophoric derivative of phenylisothiocyanate, 4-*N*,*N*-dimethylaminoazobenzene-4'-isothiocyanate (DABITC; see Figure E2.2), and analyzing the

The Edman method of NH$_2$-terminal amino acid analysis.

Phenylthiohydantoin   Intact peptide

derivatives by HPLC, it is possible to sequence peptides and proteins at the nanomolar level.

The COOH-terminal amino acid of a peptide or protein may be analyzed by either chemical or enzymatic methods. The chemical methods are similar to the procedures for NH$_2$-terminal analysis. COOH-terminal amino acids are identified by hydrazinolysis or are reduced to amino alcohols by lithium borohydride. The modified amino acids are released by acid hydrolysis and identified by chromatography. Both of these chemical methods are difficult, and clear-cut results are not readily obtained. The method of choice is peptide hydrolysis catalyzed by carboxypeptidases A and B. These two enzymes catalyze the hydrolysis of amide bonds at the COOH-terminal end of a peptide (Equation E2.3), since carboxypeptidase action requires the presence of a free α-carboxyl group in the substrate.

Carboxypeptidase A catalyzes the hydrolysis of carboxyl-terminal acidic or neutral amino acids; however, the rate of hydrolysis depends on the structure of the side chain R'. Amino acids with nonpolar aryl or alkyl side chains are cleaved more rapidly. Carboxypeptidase B is specific for the hydrolysis of basic COOH-terminal amino acids (lysine and arginine). Neither peptidase functions if proline occupies the COOH-terminal position or is the next to last amino acid.

Overview of the Experiment

In this experiment, the sequence of amino acids in a dipeptide will be determined by using some of the techniques just described. The amino acid com-

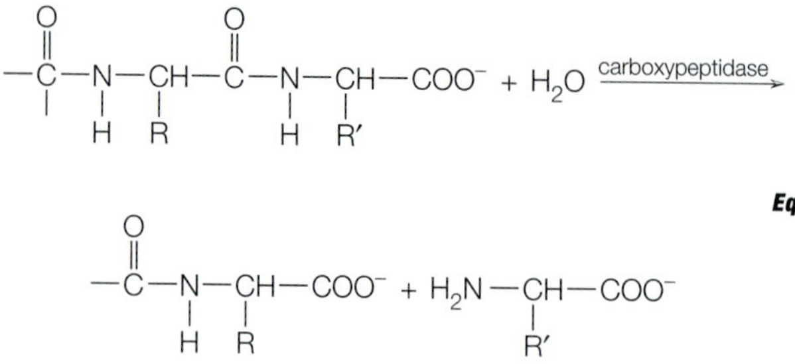

$$-\overset{\overset{\displaystyle O}{\|}}{C}-\underset{\underset{\displaystyle H}{|}}{N}-\underset{\underset{\displaystyle R}{|}}{CH}-\overset{\overset{\displaystyle O}{\|}}{C}-\underset{\underset{\displaystyle H}{|}}{N}-\underset{\underset{\displaystyle R'}{|}}{CH}-COO^- + H_2O \xrightarrow{\text{carboxypeptidase}}$$

>>

Equation E2.3

$$-\overset{\overset{\displaystyle O}{\|}}{C}-\underset{\underset{\displaystyle H}{|}}{N}-\underset{\underset{\displaystyle R}{|}}{CH}-COO^- + H_2N-\underset{\underset{\displaystyle R'}{|}}{CH}-COO^-$$

position of the unknown peptide will be found by acid-catalyzed hydrolysis followed by analysis of the amino acids by chromatography or capillary electrophoresis. Sequence analysis will then be carried out by identification of the NH_2 terminus. See Figure E2.5 for a summary of the procedure. A general experimental procedure is outlined below, but the specific analyses you will perform will depend on the directions from your instructor.

Part A. Hydrolysis of the unknown dipeptide

   1. Analysis by HPLC or CE

   2. Analysis by paper chromatography

Part B. NH_2-terminal analysis

Suggested Schedule

Period 1: Part A–Begin acid hydrolysis of peptide. Part B–Complete dansyl chloride reaction and begin acid hydrolysis of dansyl peptide. Part A.2–if applicable, prepare paper chromatogram by applying standard amino acids.

Figure E2.5

Flowchart for the analysis of a dipeptide.

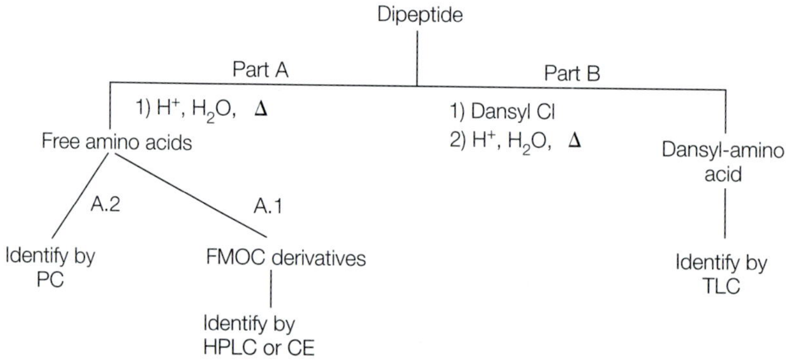

- -

Period 2: Part B—Work up the dansyl hydrolysate and spot on the TLC plate with standard dansyl amino acids. Part A.1—Work up peptide hydrolysate and prepare FMOC derivatives of amino acids for analysis by HPLC or CE. Part A.2—If applicable, develop paper chromatogram in solvent system.

II. MATERIALS AND SUPPLIES

A. Hydrolysis of Unknown Peptide

- Unknown dipeptide
- Watch glass
- 6 M HCl
- Syringe, 50 or 100 μL
- Melting point capillary tube, 1.5 × 100 mm
- Oven at 100°C
- Heat lamp

1. Analysis by HPLC or CE

- Borate buffers: 0.05 M, pH 9.0; and 0.2 M, pH 7.7
- 9-Fluorenylmethyl chloroformate (FMOC) solution in acetone, 0.015 M
- Hexane
- Speed-Vac centrifuge
- Eppendorf tubes, 1.5 mL
- FMOC derivatives of amino acids: Ala, Gly, Phe, Leu, Val
- HPLC system including two pumps, injector, controller, and UV detector operating at 254 nm. Suggested columns and conditions are:

    (a) Phenomenex HYPERSIL 5μC8(MOS-1); 250 × 4.6 mm; Mobile phase A—0.05 M acetate buffer, pH 4.2/acetonitrile/tetrahydrofuran (80:19.5:0.5). Mobile phase B—0.05 M acetate buffer, pH 4.2/acetonitrile (20:80).

    Gradient: 0–20 min, 0–50% B

    20–30 min, 50–100% B

    Flow rate: 1.0 mL/min

    Detector: UV at 254 nm

    (b) Alltech Adsorbosphere 5μC18; 250 × 4.6 mm; Mobile phase A—0.05 M acetate buffer, pH 4.2/methanol/acetonitrile (50:40:10). Mobile phase B—0.05 M acetate buffer, pH 4.2/acetonitrile (50:50).

Gradient: 0–3 min, isocratic elution with 100% A

3–12 min, 0–100% B

12–23 min, isocratic elution with 100% B

Flow rate: 1.3 mL/min

Detector: UV at 254 nm

- CE system equipped with a UV detector (248 nm), an unmodified fused silica capillary (64 cm), and computer.

Borate buffer/SDS, 0.02 M borate buffer, pH 9.2 with 0.025 M SDS.

2. Analysis by Paper Chromatography

- Whatman 3MM paper, 20 × 20 cm
- Amino acid standard solutions, 1% in water
- Chromatography chambers
- Chromatography solvent, acetonitrile and 0.1 M ammonium acetate, 60:40, pH 4.0 or pH 5.0
- Ninhydrin spray
- Ruler
- Stapler

B. NH$_2$-Terminal Analysis

Dansyl Chloride

- Unknown dipeptide, 1 mg
- Dansyl chloride solution, 5 mg/mL in acetone
- Conical centrifuge tube, 12 mL
- Sodium bicarbonate, 0.2 M
- Hydrocarbon foil
- Constant-temperature bath at 37°C
- Hydrolysis vial (½ dram with polyvinyl liner, Sargent-Welch)
- 6 M HCl
- Acetone and 6 M HCl (1:1)
- Oven at 100°C
- Thin-layer plates, polyamide, double-sided, 15 × 15 cm (Schleicher and Schuell)
- Chromatography solvents:

Solvent 1. Formic acid–H$_2$O, 1.5:100 (v/v)

Solvent 2. Toluene–acetic acid, 10:1 (v/v)

Solvent 3. Ethyl acetate–methanol–acetic acid, 2:1:1 (v/v/v)

- Standard mixture of dansyl amino acids in acetone, 1 mg/mL each: Phe, Leu, Gly, Ala, Val.
- UV detection lamp

III. EXPERIMENTAL PROCEDURE

A. Hydrolysis of Unknown Dipeptide

Place approximately 1–2 mg of your peptide sample on a small watch glass and dissolve it in 0.05 mL of 6 M HCl delivered with a syringe. With a syringe, transfer the peptide solution into a melting-point capillary, and seal the open end with a Bunsen burner. Put the sealed tube in a small labeled beaker and place in a 100° C oven for 12 to 15 hours. After hydrolysis, open the capillary by scratching one end with a sharp file, and transfer the hydrolysate to a watch glass. Evaporate the liquid by *gentle* heating under a heat lamp **(Hood!)** or by using a Speed-Vac. Add 0.1 mL of distilled water, and repeat the evaporation step.

1. Analysis by HPLC or CE

Dissolve the residue containing free amino acids in 200 μL of 0.05 M borate buffer, pH 9.0. Add 2.0 mL of 0.2 M borate buffer, pH 7.7. The final pH of the solution sould be in the range of 7.5–8.0. If not, adjust with dilute HCl or NaOH. To make the FMOC derivatives of the amino acids, transfer 500 μL of the above amino acid solution to a 15-mL centrifuge tube. Add 500 μL of the 15 mM 9-fluorenylmethyl chloroformate in acetone solution. Allow to stand at room temperature for about 45 seconds and then add 2 mL of hexane. Stopper the tube and shake for 1–2 minutes. (This extraction process removes excess FMOC reagent.) Separate the two layers by centrifuging at 1000 rpm for 5 minutes. Remove the top hexane layer with a disposable pipet and rubber bulb. Repeat the extraction step two more times using 2 mL of hexane each time. Dispose of the hexane fractions in the organic waste. Save the lower aqueous layer containing the FMOC–amino acids for HPLC or CE analysis. For HPLC analysis, use solvents, gradients and conditions as described in the Materials and Supplies section. Inject approximately 10 μL of the FMOC–amino acids from the unknown dipeptide and 10 μL of standard FMOC–amino acid derivatives provided. For CE analysis, use conditions suggested in Materials and Supplies using injection volumes of 2–5 μL.

2. Analysis by Paper Chromatography

Prepare the residue for paper chromatography by dissolving it in 0.1 mL of distilled water. Obtain a 20 $\times$ 20 cm sheet of Whatman 3MM. Wear disposable plastic gloves when you handle the paper in order to avoid transfer of amino acids from your fingers to the paper sheet. Prepare the chromatogram by applying spots of the unknown hydrolysis mixture and amino acid standards along one edge of the paper. The amino acid standards can be applied during period 1 and the hydrolysis mixture at the beginning of period 2. The spot should be no closer than 2 cm from the bottom edge or side edges. You may wish to mark the paper *lightly* 2 cm from one edge with a pencil. Prepare a "key" in your notebook to keep track of the identity of each spot (Figure E2.6).

Figure E2.6

Preparation of a thin-layer
or paper chromatogram.

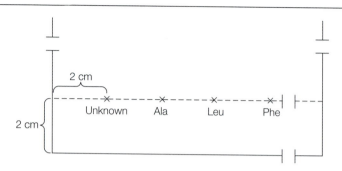

Using small capillary tubes drawn to a fine point or microcapillary pipets,
make two separate applications of your unknown solution, one containing
approximately 5 μL and the other containing two or three times as much sample. It is best to apply the larger amount in small increments. Make one application and allow it to dry before more is applied to the spot. Apply the amino
acid standards at selected spots along the 2-cm line; 5 to 10 μL of each standard should be sufficient. Applications of amino acids should be kept as small
as possible (0.2–0.3 cm) so that the developed spots do not overlap. To develop the paper chromatogram, staple the sheet into a cylinder and place it in
the chromatography jar containing a 1-cm layer of acetonitrile and 0.1 M ammonium acetate (60:40, pH 4.0 or 5.0). When preparing the cylinder, handle
the paper only with gloves and do not overlap the edges while stapling. Solvent development time is 40 to 60 minutes.

When the developing solvent reaches a few centimeters from the top of
the chromatogram, remove the paper from the jar and allow it to dry in a
hood. Remove the staples and spray the paper lightly with ninhydrin solution. Allow it to dry in a hood for 10 minutes. The color may be developed
by placing the chromatogram in an oven (100°C for 10 min) or at room temperature for 1 to 2 hours. All the spots should be circled with a pencil, as
they will fade with time. Describe in your notebook the color of each spot.

B. NH₂-Terminal Analysis

Dansyl Chloride

▶ **CAUTION**

Dansyl chloride is a corrosive organic acid halide, which should be used
only in a hood. Wear disposable gloves while using the reagent. If it is
spilled on the skin, sprinkle with sodium bicarbonate and wash with copious amounts of warm soapy water. Rinse well. For floor or bench
spills, cover with sodium bicarbonate and transfer mixture to a beaker
of water. Pour the aqueous solution down the drain with excess water.
Do not pipet solutions of dansyl chloride by mouth!

Weigh about 1 mg of the unknown peptide and transfer into a conical centrifuge tube. Dissolve the peptide in 0.5 mL of 0.2 M sodium bicarbonate. Add 0.2 mL of the solution of dansyl chloride in acetone and mix well. Cover with hydrocarbon foil and incubate for 1 hour at 37°C or 2 hours at room temperature. After completion of the reaction, evaporate the solution to dryness under vacuum or by a gentle stream of N_2 gas, with the reaction tube held in a beaker of warm water. Dissolve the residue in 0.5 mL acetone–6 M HCl solution and transfer it into the hydrolysis vial. Evaporate the acetone from the vial with a gentle stream of nitrogen gas. Seal the hydrolysis vial and heat in a 100°C oven for 10 to 12 hours. Remove the hydrolysate from the vial and evaporate it on a watch glass under a heat lamp. Dissolve the residue in a minimal amount of 50% ethanol/pyridine (about 10 μL). Using a microsyringe, spot 1–5 μL of the solution about 1 cm from the corner of a polyamide thin-layer sheet (see Figure E2.7). The diameter of the spot should not exceed 3–4 mm. Dry the spot with warm air. On the reverse side of the plate load 1 μL of the standard dansyl–amino acid mixture. This should be spotted at the same position (1 cm from corner) as the unknown. Develop the plate in solvent 1, 1.5% formic acid, in the direction shown in Figure E2.7 until the solvent reaches 2–3 cm from the top. This may take up to 1 hour. Dry the solvent from both sides of the plate in warm air. View the plate under UV light. **Be sure to wear safety goggles!** A blue fluorescent streak with faint green fluorescent spots may be seen. Develop the plate in solvent 2, toluene–acetic acid, at right angles to solvent 1 (see Figure E2.7). The fluorescent blue streak will be along the bottom of the plate. Remove the plate from the solvent, allow to air-dry, and view under UV light. There is now some separation of the blue and green spots (see Figure E2.7). If the

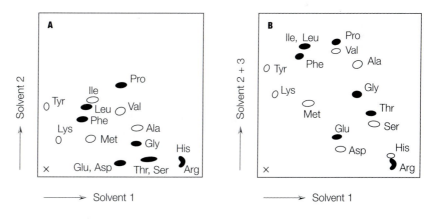

Figure E2.7

Thin-layer chromatography of dansyl–amino acids on a polyamide plate.   **A** After solvents 1 (1.5% formic acid) and 2 (toluene–acetic acid).   **B** After solvents 1, 2, and 3 (ethyl acetate–methanol–acetic acid). See text for further details.

unknown cannot be identified, run the plate in solvent 3, ethyl acetate–methanol–acetic acid, in the same direction as solvent 2.

IV. ANALYSIS OF RESULTS

A. Hydrolysis of Unknown Dipeptide

1. Analysis by HPLC and CE

Prepare a table of HPLC elution time values for each of the standard FMOC amino acids used in this experiment. FMOC amino acids usually elute from a reversed phase column in the following order (from first to last): Gly, Ala, Val, Phe, and Leu. Describe how the order of elution is related to the relative polarity of the FMOC amino acids. The FMOC reagent during the derivatization reaction often undergoes hydrolysis to a decarboxylated form of FMOC. This product, an alcohol called FMOH, usually elutes from the HPLC column between Val and Phe or between Ala and Val. Use the standard elution times to identify the unknown amino acids present in the original dipeptide.

The elution order of the standard FMOC amino acids on a CE unmodified silica capillary is usually Ala, Gly, Val, Leu, Phe. FMOH usually elutes after Phe.

2. Analysis by Paper Chromatography

Prepare a table of R_f values for the standard amino acids. You should also describe the color of the original ninhydrin spot. Since the colors vary slightly with the amino acids, this aids in the identification of an unknown amino acid. Calculate the R_f values for the constituent amino acids of the unknown peptide. From these data you should be able to identify the amino acids present in your unknown.

B. NH₂-Terminal Analysis

Dansyl Chloride

Draw a picture in your notebook of the polyamide thin-layer plate exposed under UV light after each of the two or three solvent developments. These pictures should look similar to Figure E2.7. Three fluorescent areas should be evident after solvent 2; however, better separation is achieved by solvent 3. A blue fluorescent area at the bottom of the plate is dansic acid (DNS-OH), which is a hydrolysis product of dansyl chloride. A blue-green fluorescent spot about one-third to one-half up the plate is dansyl amide (DNS-NH$_2$), which is produced by reaction of dansyl chloride with ammonia. A third spot, which usually fluoresces green, is the dansyl derivative of the NH$_2$-terminus amino acid. Note the positions of the standard dansyl amino acids and compare with the unknown. What is the identity of the NH$_2$-terminal amino acid? Are any other fluorescent spots evident on the plate? Using polarity or nonpolarity, try to explain the position of each molecule on the thin-layer plate.

Study Problems

▶ *1.* Study the details of the Edman and dansyl chloride methods for amino-terminal analysis of proteins. Discuss the advantages and disadvantages of each method.

▶ *2.* The new reagent 4-*N,N*-dimethylaminoazobenzene-4′-isothiocyanate (DABITC) is often used for the Edman method of protein sequence analysis. Write a reaction showing this use of the reagent.

 3. Describe how the HPLC and CE instruments detected the FMOC–amino acids eluting from the column.

▶ *4.* List the order of FMOC-amino acids from a typical reversed-phase HPLC column. Explain the order in terms of the relative polarity of the amino acid derivatives. Hint: See Chapter 3.

▶ *5.* Predict the order of elution from a reversed-phase HPLC column for each of the following mixtures.
(a) FMOC-Ser and FMOC-Ala
(b) FMOC-Asp and FMOC-Pro
(c) FMOC-Gly, FMOC-Pro, and FMOC-Leu

▶ *6.* Why must the layer of solvent in a chromatography jar be below the origin line containing the applied samples on a paper or thin-layer chromatogram?

▶ *7.* In the TLC analysis of dansyl amino acids, why does dansic acid move very slowly during development in solvents 2 and 3?

▶ *8.* Complete the following reaction:

$$\text{Gly--Ala} + H_2O \xrightarrow{H^+}$$

 9. Draw the structure of the FMOC derivative of the amino acid phenylalanine as it would exist at pH 7.0.

 10. The following reagents are useful for characterizing peptides and proteins. Describe how each can be used and what information is obtained. Write all appropriate reactions.
(a) Dansyl Chloride   (d) *o*-Phthalaldehyde + β-mercaptoethanol
(b) 6 *M* HCl   (e) Ninhydrin
(c) 9-Fluorenylmethyl chloroformate

Further Reading

D. Baker, *Capillary Electrophoresis* (1995), John Wiley & Sons (New York). A complete book on the theory and practice of CE.

D. Blackman, *The Logic of Biochemical Sequencing* (1993), CRC Press (Boca Raton, FL). End group analysis and sequencing of proteins.

R. Boyer, *Concepts in Biochemistry* (1999), Brooks/Cole (Pacific Grove, CA), pp. 78–105. A discussion of amino acids and protein sequencing.

C. Clapp, J. Swan, and J. Poechmann, *J. Chem. Educ.* **69,** A122–A126 (1992). Using 9-fluorenylmethyl chloroformate and HPLC to identify amino acids.

S. Einarsson, B. Josefsson, and S. Lagerkvist, *J. Chromatogr.* **282,** 609–618 (1983). FMOC derivatives of amino acids.

R. Garrett and C. Grisham, *Biochemistry,* 2nd ed. (1999), W. B. Saunders (Orlando, FL), pp. 107–152. Protein primary structure.

P. Grossman and J. Colburn, Editors, *Capillary Electrophoresis: Theory and Practice* (1992), Academic Press (San Diego). An up-to-date and complete review of CE.

E. Heimer, *J. Chem. Educ.* **49,** 547 (1972). Paper chromatography of amino acids.

F. Lai and T. Sheehan, *BioTechniques* **14,** 642–649 (1993). A comparison of reagents for amino acid derivatization.

A. Lehninger, D. Nelson, and M. Cox, *Principles of Biochemistry,* 3rd ed. (1999), Worth Publishers (New York), pp. 141–145. Protein sequencing.

C. Mathews, K. van Holde, and K. Ahern, *Biochemistry,* 3rd ed. (2000), Benjamin/Cummings (San Francisco), pp. 126–160. An introduction to amino acids and protein sequence analysis.

MIT Biology Hypertextbook
http://esg-www.mit.edu:8001/esgbio/lm/proteins/sequencing/strategies.html
An Internet discussion of protein sequencing.

C. Monnig and R. Kennedy, *Anal. Chem.* **66,** 208R–314R (1994). Review of capillary electrophoresis applications.

P. Rogers, *J. Chem. Educ.* **73,** 189 (1996). "A Simplified Procedure for Dipeptide Sequence Analysis."

T. Strein, J. Poechmann, and M. Prudenti, *J. Chem. Educ.* **76,** 820–825 (1999). Separating FMOC amino acids using capillary electrophoresis.

L. Stryer, *Biochemistry,* 4th ed. (1995), W. H. Freeman (New York), pp. 18–26; 53–60. A discussion of amino acids and protein sequence analysis.

D. Voet and J. Voet, *Biochemistry,* 2nd ed. (1995), John Wiley & Sons (New York), pp. 106–119. Sequence analysis of proteins.

D. Voet, J. Voet, and C. Pratt, *Fundamentals of Biochemistry* (1999), John Wiley & Sons (New York), pp. 77–91; 107–114. An introduction to amino acids and protein sequencing.

P. Weber and D. Buck, *J. Chem. Educ.* **71,** 609–612 (1994). Using capillary electrophoresis to separate naphthalene-2,3-dicarboxaldehyde derivatives of amino acids.

3

USING GEL FILTRATION TO STUDY LIGAND-PROTEIN INTERACTIONS

• **Recommended Reading**

Chapter 3, Section F; *Chapter 5*, Section A; *Experiment 4.*

• **Synopsis**

Serum albumin circulates in the blood stream transporting essential nutrients such as fatty acids to peripheral tissue. Transported molecules, called ligands, often have a special affinity for selected binding sites on proteins and nucleic acids. In this experiment, the dynamics of ligand-protein interactions will be explored with the binding of the dye phenol red to bovine serum albumin. The technique of gel filtration will be used to separate the dye-protein complex. Data will be analyzed in order to construct binding curves.

I. INTRODUCTION AND THEORY

Ligand Binding by Macromolecules

The technique of gel filtration is widely used in biochemical research. Chapter 3 described the theory and applications of gel filtration to the experimental procedures of desalting, separation and purification of biomolecules, and estimation of molecular weight of biomolecules. In this experiment, gel filtration procedures will be used to study the dynamic binding of small molecules by proteins. Many of the dynamic processes occurring in biological cells and organisms are the result of interactions between molecules. Often these interactions involve one or more smaller molecules binding to a macromolecule (usually a protein or nucleic acid).

You have learned about many types of complex formation between small molecules and proteins in biochemistry class. The action of hormones provides one example. A hormone response is the consequence of a weak, but specific, interaction between the hormone molecule and a receptor protein in the membrane of the target cell. Before a metabolic reaction can occur, a small substrate molecule must physically interact in a certain well-defined manner with a macromolecular catalyst, an enzyme. The biochemical action of drugs also depends on molecular interactions. A drug is first distributed throughout the body via the blood stream. Drugs in the blood stream are often bound to plasma proteins, which act as carriers. When the drug molecules are transported to their site of action, a second molecular interaction is likely to occur. Many drugs elicit their effects by interfering with biochemical processes. This may take the form of enzyme inhibition, where the drug molecule binds to a specific enzyme and prohibits its catalytic action. Table E3.1 lists several other molecular interactions that lead to some dynamic biochemical action.

All of these molecular interactions have at least two common characteristics. (1) The forces that are the basis of these interactions are usually weak and noncovalent. We usually define the interactions in terms of hydrogen bonding, hydrophobic stabilization, van der Waals forces, and electrostatic interactions. (2) The binding between molecules is selective and specific. Imagine that the interaction brings together two molecular surfaces. Where the surfaces are in contact, the forces must be complementary. If on one surface there is a nonpolar molecular group (phenyl ring, hydrophobic alkyl chain, etc.), the adjacent region on the other surface must also be hydrophobic and nonpolar. If a positive charge exists on one surface, there may be a neutralizing negative charge on the other surface. Simply stated, the two molecules must be compatible, in a chemical sense.

Quantitative Treatment of Binding

A thorough understanding of the biochemical significance of ligand binding to macromolecules comes only from a quantitative analysis of the strength of binding. (In biochemistry, a small molecule that binds to

Table E3.1

Biological Examples of Ligand-Macromolecule Interactions

Type of Interaction	Biochemical Significance
Lipids-proteins	Cell membranes
Substrate-enzyme	Metabolic reactions
Hormone-receptor protein	Regulation
Inhibitor-enzyme	Metabolic regulation
Ligand-carrier protein	Membrane transport
Antigen-antibody	Immune response
Coenzyme-enzyme	Metabolic reactions
Drug-protein	Disease treatment
Regulatory molecules–DNA	Metabolic regulation

a macromolecule is called a **ligand.**) Binding affinity between two molecules is often expressed as an equilibrium constant, the **formation constant,** K_f, which is derived from the law of mass action. Consider the specific interaction between a small molecule, L (for ligand), and a macromolecule, M (Equation E3.1). These two species combine to form a complex, LM.

>> $\text{L} + \text{M} \rightleftharpoons \text{LM}$ *Equation E3.1*

K_f, the formation constant for the complex, is defined by Equation E3.2.

>> $K_f = \dfrac{[\text{LM}]}{[\text{L}][\text{M}]}$ *Equation E3.2*

Do not confuse K_f with K_d, the dissociation constant. The relationship between K_f and K_d is defined in Equation E3.3

>> $K_d = \dfrac{[\text{L}][\text{M}]}{[\text{LM}]} = \dfrac{1}{K_f}$ *Equation E3.3*

The larger the value of K_f, the greater the strength of binding between L and M. (Large K_f implies a high concentration of LM relative to L and M.) Return to Equation E3.2 and note that in order to determine K_f, a method must be developed to measure equilibrium concentrations of L, M, and the complex LM. In a later section, we will describe experimental techniques that are applied to these measurements of binding constants, but first we must reorganize Equation E3.2 into a form that contains more readily measurable terms. We will begin with the assumption that the macromolecule, M, has several binding sites for L and that these sites do not interact with each other. That is, K_f is identical for all binding sites. The following definitions are necessary for the reorganization of Equation E3.2

$[\text{L}]$ = equilibrium concentration of free or unbound ligand

$[\text{M}]$ = equilibrium concentration of macromolecule with no bound L, or the concentration of unoccupied binding sites

$[\text{LM}]$ = equilibrium concentration of ligand-macromolecule complex, or the concentration of occupied sites

$[\text{M}]_0$ = total or initial concentration of macromolecule, or total concentration of available binding sites

$[\text{L}]_0$ = total concentration of bound and unbound ligand; or initial concentration of ligand

v = fraction of available sites on M that are occupied, or the fraction of M that has L in binding site:

$$v = \frac{[LM]}{[M]_0}$$

Equation E3.4

The term v is particularly significant because it can be considered a ratio of the number of occupied sites to the total number of potential binding sites on M. It can be measured experimentally, but first it must be redefined in the following manner: Since $[M]_0 = [LM] + [M]$, then

$$v = \frac{[LM]}{[LM] + [M]}$$

From Equation E3.2, $[LM] = K_f[L][M]$. Therefore,

$$v = \frac{K_f[L][M]}{K_f[L][M] + [M]}$$

Simplifying,

$$v = \frac{K_f[L]}{K_f[L] + 1}$$

Equation E3.5

You should recognize the similarity of Equation E3.5 to the Michaelis-Menten equation for enzyme catalysis. A graph of v vs. [L] yields a hyperbolic curve (see Figure E3.1) that approaches a limiting value or saturation level. At this point, all binding sites on M are occupied. Because of the difficulty of measuring the exact point of saturation, this nonlinear curve is seldom used to determine K_f. Linear plots are more desirable, so Equation E3.5 is converted to an equation for a straight line. The equation will now be put into a more general form to account for any number of potential binding sites on M. The symbol $\bar{v}$ will be used to represent the average number of occupied sites per M, and n will represent the number of potential binding sites per M molecule. Assuming that all the binding sites on M are equivalent, Equation E3.5 becomes

$$\bar{v} = \frac{nK_f[L]}{K_f[L] + 1}$$

Equation E3.6

Equation E3.6 contains terms such as [L] that are difficult to determine, so it must be converted to a form amenable to experimental measurements.

If $\bar{v}$ is the average number of occupied sites per M molecule, then $n - \bar{v}$ is the average number of unoccupied sites per M molecule.

$$n - \bar{v} = n - \frac{nK_f[L]}{K_f[L] + 1}$$

A plot illustrating saturation of binding sites with ligand. The *n* is estimated at the point when all binding sites are occupied by ligand. K_f is represented by the [L] when one-half of the sites are occupied by ligand.

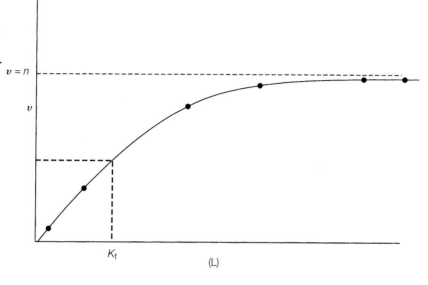

Simplifying,

$$(n - \bar{v})\,(K_f[L] + 1) = n(K_f[L] + 1) - n(K_f[L])$$

>>
$$(n - \bar{v}) = \frac{n}{K_f[L] + 1}$$

Equation E3.7

To further simplify Equation E3.7, let the term $\bar{v}/n - \bar{v}$ represent the ratio of occupied sites to nonoccupied sites on M; it can be mathematically represented as

$$\frac{\bar{v}}{n - \bar{v}} = \left(\frac{nK_f[L]}{K_f[L] + 1}\right)\left(\frac{K_f[L] + 1}{v}\right)$$

or

>>
$$\frac{\bar{v}}{n - \bar{v}} = K_f[L]$$

Equation E3.8

A more desirable form for graphical use is known as **Scatchard's equation:**

>>
$$\frac{\bar{v}}{[L]} = K_f(n - \bar{v})$$

Equation E3.9

If a plot of $\bar{v}/[L]$ versus $\bar{v}$ yields a straight line, shown by the solid line in Figure E3.2, then all the binding sites on M are identical and independent, and K_f and n are estimated as shown in the figure.

The derivation of Equation E3.9 assumes that K_f is identical for all binding sites; that is, the binding of one molecule of L does not influence the binding of other L molecules to binding sites on M. However, it is common for ligand-macromolecule interactions to display such influences. The binding of one L molecule to M may encourage or inhibit the binding of a second L molecule to M. For example, the binding of dioxygen to one of the four subunits of hemoglobin increases the affinity of the other subunits for oxygen. There is said to be **cooperativity** of sequential binding. If the sites do show cooperative binding, the plot is nonlinear, as shown by the dashed line in Figure E3.2. The shape of the nonlinear curve may be used to determine the number of types of binding sites. The dashed line in Figure E3.2 can be resolved into two lines, indicating that two types of

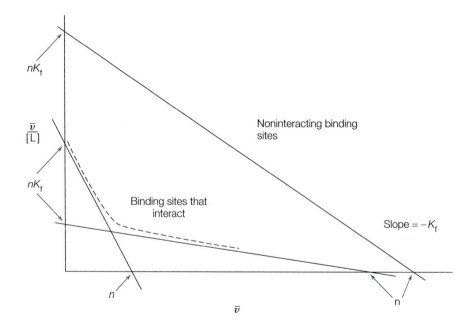

Figure E3.2

Scatchard plot. Two types of Scatchard curves are illustrated. The upper plot (solid line) represents binding to a macromolecule with noninteracting sites. The binding of a ligand molecule at one site is independent of the binding of a second ligand molecule at another site. The plot can be extrapolated to each axis and the n and K_f calculated. The lower, dashed line represents binding of ligand molecules to binding sites that interact. One ligand molecule bound to the macromolecule influences the rate of binding of other ligand molecules. The dashed line is evaluated as shown in the figure. The curved line has two distinct slopes. It can be resolved into two straight lines, each of which may be evaluated for n and K_f. This would indicate that there are two types of binding sites, each with a unique n and K_f.

binding sites are present on M. The n and K_f for each type of binding site may be estimated by resolving the smooth curve into straight lines as shown in figure E3.2. K_f and n can be estimated by extrapolating the two straight lines to the axes.

Experimental Measurement of Ligand-Binding Interactions

In order to analyze ligand-protein interactions quantitatively (to use Equation E3.9), one must be able to distinguish experimentally between bound ligand (LM) and free ligand (L). Many techniques have been developed for measuring the dynamics of ligand-macromolecular interactions. Widely used techniques include equilibrium dialysis, ultrafiltration, and spectroscopic (especially fluorescence) measurements. Most of these techniques require specialized and expensive equipment, sometimes cumbersome procedures, and the use of radiolabeled ligands. Since these methods are not practical for a teaching laboratory, other procedures have been developed.

One of the simplest and most convenient methods for monitoring ligand binding is the differential method, which detects and quantifies some measurable change in spectral absorption in the UV-VIS regions. Many binding ligands absorb strongly in the UV-VIS region; however, there may be no significant spectral changes when the ligand binds to a protein. Therefore, before the extent of complex formation can be measured, the LM complex must be separated from free, unbound ligand. Gel filtration is the ideal method for this molecular separation because the ligand-protein complex often has a molecular weight much greater than that of the free ligand, which is usually a smaller molecule. After separation, absorbance measurements may be used to quantify the ligand-protein complex and free ligand concentrations. The assumption must be made that there is not significant dissociation of the protein-ligand complex during separation.

Transport Properties of Serum Albumin

Albumin is the most abundant protein in human and other animal plasma. It is estimated that up to 40% of the total albumin in humans is in circulation transporting essential nutrients, especially those that are sparingly soluble in aqueous-based plasma. For example, the fatty acids, which are important fuel molecules for the peripheral tissue, are distributed by albumin. In addition, albumin is the plasma transport protein for other substances including bilirubin, thyroxine, and steroid hormones. Also, many drugs including aspirin, sulfanilamides, clofibrate, and digitalis bind to albumin and are most likely carried to their sites of action by the protein.

Extensive research on albumin has led to an increasingly clear picture of ligand binding. The dye phenol red has been widely used as a model for the binding of natural ligands to proteins. Experimental results have shown that each molecule of albumin binds at least six molecules of phenol red. The presence of fatty acids such as decanoate, palmitate, stearate, and oleate inhibits phenol red binding, so these natural ligands compete for the same binding sites on albumin. The binding of phenol red to serum albumin also is pH sensitive with a binding maximum in the range of pH 3.0 to 5.0.

Overview of the Experiment

In this experiment, you will evaluate the binding of the dye phenol red to bovine serum albumin. Increasing amounts of dye will be added to fixed amounts of protein in buffered solutions. The reaction mixture will be subjected to gel filtration to separate the dye-protein complex from free, unbound dye. Two types of binding studies can be completed:

1. Collection of data for binding graphs. Student groups may be assigned different concentrations of dye and binding data are collected for a class project.

2. Variation of pH. Student groups may do the binding studies at pH 4.0, 4.5, 5.0, 6.0, 7.0, and 8.0.

Approximately 1 hour is required to pour and equilibrate the Sephadex column. Each packed column may be used several times as long as it is equilibrated with the proper buffer for each run. The time required for each reaction and column run is about 1 hour with an additional ½ hour needed to measure absorbances of collected fractions.

II.   MATERIALS AND SUPPLIES

- Bovine serum albumin (BSA); each column run requires 20 mg of protein. The BSA must be essentially fatty acid free.
- Acetate buffer; 0.1 M at various pH values: 4.0, 4.5, 5.0
- Sodium phosphate buffer; 0.1 M at various pH values: 6.0, 7.0, 8.0
- Phenol red in buffer; it is most efficient to have a solution of phenol red prepared in each pH buffer to be used. Concentration = 1.0 g of phenol red/100 mL of buffer.
- Sephadex G-25-150; slurry in buffer
- Chromatography columns; 1.5 × 15 cm
- Fraction collector
- Spectrometer for A_{520} measurements and cuvettes

III.   EXPERIMENTAL PROCEDURE

Preparing the Sephadex Column

Experimental studies to construct a binding curve will be described at pH 4.5. To prepare the gel filtration column, obtain a chromatography column (about 1.5 × 15 cm). It should be equipped with a porous glass disk and a stopcock at the bottom to control flow rate. Clamp the column to a ring stand and connect the bottom tubing to a fraction collector containing 50 test tubes. If a fraction collector is not available, place a rack of test tubes (approximately 15 × 85 mm) under the column for collecting fractions. Sephadex G-25 has been prepared for your use. The dry beads were preswollen in water for several hours, fine particles that may slow column flow have been removed by defining, and the gel has been equilibrated in 0.1 M acetate buffer, pH 4.5.

The column is poured using between 20 and 25 mL of Sephadex slurry. A flow rate of approximately 1 drop per second is set with the stopcock. Allow the gel to settle into a column about 12 cm in height. Never allow the buffer level to fall below the top of the gel column. More buffer may have to be carefully added to pack the Sephadex completely into a tight column. Stop the column flow when there is about 1 cm of buffer on top of the packed gel. Cut a piece of filter paper into a circle with a diameter slightly less than the inside of the column. Carefully place the paper in the buffer and allow it to settle on the top of the gel. This protects the gel from disturbances while you are loading sample and eluting with buffer.

Eluting Phenol Red–BSA Mixtures on Sephadex Columns

Samples assigned by your instructor are prepared as described in Table E3.2. They must be made just before column elution. To prepare a sample, carefully weigh 20 mg of fatty acid–free bovine serum albumin onto a small piece of creased weighing paper and transfer to a small test tube. Add the appropriate amount of buffer and gently mix to dissolve protein. Add the indicated amount of phenol red solution. Mix gently but well for about 3–4 minutes (shaking will denature protein). Use an automatic pipettor to apply 250 μL of the reaction mixture to the top of the Sephadex column. To do this, remove all buffer from the top of the gel and carefully drop the sample directly onto the top of the paper disk. As soon as the colored solution has entered the gel, add a few drops of buffer to wash all reaction mixture into the gel. Then carefully add buffer to the top of the gel, begin to collect 1-mL fractions, and set a column flow rate of about 1 drop per 1–2 seconds. Add buffer continuously to the top of the column as you collect fractions. Continue to collect 1-mL fractions and add buffer until all yellow dye color has eluted from the column (about 50 fractions). Turn off the stopcock of the column. The yellow color in the fractions is due to the presence of phenol red. Phenol red is an indicator dye that changes color from yellow (acidic) to

Table E3.2

Preparation of Samples for Binding Studies

Reagent	Reaction No.					
	1	2	3	4	5	6
Bovine serum albumin (mg)	20	20	20	20	20	20
Acetate buffer (mL)	0.95	0.90	0.80	0.70	0.60	0.40
Phenol red solution (mL)	0.05	0.10	0.20	0.30	0.40	0.60

red (basic) depending on the pH. The red color is more intense, so the fractions are made basic by the addition of 200 μL of 1 M NaOH solution with an automatic pipettor. Measure and record the absorbance at 520 nm for all of the fractions. Use acetate buffer as reference in the spectrometer.

The Sephadex column may be prepared for another reaction mixture by eluting an additional 15–20 mL of buffer to be sure all components are washed from the column. If you are finished using the column, pour the gel into a container labeled "Used Sephadex G-25". The gel may be recycled and used again.

Reaction Mixtures at Different pH Values

A study of phenol red binding under different pH conditions may be completed by changing the pH of the reaction mixtures and Sephadex gel column. For each pH to be studied, the column must first be equilibrated with the proper buffer. Several buffers are available including acetate buffer (pH 4.0, 4.5, and 5.0) and phosphate buffer (pH 6.0, 7.0, and 8.0). Equilibrate the column with approximately 20–25 mL of the new buffer. To be assured of the proper pH, check the column eluent with a pH indicator strip or collect a fraction for measurement with a pH meter. Prepare the reaction mixtures by mixing the protein with the appropriate buffer and phenol red. Be sure to note that solutions of phenol red are prepared in different buffers. Load the reaction mixture on the equilibrated Sephadex G-25 column and develop and analyze the column fractions as described above.

IV. ANALYSIS OF RESULTS

Eluting Phenol Red-BSA Mixtures on Sephadex Columns

Describe the color of the reaction mixtures just before loading onto the column. Make observations about the column development process. Analyze the absorbance data by preparing an elution curve. On graph paper or a computer graphics program, plot A_{520} (y-axis) versus fraction number. How many peaks were obtained? What are the components represented by each peak? Use this elution curve to explain the action of a gel filtration column. Explain the fact that phenol red is present in both colored fractions. In what fraction is the bovine serum albumin?

Preparation of Binding Graphs

Several different methods may be used to plot data from this experiment, but two will be described here:

> Curve 1: Direct plot of v versus [L].
> Curve 2: Scatchard plot of v/[L] versus v.

The value of v, the amount of phenol red ligand bound by the protein, may be calculated as the percentage of the total amount of phenol red. Total absorbance values for each peak may be determined, or the areas under each peak may be used to represent concentrations of bound and free phenol red. The percent of bound phenol red is determined by:

$$\% \text{ bound dye} = \frac{\text{total } A_{520} \text{ or area of peak 1}}{\text{total } A_{520} \text{ or area of peak 2} + \text{total } A_{520} \text{ or area of peak 1}} \times 100$$

Each reaction mixture provides data for one point on the binding graph. Collect data from other students to complete the graph. What are the shapes of curve 1 and curve 2? Does the Scatchard plot show cooperativity?

Effect of pH on Ligand Binding

For each reaction mixture, calculate the bound phenol red as the percentage of total phenol red as above. Prepare a plot of % bound (y-axis) versus pH. In previous studies it has been shown that optimum binding of phenol red to BSA occurs in the pH range of 3 to 5. Binding affinity gradually declines between pH 5.0 and 8.0 and is insignificant above pH 8.

Study Problems

1. Equation E3.5 in this experiment can be used to determine K_f values, but hyperbolic plots are obtained. Can you convert Equation E3.5 into an equation that will yield a linear plot without going through all the changes necessary for the Scatchard equation? Hint: Study the conversion of the Michaelis-Menten equation to the Lineweaver-Burk equation.

▶ *2.* The binding of a ligand, L, to a protein is studied as described in this experiment. The following data are collected using 7.5×10^{-6} molar protein.

Reaction No.	L added (μM)	L bound (μM)
1	20	11
2	50	26
3	100	44
4	150	55
5	200	60
6	400	70

Use a graphical analysis to determine n and K_f.

▶ *3.* A research project you are working on involves the study of sugar binding to human albumin. The sugars to be tested are not fluorescent, and you do not wish to use a secondary fluorescent probe. Human albumin has only one tryptophan residue, and you know that this amino acid is fluorescent. You find that the tryptophan fluorescence spectrum of human albumin undergoes changes when various sugars are added. Can you explain the results of this experiment and discuss the significance of the finding?

> 4. A drug, X, has a strong affinity for serum albumin. When X was bound to albumin, an increase in absorbance was noted. The $\bar{v}$ values were determined from these absorbance measurements. Use the data below to determine K_f and n for the interaction between X and albumin. Prepare two types of graphs and compare the results. In one graph plot $\bar{v}$ versus [X], and in the second plot $\bar{v}/[X]$ versus $\bar{v}$. Which is the better method for calculating binding constants? Why?

[X]	$\bar{v}$
0.36	0.43
0.60	0.68
1.2	1.08
2.4	1.63
3.6	1.83
4.8	1.95
6.0	1.98

5. In your biochemistry research project you find that the binding of a ligand to a protein decreases if the ionic strength of the buffer solvent is increased. What type of noncovalent bonding might be involved in the ligand-protein complex?

6. Explain how your experimental results would change if you did the phenol red–BSA binding studies with added stearic acid.

7. The Bradford protein assay as described in Chapter 2 is based on the absorbance change that occurs upon binding of Coomassie Blue dye to proteins. Explain how you would study the dynamics of this binding process and experimentally determine the number of binding sites on a protein.

8. Without looking back in the experiment, write a list of four natural or synthetic molecules that may be transported by human serum albumin.

9. Explain how you could use a gel filtration column to remove sodium chloride from a solution of bovine serum albumin. Draw a graph representing the elution curve for separation. Assume that two measurements were made on each fraction, a test for the presence of chloride and absorbance at 280 nm.

10. Below is a table showing the % binding of phenol red to human serum albumin versus pH of the reaction mixture. Draw a graph showing the results and explain the shape of the curve.

pH	% of Phenol Red Bound
3.0	65
4.0	70
5.0	50
6.0	30
7.0	30
8.0	20
9.0	5

Further Reading

G. Ackers, *Methods Enzymol.* **27,** 441–455 (1973), Academic Press (New York). Studies of protein ligand binding by gel filtration techniques.

A. Attie and R. Raines, *J. Chem. Educ.* **72,** 119–124 (1995). "Analysis of Receptor-Ligand Interactions."

A. Bordbar, A. Saboury, and A. Moosavi-Movahedi, *Biochem. Educ.* **24,** 172–175 (1996). "The Shapes of Scatchard Plots for Systems with Two Sets of Binding Sites."

R. Boyer, *Concepts in Biochemistry* (1999), Brooks/Cole (Pacific Grove, CA), pp. 102–104. An introduction to gel filtration.

C. Cantor and P. Schimmel, *Biophysical Chemistry,* Part III, W. H. Freeman (San Francisco), pp. 849–886. "Ligand Interactions at Equilibrium."

D. Freifelder, *Physical Biochemistry,* 2nd ed. (1982), W. H. Freeman (San Francisco), pp. 654–684. Introduction to ligand binding.

U. Kragh-Hansen, *Biochem J.* **195,** 603–613 (1981). "Effects of Aliphatic Fatty Acids on the Binding of Phenol Red to Human Serum Albumin."

M. Lee and J. Debro, *J. Chromatog.* **10,** 68–72 (1963). "The Application of Gel Filtration to the Measurement of the Binding of Phenol Red to Human Serum Proteins."

A. Lehninger, D. Nelson, and M. Cox, *Principles of Biochemistry,* 3rd ed. (1999), Worth Publishers (New York), p. 439. Scatchard analysis.

C. Mathews, K. van Holde, and K. Ahern, *Biochemistry,* 3rd ed. (2000), Benjamin/Cummings (San Francisco), pp. 151–152. Introduction to gel filtration chromatography.

Pharmacia Biotechnology, *Gel Filtration: Principles and Methods,* 5th ed. (1991), Pharmacia Biotechnology, S-751 82 Uppsala, Sweden, or 800 Centennial Avenue, Piscataway, NJ 08854. An excellent guide on theory, practice, and applications of gel filtration.

J. Robyt and B. White, *Biochemical Techniques: Theory and Practice* (1987), Brooks/Cole (Monterey, CA), pp. 88–95. An introduction to gel filtration techniques.

H. Rowe, *J. Chem. Educ.* **70,** 415–416 (1993). "An Inexpensive Gel-Filtration Chromatography Experiment."

L. Stryer, *Biochemistry,* 4th ed. (1995), W. H. Freeman (New York), pp. 48–50. Introduction to gel filtration chromatography.

D. Voet and J. Voet, *Biochemistry,* 2nd ed. (1995), John Wiley & Sons (New York), pp. 82–85; 1274–1276. Introduction to gel filtration and hormone binding to receptors.

D. Voet, J. Voet, and C. Pratt, *Fundamentals of Biochemistry* (1999), John Wiley & Sons (New York), pp. 101–102. Use of gel filtration.

Ligand Binding and Gel Filtration on the Web

http://www.scripps.edu/pub/olson-web/people/gmm/
    Study Ligand-Protein Docking for information about drug design. Review movies showing interactions with a variety of ligand-protein systems.

http://www.science.smith.edu/Biochem/Chm_357/design.htm
  Introduction to Computational Drug Design.

http://www.graphpad.com/www/radiolig/radiolig1.htm
  GraphPad Software for analyzing binding data. Provides an introduction to
  analysis of ligand-protein interactions.

http://www.chem.umd.edu/biochem/jollie/prob_sets/set2_sols.htm
  For experience in analysis of ligand binding data, work Problems No. 4 and 5.
  The solutions are provided.

http://laxmi.nuc.ucla.edu:8241/Pharm241_97//Nonlin_comp/Section3.html
  Read notes for Pharmacology 241. Introduction to ligand protein binding.
  Topics to review include Law of Mass Action, Michaelis-Menten Kinetics,
  Details of M-M Kinetics, Single-Binding Site, and Classical Scatchard Plot.

http://ntri.tamuk.edu/fplc/siz.html
  Introduction to gel filtration chromatography.

4

ISOLATION AND CHARACTERIZATION OF BOVINE MILK α-LACTALBUMIN

- **Recommended Reading**

Chapter 2; Chapter 3, Sections A, D, E, F, G, H; *Chapter 4,* Sections A, B; *Chapter 5,* Section A; *Chapter 7,* Sections A, B; *Experiment 7.*

- **Synopsis**

α-Lactalbumin is one of the major proteins found in milk. Its function is to regulate the synthesis of the milk sugar lactose. It can be isolated readily from milk and purified by Sephadex gel filtration or affinity chromatography. This experiment introduces the student to a series of protein purification steps that yield α-lactalbumin in sufficient amounts and purity for further physical characterization by UV spectroscopy and SDS-polyacrylamide gel electrophoresis.

I. INTRODUCTION AND THEORY

Protein purification is an activity that has occupied the time of biochemists throughout the history of biochemistry. In fact, a large percentage of the biochemical literature is a description of how specific proteins have been separated from the thousands of other proteins and biomolecules in tissues, cells, and biological fluids. Biochemical investigations of all biological processes require, at some time, the isolation, purification, and characterization of a protein. Of course, there is no single technique or sequence of techniques that can be followed to purify all proteins. Most investigators approach the problem by trial and error. Fortunately, the experiences and discoveries of hundreds of biochemists have been combined so that, today, a general and somewhat systematic approach is used for protein purification.

The best purification procedure is one that yields a maximum amount of the purified protein in a minimum amount of time. The following discussion outlines the basic steps that must be considered in order to develop a protein purification scheme (Table E4.1).

Development of Protein Assay

First and foremost in any protein purification scheme is the development of an assay for the protein. This procedure, which may have a physical, chemical, or biological basis, is necessary in order to determine quantitatively and/or qualitatively the presence of the specific protein. During the early stages of purification, the particular protein desired must be distinguished from thousands of other proteins present in crude cell homogenates. The desired protein may be less than 0.1% of the total protein content of the crude extract. If the desired protein is an enzyme, the obvious assay will be based on biological function, that is, a measurement of the enzymatic activity after each isolation-purification step. The development of enzymatic assays is considered in greater detail in Experiment 5. If the protein to be isolated is not an enzyme or if the biological activity of the protein is unknown, physical or chemical methods must be used. One of the most useful analytical methods is electrophoresis (Chapter 4).

Source of the Protein

The selection of a source from which the desired protein is to be isolated should be considered. If the objective is simply to obtain a certain quantity of a protein for further study, you would choose a source that contains large amounts of the protein. A possible choice is an organ from a large animal that can be obtained from a local slaughterhouse. Microorganisms are also good sources because they can be harvested in large quantities. If, however, you desire a specific protein from a specific type of cell, tissue, cell organelle, or biological fluid, the source is limited. If the desired protein is known to be located in an organelle or subcompartment of the cell, partial purification of the protein is achieved by isolating the organelle. (See Chapter 7,

---

Table E4.1

Typical Sequence for the Purification of a Protein

1. Develop an assay for the desired protein.
2. Select the biological source of the protein.
3. Release the protein from the source and solubilize it in an aqueous buffer system.
4. Fractionate the cell components by physical methods (centrifugation).
5. Fractionate the cell components by differential solubility.
6. Ion-exchange chromatography.
7. Gel filtration.
8. Affinity chromatography.
9. Isoelectric focusing.
10. Determine purity by electrophoresis or HPLC.

Figure 7. 11.) If the protein is, instead, in the soluble cytoplasm of the cell, it will remain dissolved in the final supernatant obtained after centrifugation at 100,000 × *g*.

With the introduction of new technology in recombinant DNA research, it is possible to transfer via a plasmid the specific gene for a protein into another organism (usually *Escherichia coli* cells) and allow that organism to synthesize the desired protein (see Experiments 14 and 15).

Preparation of Crude Extract

Once the protein source has been selected, the next step is to release the desired protein from its natural cellular environment and solubilize it in aqueous solution. This calls for disruption of the cell membrane without damage to the cell contents. Proteins are relatively fragile molecules and only gentle procedures are allowed at this stage. The gentlest methods for cell breakage are osmotic lysis, gentle grinding in hand-operated glass homogenizers, and disruption by ultrasonic waves. These methods are useful for "soft tissue" as found in green plants and animals. When dealing with bacterial cells, where rigid cell walls are present, the most effective methods are grinding in a mortar with an inert abrasive such as sand or alumina, treatment with lysozyme, or both. Lysozyme is an enzyme that catalyzes the hydrolysis of polysaccharide moieties present in cell walls. When very resistant cell walls are encountered (yeast, for example), the French press must be used. Here the cells are disrupted by passage, under high pressure, through a small hole.

Osmotic lysis consists of suspending cells in a solution of relatively high ionic strength. This causes water inside the cell to diffuse out through the membrane. The cells are then isolated by centrifugation and transferred to pure water. Water rapidly diffuses into the cell, bursting the membrane.

Another alternative, gentle grinding, is best accomplished with a glass or Teflon homogenizer. This consists of a glass tube with a close-fitting piston. Several varieties are shown in Figure E4.1A, B. The cells are forced against the glass walls under the pressure of the piston, and the cell components are released into an aqueous solution.

Ultrasonic waves, produced by a sonicator, are transmitted into a suspension of cells by a metal probe (Figure E4.1C). The vibration set up by the ultrasonic waves disrupts the cell membrane, releasing the cell components into the surrounding aqueous solution.

A less gentle, but widely used, solubilization device is the common electric blender. This method may be used for plant or animal tissue, but it is not effective for disruption of bacterial cell walls. A blender that is more scientifically designed is the rotor stator homogenizer. The stator is a hollow tube and the rotor, attached to the stator, is a rapidly turning knife blade. Cells are torn apart by the turbulence and shear generated by the rotor.

The most recently developed homogenizer is the so-called cell bomb, which makes use of high pressure and decompression to disrupt cells. In this

Figure E4.1

Tools for the preparation of a crude cell extract.　**A** Hand-operated homogenizers, courtesy of Ace Glass, Inc., Vineland, NJ. **B** Homogenizer with electric motor, courtesy of VWR Scientific, Division of Univar.　**C** Sonicator, courtesy of *Curtin Matheson Scientific, Inc.*

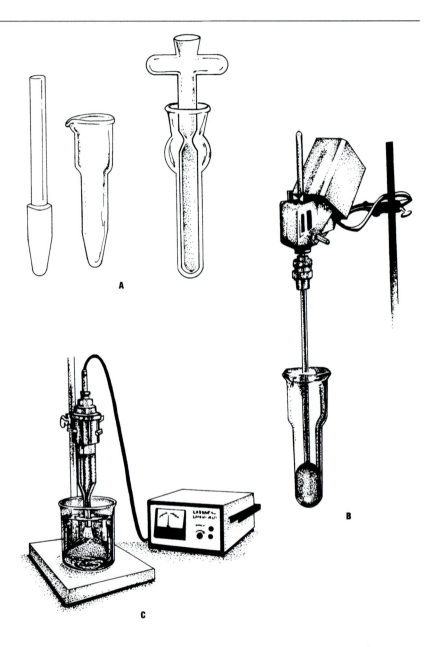

A

B

C

technique, a gas, usually nitrogen, helium, or air, is forced into cells under high pressure. When the pressure is released, expanding bubbles of the gas (nebulization) rupture the cell membrane. Cell bombs are available with pressure ranges of 250 to 25,000 psi. Cell bombs have an advantage over sonicators in that they cause no temperature increase that can denature proteins. A disadvantage of cell bombs is the potential for explosions because of the high-pressure conditions.

Since so many cell disruption methods are available, the experimenter must, by trial and error, find a convenient method that yields the maximum quantity of the protein with minimal damage to the molecules.

Stabilization of Proteins in a Crude Extract

Continued stabilization of the protein must always be considered in protein purification. While the protein is inside the cell, it is in a highly regulated environment. Cell components in these surroundings are protected against sudden changes in pH, temperature, or ionic strength and against oxidation and enzymatic degradation. Once the cell wall barrier is destroyed, the protective processes are no longer functional and degradation of the desired protein is likely to begin. An artificial environment that mimics the natural one must be maintained so that the protein retains its chemical integrity and biological function throughout the purification procedure. What factors are important in maintaining an environment in which proteins are stable? Although numerous factors must be considered, the most critical are (1) ionic strength and polarity, (2) pH, (3) the presence or absence of metal ions, (4) oxidation, (5) endogenous proteases, and (6) temperature.

The standard cellular environment is, of course, aqueous; because of the presence of inorganic salts, though, its ionic strength is relatively high. Addition of KCl, NaCl, or $MgCl_2$ to the cell extract may be necessary to maintain this condition. Proteins that are normally found in the hydrophobic regions of cells (i.e., membranes) are generally more stable in aqueous environments in which the polarity has been reduced by addition of 1 to 10% glycerol or sucrose.

The pH of a biological cell is controlled by the presence of natural buffers. Since protein structure is often irreversibly altered by extremes in pH, a buffer system must be maintained for protein stabilization. The importance of proper selection of a buffer system cannot be overemphasized. The criteria that must be considered in selecting a buffer have been discussed in Chapter 2. For most cell homogenates at physiological pH values, Tris and phosphate buffers are widely used.

The presence of metal ions in mixtures of biomolecules can be both beneficial and harmful. Metal ions such as Na^+, K^+, Ca^{2+}, Mg^{2+}, and Fe^{3+} may actually increase the stability of dissolved proteins. Many enzymes require specific metal ions for activity. In contrast, heavy metal ions such as Ag^+, Cu^{2+}, Pb^{2+}, and Hg^{2+} are deleterious, particularly to proteins that depend on sulfhydryl groups for structural and functional integrity. The main sources of contaminating metals are buffer salts, water used to make buffer solutions, and metal containers and equipment. To avoid heavy metal contamination, you should use high-purity buffers, glass-distilled water, and glassware specifically cleaned to remove extraneous metal ions (Chapter 1). If metal contamination still persists, a chelating agent such as ethylenediaminetetraacetic acid (EDTA, 1×10^{-4} M) may be added to the buffer.

Many proteins are susceptible to oxidation. This is especially a problem with proteins having free sulfhydryl groups, which are easily oxidized and converted to disulfide bonds. A reducing environment can be maintained by adding mercaptoethanol, cysteine, or dithiothreitol (1×10^{-3} M) to the buffer system.

Many biological cells contain degradative enzymes (proteases) that catalyze the hydrolysis of peptide linkages. In the intact cell, functional proteins are protected from these destructive enzymes because the enzymes are stored in cell organelles (lysosomes, etc.) and released only when needed. The proteases are freed upon cell disruption and immediately begin to catalyze the degradation of protein material. This detrimental action can be slowed by the addition of specific protease inhibitors such as phenylmethylsulfonyl fluoride or certain bioactive peptides. These inhibitors are to be used with extreme caution because they are potentially toxic.

Many of the above conditions that affect the stability of proteins in solution are dependent on chemical reactions. In particular, metal ions, oxidative processes, and proteases bring about chemical changes in proteins. It is a well-accepted tenet in chemistry that lower temperatures slow down chemical processes. We generally assume that proteins are more stable at low temperatures. Although there are a few exceptions to this, it is fairly common practice to carry out all procedures of protein isolation under reduced-temperature conditions (0 to 4°C).

After cell disruption, gross fractionation of the properly stabilized, crude cell homogenate may be achieved by physical methods, specifically centrifugation. Figure 7.11, Chapter 7, outlines the stepwise procedure commonly used to separate subcellular organelles such as nuclei, mitochondria, lysosomes, and microsomes.

Separation of Proteins Based on Solubility Differences

Proteins are soluble in aqueous solutions primarily because their charged and polar amino acid residues are solvated by water. Any agent that disrupts these protein-water interactions decreases protein solubility because the protein-protein interactions become more important. Protein-protein aggregates are no longer sufficiently solvated, and they precipitate from solution. Because each specific type of protein has a unique amino acid composition and sequence, the degree and importance of water solvation vary from protein to protein. Therefore, different proteins precipitate at different concentrations of precipitating agent. The agents most often used for protein precipitation are (1) inorganic salts, (2) organic solvents, (3) polyethylene glycol (PEG), (4) pH, and (5) temperature. The most commonly used inorganic salt, ammonium sulfate, is highly solvated in water and actually reduces the water available for interaction with protein. As ammonium sulfate is added to a protein solution, a concentration of salt is reached at which there is no longer sufficient water present to maintain a particular type of protein in solution. The protein precipitates or is "salted out" of solution. The concentration of ammonium sulfate at which the desired protein precipitates from solution cannot be calculated but must be established by trial and error. In practice, ammonium sulfate precipitation is carried out in stepwise intervals.

For example, a crude cell extract is treated by slow addition of dry, solid, high-purity ammonium sulfate in order to achieve a change in salt concentration from 0 to 25% in ammonium sulfate, is gently stirred for up to 60 minutes, and is subjected to centrifugation at $20,000 \times g$. The precipitate that separates upon centrifugation and the supernatant are analyzed for the desired protein. If the protein is still predominantly present in the supernatant, the salt concentration is increased from 25 to 35%. This process of ammonium sulfate addition and centrifugation is continued until the desired protein is salted out.

Organic solvents also decrease protein solubility, but they are not as widely used as ammonium sulfate. They are thought to function as precipitating agents in two ways: (1) by dehydrating proteins, much as ammonium sulfate does, and (2) by decreasing the dielectric constant of the solution. The organic solvents used (which, of course, must be miscible with water) include methanol, ethanol, and acetone.

A relatively new method of selective protein precipitation involves the use of nonionic polymers. The most widely used agent in this category is polyethylene glycol. The polymer is available in a variety of molecular weights ranging from 400 to 7500. The biochemical literature reports successful use of different sizes, but lower-molecular-weight polymer has been shown to be the most specific. The principles behind the action of PEG as a protein precipitating agent are not completely understood. Possible modes of action include (1) complex formation between protein and polymer and (2) exclusion of the protein from part of the solvent (dehydration) followed by protein aggregation and precipitation. The most attractive advantage of PEG is its ability to fractionate proteins on the basis of size and shape as in gel filtration.

Finally, changes in pH and temperature have been used effectively to promote selective protein precipitation. A change in the pH of the solution alters the ionic state of a protein and may even bring some proteins to a state of charge neutrality. Charged protein molecules tend to repel each other and remain in solution; however, neutral protein molecules do not repel each other, so they tend to aggregate and precipitate from solution. A protein is least soluble in aqueous solution when it has no net charge, that is, when it is isoelectric. This characteristic can be used in protein purification, since different proteins usually have different isoelectric pH values.

An increase in temperature generally causes an increase in the solubility of solutes. This general rule is followed by most proteins up to about 40°C. Above this temperature, however, many proteins aggregate and precipitate from solution. If the protein of interest is heat stable and still water soluble above 40°C, a major step in protein purification can be achieved because most other proteins precipitate at these temperatures and can be removed by centrifugation.

Selective Techniques in Protein Purification

After gross fractionation of proteins, as discussed above, more refined methods with greater resolution can be attempted. These methods, in order of increasing resolution, are gel filtration, ion-exchange chromatography, affinity chromatography, and isoelectric focusing. Since the basis of protein separation is different for each of these techniques, it is often most effective and appropriate to use all of the techniques in the order given.

Chromatographic methods for protein purification have been discussed in Chapter 3. However, we must consider another important topic, preparation of protein solutions for chromatography. Fractionation of heterogeneous protein mixtures by inorganic salts, organic solvents, or PEG usually precedes ion-exchange chromatography. The presence of the precipitating agents will interfere with the later chromatographic steps. In particular, the presence of ammonium sulfate increases the ionic strength of the protein solution and damps the ionic protein–ion exchange resin interactions. Procedures that are in current use to remove undesirable small molecules from protein solutions include ultrafiltration (Chapter 2), dialysis (Chapter 2), and gel filtration (Chapter 3).

Chromatography is now and will continue to be the most effective method for selective protein purification. The more conventional methods (ion exchange and gel filtration) rely on rather nonspecific physicochemical interactions between a stationary support and protein molecule. These techniques, which separate proteins on the basis of net charge, size, and polarity, do not have a high degree of specificity.

The highest level of selectivity in protein purification is offered by affinity chromatography—the separation of proteins on the basis of specific biological interactions (Chapter 3). A modified form of affinity chromatography, immobilized metal-ion affinity chromatography (IMAC), was introduced by Porath in 1975. In this method, the bioactive support consists of metal ions chelated to an insoluble matrix. A typical affinity support for IMAC, iminodiacetic acid (IDA) covalently linked to agarose, is shown in Figure E4.2. Proteins are separated according to their individual abilities to bind to the immobilized metal ions. Electron-donating amino acid side chains on the surfaces of protein molecules interact with the metal ions. Important electron-rich groups on the protein include the imidazole ring of histidine, the indole ring of tryptophan, and the sulfhydryl group of cysteine. These amino acid side chains are able to displace weakly bound ligands such as water. A typical immobilized metal ion as shown in Figure E4.2 may have up to three adsorption sites for protein binding. Proteins bound to affinity supports may be displaced by free ligands, such as imidazole, ammonia, pyrophosphate, or amino acids, that bind to the metal ion.

To some individuals, especially those who recall their experiences in organic chemistry laboratory, the ultimate step in purification of a molecule is crystallization. The desire to obtain crystalline protein has long been strong and many proteins have been crystallized. However, there is a common misconception that the ability to form crystals of a protein ensures that the protein is homogeneous. For many reasons (entrapment of contaminants within crystals, aggregation of protein molecules, etc.), the ability to crystallize a protein should not be used as a criterion of purity. The interest in pro-

Figure E4.2

Structure of metal ion—
IDA-agarose affinity support.
Me = metal ion such as Cu(II),
Zn(II), Ni(II), or Fe(III).

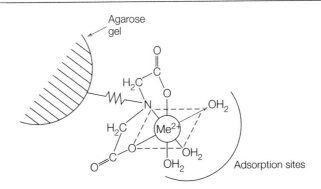

tein crystals today has its origin in the demand for X-ray crystallographic analysis of protein structure.

Purification of α-Lactalbumin

Milk is composed of several proteins, including the caseins (phosphoproteins), β-lactoglobulin, α-lactalbumin, albumin, and immunoglobulins. A crude preparation of α-lactalbumin was first obtained from milk in 1899. Since that time, several investigators have developed procedures for isolating and purifying the protein. α-Lactalbumin has a molecular weight of 14,200 and contains 129 amino acid residues. It is present in milk at concentrations averaging 1 mg/mL and was at first thought to serve a nutritional function. We now know that it is an essential component of the lactose synthase system. Thorough discussions of the history and biological role of α-lactalbumin are available in the literature (Ebner, 1970; Hall and Campbell, 1988).

The primary proteins in milk are the caseins. These may be removed from milk by acid and heat-induced precipitation at pH 4.6 followed by centrifugation. The remaining supernatant (called whey) contains primarily α-lactalbumin and β-lactoglobulin. These two proteins may be separated by gel filtration or with a Cu(II)–IDA-agarose affinity column.

The specific amino acid side chains on α-lactalbumin responsible for binding to the metal support are not known; however, α-lactalbumin is a metalloprotein. Under physiological conditions, it carries one Ca(II) per molecule; hence, there are metal binding sites on the protein. Column-bound α-lactalbumin is eluted by a solution of the free ligand imidazole. A flowchart outlining these procedures is shown in Figure E4.3.

Analysis and Characterization of α-Lactalbumin

Recall from the previous discussion of protein purification that an important step is the development of a specific assay for the desired protein. The assay must be specific so that the desired protein can be detected both

EXPERIMENT 4

Figure E4.3

Flowchart for isolation and purification of α-lactalbumin from milk.

- -

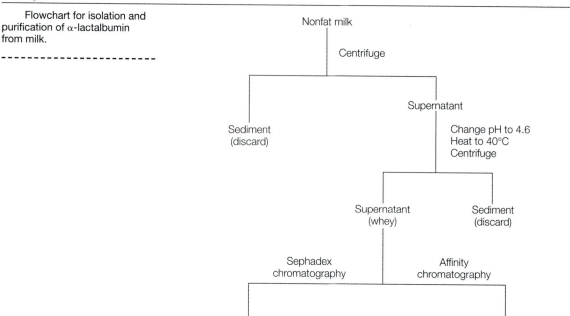

quantitatively and qualitatively in a solution containing thousands of other proteins. This is difficult for α-lactalbumin because it has no unique physical, chemical, or biological property that can be readily measured. Measuring its biological activity as a modifier protein in the lactose synthase system is the most specific way to analyze α-lactalbumin; however, the assay, which requires a coupled enzyme system and preparation of numerous solutions, is too time-consuming to complete in this laboratory period.

Most proteins have a broad characteristic absorption spectrum centered at about 280 nm. The major absorption is due to the presence of aromatic moieties in the amino acids phenylalanine, tyrosine, and tryptophan. During the α-lactalbumin purification described in this experiment, you will monitor the process by measuring the absorption at 280 nm (A_{280}) of column fractions to be sure the experiment is proceeding correctly. You must recognize that you are measuring not the concentration or presence of α-lactalbumin specifically but the total amount of all proteins present.

The procedure outlined here will provide a final product that has an ultraviolet absorption at 280 nm, indicating that protein material is likely present in solution. However several questions remain unanswered:

1. How much protein was isolated?
2. What is the purity of the protein sample?
3. What is the identity of the protein?
4. What is the molecular weight of the protein?

Several useful methods for the quantitative determination of protein solutions were discussed in Chapter 2. Two of those methods, the Bradford protein assay and the direct spectrophotometric assay, will be applied to the α-lactalbumin solutions. Neither of these assays is specific for a certain type of protein; rather they both estimate total protein content.

It is unlikely that the protein fractions from this experiment contain a single type of protein. How many different proteins are present? What is the relative abundance of each protein? Is α-lactalbumin the predominant protein in the isolated fractions? What are the approximate molecular weights of α-lactalbumin and other proteins? These questions may be answered by analysis of the isolated fractions by sodium dodecyl sulfate–polyacrylamide gel electrophoresis (SDS-PAGE) (see Chapter 4). The technique of SDS-PAGE will be introduced and applied to the column-purified fractions, crude whey fraction, and standard α-lactalbumin.

The UV-VIS absorption spectrum of a protein is a characteristic property of that specific protein and can aid in the identification of an unknown. The α-lactalbumin fraction will first be characterized by measurement of the spectrum from 240 to 340 nm. From this, the absorption coefficient is calculated using the Beer-Lambert relationship. Finally, the A_{280}/A_{290} ratio is calculated; this may be used as an indicator of the purity of the isolated α-lactalbumin.

Overview of the Experiment

The complete isolation, purification, and characterization of α-lactalbumin as described here require about 9 hours. The preparation of whey is completed in approximately 3 hours and the chromatographic step (Sephadex or affinity) requires another 3 hours. The analysis procedures require 3 hours. The whey fraction may be stored in a freezer for several weeks if desired. As an alternative to isolation, students may be provided commercial α-lactalbumin, which is then further purified by chromatography or analyzed directly by SDS-PAGE. The time for gel electrophoresis can be greatly reduced if commercially prepared gel plates are used.

II. MATERIALS AND SUPPLIES

A. Preparation of Milk Whey

– Nonfat milk, raw or pasteurized, 100 mL
– HCl, 12 M and 0.5 M
– Refrigerated centrifuge and tubes, capable of 16,000 $\times$ g
– pH meter

–   Heater/stirrer
–   Syringe cartridge filter, 0.45 μm

B. Chromatography

Sephadex Chromatography

–   Sephadex G-50, fine mesh, in 0.02 M Tris, pH 7.0 (100 mL slurry)
–   Glass column, 2.5 × 40 cm
–   UV monitor and fraction collector

Affinity Chromatography

–   IDA-agarose, prepacked in column or provided in slurry form
–   Buffer A: 0.020 M Tris, 0.5 M NaCl, pH 7.0
–   Buffer B: 0.020 M Tris, 0.5 M NaCl, 0.020 M imidazole, pH 7.0
– $CuSO_4$ in H_2O, 0.1 M
–   UV monitor and fraction collector
–   Quartz cuvettes
–   UV-VIS spectrometer

C. SDS-PAGE of α-Lactalbumin

It is recommended that precast gels be obtained commercially (Jule or Novex). They are cost- and time-effective and they come in a variety of sizes and types. If students are to make their own gels, the following items, reagents, and solutions should be provided. Many types of gel casters are commercially available.

–   Glass plates, spacers, comb, electrical tape (2 inch diameter), and clamps or precast gels of 12% acrylamide; 0.75 mm thickness, 8 × 10 cm.
–   Stock solutions:

    1.   Tris-glycine SDS buffer, pH 8.2
    2.   30% Acrylamide plus 0.74% bisacrylamide
    3.   10% Sodium dodecyl sulfate
    4.   5% Ammonium persulfate
    5.   TEMED (N,N,N', N'-tetramethylethylenediamine)

–   Electrophoresis apparatus for vertical slab gels and power supply
–   Sample application buffer: Tris, glycerol, bromophenol blue, pH 6.8
–   Protein solution buffer: Tris, pH 6.8
–   α-Lactalbumin samples (standard and fractions from parts A and B)
–   Molecular weight standards; use a mixture of at least six proteins in the molecular weight range of 10,000–70,000. These are commercially available.

– 2-Mercaptoethanol (STENCH!)

– Boiling-water bath

– Staining with Coomassie Blue:

Dye solution, 0.25% Coomassie Blue in methanol–acetic acid–water (5:1:5)

Destaining solution, acetic acid–methanol–water (7:7:86)

D. Bradford Protein Assay

– α-Lactalbumin from parts A and B
– Bovine gamma globulin standard, 0.1 mg/mL in H_2O
– Bradford dye reagent. This is a commercially available mixture of Coomassie Brilliant Blue G-250 dye, phosphoric acid, and methanol.
– Spectrophotometer for reading A_{595} with glass cuvettes (1 or 3 mL)
– Test tubes, 10 × 100 mm

III. EXPERIMENTAL PROCEDURE

A. Preparation of Milk Whey

Obtain 10 mL of nonfat milk and centrifuge at 16,000 × g for 45 minutes in a refrigerated centrifuge. (During this step, proceed directly to the chromatography step.) Decant the supernatant into a small beaker. The sediment and floating lipid layer should remain in the centrifuge tube and be discarded. Adjust the pH of the supernatant to 4.6 in two steps:

1. Adjust to approximately pH 5 with dropwise addition of 12 M HCl.
2. Adjust to the desired pH (4.5) with dropwise addition of 0.5 M HCl.

Heat the coagulated solution with constant stirring at 40°C for 30 minutes. Centrifuge at 16,000 × g for 30 minutes. Collect the whey (supernatant) and clarify by filtering through a 0.45-μm syringe cartridge filter. The whey is stored for short periods of time (hours) in crushed ice or for longer periods in a freezer.

B. Chromatography

Sephadex Chromatography

Packing the Sephadex Column

Sephadex G-50 has been prepared for your use. The dry beads have been preswollen in water for several hours, fine particles that may slow column flow have been removed, and the gel has been equilibrated in Tris buffer. If chromatography columns are not available, one can easily be constructed from a glass tube (2.5 × 40 cm), cork stoppers, Tygon tubing, a screw clamp, and small glass tubing as shown in Figure E4.4. With the screw clamp closed, add 10 to 15 mL of Tris buffer and insert a small piece of

Figure E4.4

A column for
chromatography.

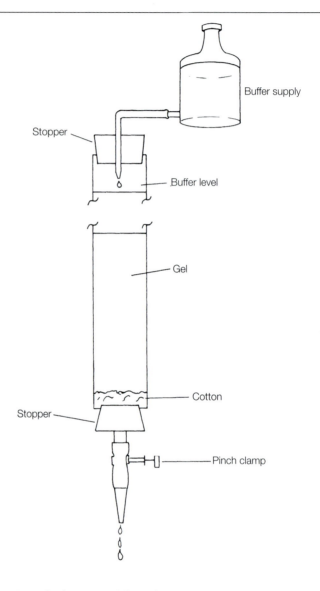

cotton into the bottom of the column. Force air bubbles out of the cotton
with a long glass rod. Pour a well-mixed slurry of Sephadex G-50 into the
column. Open the bottom clamp to allow a slow column flow. As the gel
settles into a bed, add more slurried G-50 until the settled bed reaches a
height of about 30 cm. During this packing process, never allow the column
to run dry. If more buffer solution is necessary, add it carefully without disturbing the gel. After a bed of desired height has been prepared, continue
eluting the column with buffer. Cut a piece of filter paper into a circle with
a diameter of 2 to 2.3 cm and put it on top of the Sephadex bed. Be careful
not to disturb the gel at the top of the column.

Sample Application

Measure out 4–5 mL of whey for application to the column. If you are using commercial α-lactalbumin, prepare a solution of 3–4 mg of protein in 1–2 mL of Tris buffer. After the eluting solvent has just drained into the top of the G-50 column, add the α-lactalbumin solution dropwise to the top of the gel. Use a pipet and avoid disturbing the top surface of the gel. After the protein solution has just entered the gel, slowly and carefully add 10 mL of buffer to the top of the column without disturbing the gel surface and allow it to penetrate the surface of the column. Then, immediately add buffer to fill the column. Set up an eluting solvent supply as shown in Figure E4.4. This will allow solvent to enter the column at the same rate as solvent elutes from the column.

Column Development, Fraction Collection, and Analysis of Fractions

The stopper in the top of the column must be airtight. Adjust the screw clamp so that the flow rate of the column is about 5 to 10 mL/hr. Collect 2-mL fractions in test tubes in a fraction collector. Transfer the contents of every other fraction to a quartz cuvette and measure the absorbance at 280 nm or use a UV monitor if available. For a blank, use Tris buffer. Record the A_{280} values in your notebook. When you begin to detect protein eluting from the column ($A_{280} > 0$), measure the A_{280} of every fraction. On a single sheet of graph paper, plot A_{280} vs. fraction number. Continue collecting fractions until no more protein is eluted (about 30 fractions). Save all fractions that have A_{280} over 0.2.

When you are finished with the Sephadex column, pour the gel into a container labeled "used Sephadex G-50." The gel may be recycled and used later.

Affinity Chromatography

Obtain a prepacked column and clamp it to a ring stand. If you must prepare your own column of IDA-agarose, use a 1×6–8 cm column. Pour in about 2 mL of the IDA-agarose slurry. (Be sure the column outlet is closed.) Allow the gel to settle to a column 1–2 cm high. Protect the surface of the gel by allowing a small circle of filter paper to settle onto the top. Allow most of the solution to pass through the column, close the outlet, and add buffer A to fill the column.

Wash the column (prepacked or handmade) with 10 mL of buffer A. Any flow rate up to 2 mL/min is permissible. Since this eluent will be discarded, it may be collected in a single container. *Do not allow the column to go dry*. To load the column with copper, drain buffer A to a level just slightly above the gel top. Add 0.5 mL of 0.1 M CuSO$_4$ solution to the column, and allow it to enter the gel. Once all the CuSO$_4$ is in the gel, immediately add buffer A, adjust the flow rate to 2 mL/min, and continue to wash the column until all excess Cu(II) has been eluted (about 25 mL). The resin will change from white to a pale blue color.

Attach the column outlet to a UV flow cell, fraction collector, and recorder, if available. Alternatively, collect fractions by hand. Allow buffer A to drain just to the top of the gel, and add 0.5 mL of whey directly to the top of the gel. Begin to collect 1-mL fractions in individual test tubes. Let the whey enter the gel, and add 1.0 mL of buffer A to rinse the inside of the column. Allow this to enter the gel; then fill the column with buffer A, and continue to collect 1-ml fractions. Begin to take A_{280} measurements of each fraction by transferring the contents of each fraction into a quartz cuvette. Buffer A should be used to zero the spectrophotometer. Begin the preparation of a graph of A_{280} (y axis) vs. fraction number (x axis). Continue to wash the column with buffer A until the A_{280} is approximately 0 or until the recorder has returned to the baseline. This will require 20–25 mL of buffer A. To elute α-lactalbumin, pass buffer B through the column. Continue to collect 1-mL fractions, measure the A_{280} of each, and add to the graph. This time use buffer B to zero the spectrophotometer. Elute with a total of 10 mL of buffer B. The gel may be reused after removing Cu(II) with 0.2 M EDTA.

C. SDS-PAGE of α-Lactalbumin

Preparation of the Slab Polyacrylamide Gel

CAUTION

Acrylamide in the unpolymerized form is a skin irritant and a potential neurotoxin. Wear gloves and a mask while weighing the dry powder. Do not breathe the dust. Prepare all acrylamide solutions in the hood. Do not mouth pipet any solutions used for gel formation or staining.

Do not touch the electrophoresis chamber or wires while the electrophoretic operation is in progress. Voltages may be as high as 300 V and shocks may be fatal.

If precast gels are provided, move ahead to the preparation of protein samples for electrophoresis. If gels are to be made, follow these instructions for preparation of an SDS gel of 12% acrylamide.

Wash the glass plates thoroughly with detergent and warm water. Rinse well with distilled water and then 95% ethanol. Wipe dry with tissue. Prepare gels 0.7 mm thick by following the directions provided with the electrophoresis apparatus (see Chapter 4).

To prepare a 12% acrylamide gel, mix the following reagents and solutions in a 50-mL Erlenmeyer flask:

Reagent	Amount (mL)
H_2O	6.4
30% Acrylamide/bis	8.0
Tris-glycine SDS, pH 8.2	5.0
10% SDS	0.2
5% Ammonium persulfate	0.4

Mix all the components well. Add 8 μL of pure TEMED. Mix well and immediately pour between glass plates. Fill the space to the top of the plates. Immediately insert the comb to make sample wells in the gel. Allow the gel to polymerize (about 30 min). Remove the comb and carefully rinse the sample wells with a 1:10 dilution of Tris-glycine SDS buffer. Prepare each protein sample including the molecular weight standards as follows:

Protein samples (1–2 mg/mL in 62.5 mM Tris-HCl, pH 6.8)	50 μL
10% SDS	10 μL
2-Mercaptoethanol (STENCH!)	3 μL
Sample application buffer, pH 6.8	40 μL

Mix components in a 0.5-mL Eppendorf tube and heat in a boiling water bath for 3 minutes to denature the proteins. Cool to room temperature and apply 10 μL of a sample to a well on the gel. Use gel-loading tips on an automatic pipettor to deliver these small samples. One or two wells should contain molecular weight standards. Be sure to record placement of each protein and standards in a sample well in your notebook. Follow the instructions provided with your electrophoresis apparatus and by the supplier of precast gels.

Run the slab gel in a vertical dimension. The current should be set at 25–30 mA per slab gel. Allow the electrophoresis to proceed until the tracking dye is at the bottom of the gel. *Turn the power off* and remove the gels from the buffer chambers. Carefully separate the sandwiched plates with a thin spatula and transfer the gel in one piece to a tray containing staining solution.

For Coomassie dye staining, allow the slab to soak for about 30 minutes. Remove the dye solution by decanting and cover the gel with destaining solution. Replace the destaining solution with fresh solution about every 4–6 hours until background color is removed. The destaining procedure may require several days. Proteins will show up as dark blue bands on a nearly colorless background.

Measure the total length of the slab gel and the distance migrated by each protein and the tracking dye from the top. In your notebook, draw a picture of the gel showing protein bands. A photograph may be taken as a permanent record.

D. Bradford Protein Assay

Set up 12 test tubes (10 × 100 mm, colorimetric tubes) and add water and proteins according to the top three rows of Table E4.2. Tube 1 is used as a blank and tubes 2 through 6 are for construction of a standard calibration curve. Tubes 7 to 10 are duplicates of two different concentrations of the isolated α-lactalbumin solution, and tubes 11 and 12 are two concentrations of whey. Water is added to give a final volume of 1.0 mL in each tube. Add 5.0 mL of dilute Bradford dye reagent to each tube and mix well by gentle inversion. After a period of at least 5 minutes, read A_{595} for each tube, using tube 1 as a blank. The tubes should be read within an hour after adding the

EXPERIMENT 4

Table E4.2

Procedure for the Bradford Protein Assay[1]

Reagents	Tube Number											
	1	2	3	4	5	6	7	8	9	10	11	12
Water	1.0	0.9	0.8	0.6	0.4	0.2	0.7	0.7	0.4	0.4	0.95	0.90
Standard												
gamma globulin	—	0.1	0.2	0.4	0.6	0.8	—	—	—	—	—	—
α-Lactalbumin	—	—	—	—	—	—	0.3	0.3	0.6	0.6	—	—
Whey	—	—	—	—	—	—	—	—	—	—	0.05	0.10
Dye solution	5.0 ⟶											

Mix well, allow to sit at least 5 min, and read A_{595}

[1] Units are mL.

dye. If absorbance readings are too low (<0.05) or too high (>1.0), repeat the assay with more or less protein fraction.

E. Ultraviolet Spectrum of α-Lactalbumin

Turn on the recording spectrophotometer and the UV lamp. Allow the instrument to warm up for about 15 minutes. Set the wavelength to 260 nm. During the warm-up period, prepare samples by transferring 1 or 3 mL of Tris buffer into a quartz cuvette and an equal volume of purified α-lactalbumin solution into a matched cell. Place the cuvette with buffer only in the sample beam and adjust the absorbance to zero. Remove the blank cuvette and replace with the cuvette containing α-lactalbumin. Read and record the absorbance at 260 nm. If the A_{260} is greater than 1.0, dilute the sample with a known volume of Tris buffer and repeat the reading.

Using the above procedure, read and record the absorbances at 280 and 290 nm.

Obtain the continuous spectrum of the purified α-lactalbumin sample from 240 to 340 nm. Do this by placing a cuvette containing Tris buffer in the reference beam and α-lactalbumin in the sample beam. Record the UV spectrum. Dilute the solution with a known amount of buffer if the recorder runs off scale. Record the spectrum with standard α-lactalbumin.

IV.   ANALYSIS OF RESULTS

A. and B. Isolation of α-Lactalbumin

Describe the appearance of the whey before chromatography. Study your experimentally derived plot of A_{280} vs. Sephadex or affinity column fraction number. How many major protein peaks are present? Which peak has the greatest amount of protein? Do you believe that each peak contains a single type of protein, or does it contain a mixture? If each peak represents a mixture of proteins, what one feature does each protein have in common?

The presence of α-lactalbumin in a fraction cannot be confirmed until further characterization. However, if you know that α-lactalbumin is among the smallest proteins in milk (molecular weight is 14,200), can you determine which protein peak contains α-lactalbumin?

C. Electrophoresis

Study the destained gel or diagram and estimate the number of protein components in each sample. Calculate the **relative mobility** of each protein band using Equation E4.1.

>> $$\text{Relative mobility} = \frac{\text{distance moved by protein}}{\text{distance moved by dye}}$$ *Equation E4.1*

Prepare a table of relative mobilities of all bands in the gel. Compare the profile for isolated α-lactalbumin to that of standard α-lactalbumin. Is α-lactalbumin the predominant protein in your preparation from milk? In whey? Try to estimate the percentage of the isolated sample that is α-lactalbumin. Assume that all proteins on the gel stain to the same extent with the dye, even though this is probably not true. Approximately what percent of total whey proteins is α-lactalbumin? Use the molecular weight standards to estimate the molecular weight of α-lactalbumin and other proteins.

D. The Bradford Assay

Prepare a standard calibration curve using the data obtained for standard gamma globulin (Table E4.2, tubes 2 to 6). Plot the absorbance on the y axis and mg of standard protein per assay on the x axis. Use the standard curve to calculate the protein concentration in the isolated α-lactalbumin fractions (in units of mg/mL.) Compare your graph with Figure 2.6.

E. Ultraviolet Spectrum of α-Lactalbumin

Calculation of Protein Concentration

Calculate the ratio A_{280}/A_{260} and estimate the concentration of α-lactalbumin.

>> Protein concentration (mg/mL) $= 1.55A_{280} - 0.76A_{260}$ *Equation E4.2*

Compare these results with the results from the Bradford assay. Explain any differences.

Calculation of Absorption Coefficient

Use the Beer-Lambert relationship (Equation 4.3) to calculate the absorption coefficient in terms of $E^{1\%}$ at 280 nm.

>> $A = Elc$ *Equation E4.3*

where

A = absorbance at 280 nm

E = absorption coefficient, $E^{1\%}$

l = path length (usually 1 cm)

c = concentration of protein, g/100 mL

The literature value for $E^{1\%}_{280}$ of purified bovine α-lactalbumin is 20.1.

Calculate the ratio A_{280}/A_{290}, and compare to the standard value of 1.30 for purified α-lactalbumin.

Ultraviolet Spectrum

Compare the spectrum of isolated α-lactalbumin with that of standard α-lactalbumin. Describe and explain any differences.

Study Problems

1. What other purification techniques could be applied to the purification of α-lactalbumin? Use a MEDLINE literature search. Study the following references for methods that have been applied to the isolation and purification of α-lactalbumin: W. G. Gordon and W. F. Semmett, *J. Amer. Chem. Soc.* **75,** 328 (1953). W. G. Gordon and J. Ziegler, *Biochem. Prep.* **4,** 16 (1955). U. Brodbeck, W. L. Denton, N. Tanahashi, and K. E. Ebner, *J. Biol. Chem.* **242,** 1391 (1967). F. J. Castellino and R. L. Hill, *J. Biol. Chem.* **245,** 417 (1970). L. Lindahl and H. Vogel, *Anal. Biochem.* **140,** 394 (1984).

2. What methods of analysis could you perform on the purified α-lactalbumin in order to determine the extent of purity?

3. What assumptions must be made about the relative mobility of bromophenol blue when used as a tracking dye?

4. What would be the effect of each of the following changes on the relative mobility of α-lactalbumin in SDS-PAGE?
   (a) Lower the pH of all buffers.
   (b) Increase the ionic strength of the buffers.
   (c) Change the temperature of the electrophoretic operation.
   (d) Increase the concentration of α-lactalbumin.
   (e) Increase the concentration of bis-acrylamide in the gels.
   (f) Decrease the electrophoretic running time.

5. Describe the chemistry involved in the Bradford assay.

6. Describe the procedure for obtaining a "difference spectrum" of isolated α-lactalbumin vs. standard α-lactalbumin.

> 7. Why do most protein solutions absorb light of 280 nm wavelength?

> 8. Comment on the validity of the following statement: "Ideally, the best Bradford calibration curve for this experiment would use purified α-lactalbumin rather than gamma globulin."

> 9. Absorbance measurements of column fractions at 280 nm were used in this experiment to detect the presence of protein material. Do you think this method of analysis could lead to a quantitative determination of protein concentration? What other biomolecules might interfere with this measurement?

> 10. What amino acid residues in proteins bind to chromatography gels containing immobilized metal ions?

Further Reading

R. Boyer, *J. Chem. Educ.* **68,** 430–432 (1991). "Purification of Milk Whey α-Lactalbumin by Immobilized Metal-Ion Affinity Chromatography."

R. Boyer, *Concepts in Biochemistry,* (1999), Brooks/Cole (Pacific Grove, CA), pp. 102–105. Protein purification and analysis.

K. Ebner, *Acct. Chem. Res.* **3,** 41–47 (1970). Biological role of α-lactalbumin.

R. Garrett and C. Grisham, *Biochemistry,* 2nd ed. (1999), W. B. Saunders (Orlando, FL), pp. 153–157. Protein techniques.

L. Hall and P. Campbell, in *Essays in Biochemistry,* Vol. 22, R. Marshall and K. Tipton, Editors (1988), Academic Press (London), pp. 1–26. A review on α-lactalbumin.

E. Harris and S. Angal, Editors, *Protein Purification Applications: A Practical Approach,* Vol. 2 (1990), IRL Press (Oxford). An excellent book for undergraduates.

B. Johnson, *The Scientist,* November 9, 1998. A review of cell homogenizers.

A. Lehninger, D. Nelson, and M. Cox, *Principles of Biochemistry,* 3rd ed. (1999), Worth Publishers (New York), pp. 130–133. Working with proteins.

C. Mathews, K. van Holde, and K. Ahern, *Biochemistry,* 3rd ed. (2000), Benjamin/Cummings (San Francisco), pp. 148–152. Isolation and purification of proteins.

J. Porath, J. Carlsson, I. Olsson, and G. Belfrage, *Nature* **258,** 598–599 (1975). "Metal Chelate Affinity Chromatography, a New Approach to Protein Fractionation."

L. Stryer, *Biochemistry,* 4th ed. (1995), W. H. Freeman (New York), pp. 45–52. Isolation, purification, and characterization of proteins.

D. Voet and J. Voet, *Biochemistry,* 2nd ed. (1995), John Wiley & Sons (New York), pp. 72–104. Study of proteins.

D. Voet, J. Voet, and C. Pratt, *Fundamentals of Biochemistry* (1999), John Wiley & Sons (New York), pp. 96–107. An introduction to protein purification and analysis.

On the Web

Worthington Biochemical Corp.
  http://www.worthington-biochem.com/pricelist/G/Galactosyltransferase.html
  A discussion of the biological role of α-lactalbumin.

5

KINETIC ANALYSIS OF TYROSINASE

- **Recommended Reading**

Chapter 5, Section A.

- **Synopsis**

Tyrosinase, a copper-containing oxidoreductase, catalyzes the orthohydroxylation of monophenols and the aerobic oxidation of catechols. The enzyme activity will be assayed by monitoring the oxidation of 3,4-dihydroxyphenylalanine (dopa) to the red-colored dopachrome. The kinetic parameters K_M and V_{max} will be evaluated using Lineweaver-Burk or direct linear plots. Inhibition of tyrosinase by thiourea and cinnamate will also be studied. Two stereoisomers, L-dopa and D-dopa, will be tested and compared as substrates.

I. INTRODUCTION AND THEORY

Enzymes are biological catalysts. Without their presence in a cell, most biochemical reactions would not proceed at the required rate. The physicochemical and biological properties of enzymes have been investigated since the early 1800s. The unrelenting interest in enzymes is due to several factors—their dynamic and essential role in the cell, their extraordinary catalytic power, and their selectivity. Two of these dynamic characteristics will be evaluated in this experiment, namely a kinetic description of enzyme activity and molecular selectivity.

Enzyme-catalyzed reactions proceed through an ES complex as shown in Equation 5.1. The individual rate constants, k_n, are placed with each arrow.

>> $$E + S \overset{k_1}{\underset{k_2}{\rightleftharpoons}} ES \overset{k_3}{\underset{k_4}{\rightleftharpoons}} E + P$$ *Equation E5. 1*

E represents the enzyme, S the substrate or reactant, and P the product. For a specific enzyme, only one or a few different substrate molecules can bind in the proper manner and produce a functional ES complex. The substrate must have a size, shape, and polarity compatible with the active site of the enzyme. Some enzymes catalyze the transformation of many different molecules as long as there is a common type of chemical linkage in the substrate. Others have absolute specificity and can form reactive ES complexes with only one molecular structure. In fact, some enzymes are able to differentiate between D and L isomers of substrates.

Kinetic Properties of Enzymes

The initial reaction velocity, v_0, of an enzyme-catalyzed reaction varies with the substrate concentration, [S], as shown in Figure E5.1. The Michaelis-Menten equation has been derived to account for the kinetic properties of enzymes. (Consult a biochemistry textbook for a derivation of this equation and for a discussion of the conditions under which the equation is valid.) The common form of the equation is

>> $$v_0 = \frac{V_{max}[S]}{K_M + [S]}$$ *Equation E5.2*

Figure E5.1

Michaelis-Menten plot for an enzyme-catalyzed reaction. From *Biochemistry,* 3rd ed., p. 377, by C. Mathews, K. van Holde, and K. Ahern, Benjamin/ Cummings, Redwood City, CA, 2000.

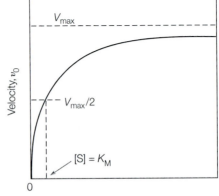

where

v_0 = initial reaction velocity

V_{max} = maximal reaction velocity; attained when all enzyme active sites are filled with substrate molecules

[S] = substrate concentration

K_M = Michaelis constant = $\dfrac{k_2 + k_3}{k_1}$

The important kinetic constants, V_{max} and K_M, can be graphically determined as shown in Figure E5.1. Equation E5.2 and Figure E5.1 have all of the disadvantages of nonlinear kinetic analysis. V_{max} can be estimated only because of the asymptotic nature of the line. The value of K_M, the substrate concentration that results in a reaction velocity of $V_{max}/2$, depends on V_{max}, so both are in error. By taking the reciprocal of both sides of the Michaelis-Menten equation, however, it is converted into the Lineweaver-Burk relationship (Equation E5.3).

>>

$$\frac{1}{v_0} = \frac{K_M}{V_{max}} \cdot \frac{1}{[S]} + \frac{1}{V_{max}}$$

Equation E5. 3

This equation, which is in the form $y = mx + b$, gives a straight line when $1/v_0$ is plotted against $1/[S]$ (Figure E5.2). The intercept on the $1/v_0$ axis is $1/V_{max}$ and the intercept on the $1/[S]$ axis is $-1/K_M$. A disadvantage of the Lineweaver-Burk plot is that the data points are compressed in the high substrate concentration region. A third type of graphical analysis results from the Eisenthal and Cornish-Bowden modification (Equation E5.4).

>>

$$V_{max} = v_0 + \frac{v_0}{[S]} K_M$$

Equation E5. 4

This modification is a direct linear graph that results from plotting each pair of experimental values of v_0 and [S]. Figure E5.3 is an example of the Eisenthal and Cornish-Bowden analysis. K_M and V_{max} values are read directly from the graph.

Significance of Kinetic Constants

The **Michaelis constant,** K_M, for an enzyme-substrate interaction has two meanings: (1) K_M is the substrate concentration that leads to an initial reaction velocity of $V_{max}/2$ or, in other words, the substrate concentration that results in the filling of one-half of the enzyme active sites, and (2) $K_M = (k_2 + k_3)/k_1$. The second definition of K_M has special significance in certain

Lineweaver-Burk plot for an
enzyme-catalyzed reaction.

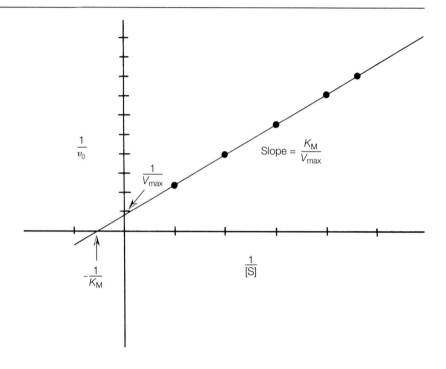

cases. When $k_2 \gg k_3$, then $K_M = k_2/k_1$ so K_M is equivalent to the dissociation constant of the ES complex. When $k_2 \gg k_3$, a large K_M implies weak interaction between E and S, whereas a small K_M indicates strong binding between E and S.

V_{max} is important because it leads to the analysis of another kinetic constant, k_3, **turnover number.** The analysis of k_3 begins with the basic rate law for an enzyme-catalyzed process (Equation E5.5), which is derived from Equation E5.1:

$$v_0 = k_3[ES]$$

Equation E5. 5

If all of the enzyme active sites are saturated with substrate, then [ES] in Equation E5.5 is equivalent to [E_T], the total concentration of enzyme, and v_0 becomes V_{max}; hence

$$V_{max} = k_3[E_T]$$

Equation E5. 6

or

$$k_3 = \frac{V_{max}}{[E_T]}$$

Figure E5.3

Eisenthal and Cornish-Bowden (direct linear) plot for an enzyme-catalyzed reaction.

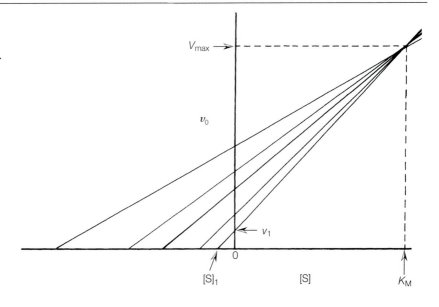

For an enzyme with one active site per molecule, the turnover number, k_3, is the number of substrate molecules transformed to product by one enzyme molecule per unit time, usually in minutes or seconds. The turnover number is a measure of the efficiency of an enzyme.

Inhibition of Enzyme Activity

Nonsubstrate molecules may interact with enzymes, leading to a decrease in enzymatic activity. The study of enzyme inhibition is of interest because it often reveals information about the mechanism of enzyme action. Also, many toxic substances, including drugs, express their action by enzyme inhibition.

The process of reversible inhibition is described by an equilibrium interaction between enzyme and inhibitor. Most inhibition processes can be classified as **competitive** or **noncompetitive,** depending on how the inhibitor impairs enzyme action. A competitive inhibitor is usually similar in structure to the substrate and is capable of reversible binding to the enzyme active site. In contrast to the substrate molecule, the inhibitor molecule cannot undergo chemical transformation to a product; however, it does interfere with substrate binding. A noncompetitive inhibitor does not bind in the active site of an enzyme but binds at some other region of the enzyme molecule. Upon binding of the noncompetitive inhibitor, the enzyme is reversibly converted to a nonfunctional conformational state, and the substrate, which is fully capable of binding to the active site, is not converted to product.

Any complete inhibition study requires the experimental differentiation between competitive and noncompetitive inhibition. The two inhibitory processes are kinetically distinguishable by application of the Lineweaver-Burk or direct linear plot. For each inhibitor studied, at least two sets of rate experiments are completed. In all sets, the enzyme concentration is identical. In set 1, the substrate concentration is varied and no inhibitor is added. In set 2, the same variable substrate concentrations are used as in set 1, and a constant amount of inhibitor is added to each assay. If additional data are desired, more sets are prepared with variable substrate concentrations as in set 2; but a constant, and different, concentration of inhibitor is present. These data, when plotted on a Lineweaver-Burk graph, lead to three lines as shown in Figure E5.4 and E5.5. The Lineweaver-Burk plot of a competitively inhibited process is shown in Figure E5.4. All three lines intersect at the same point on the $1/v_0$ axis; hence, V_{max} is not altered by the competitive inhibitor. If enough substrate is present, the competitive inhibition can be overcome. The apparent K_M value (measured on the $1/[S]$ axis) changes with each change in inhibitor concentration. Figure E5.4 reveals several kinetic characteristics of the inhibition process: K_M, V_{max}, and K_I. The quantity K_I is the dissociation constant for the EI complex, and its value indicates the strength of binding between enzyme and inhibitor. A large K_I implies relatively weak interaction between E and I.

Figure E5.4

Lineweaver-Burk plot for competitive inhibition.

- -

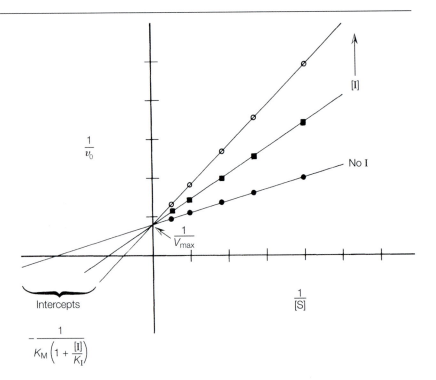

A Lineweaver-Burk plot of enzyme kinetics in the presence and absence of a noncompetitive inhibitor is shown in Figure E5.5. V_{max} in the presence of a noncompetitive inhibitor is decreased, but K_M is unaffected. The effect of a competitive inhibitor on the direct linear plot is shown in Figure E5.6.

Units of Enzyme Activity

The actual molar concentration of an enzyme in a cell-free extract or purified preparation is seldom known. Only if the enzyme is available in a pure crystalline form, carefully weighed, and dissolved in a solvent can the actual molar concentration be accurately known. It is, however, possible to develop a precise and accurate assay for enzyme activity. Consequently, the amount of a specific enzyme present in solution is most often expressed in units of activity. Three units are in common use, the *international unit* (IU), the *katal,* and *specific activity.* The International Union of Biochemistry Commission on Enzymes has recommended the use of a standard unit, the **international unit,** or just **unit,** of enzyme activity. One IU of enzyme corresponds to the amount that catalyzes the transformation of 1 μmole of substrate to product per minute under specified conditions of pH, temperature, ionic strength, and substrate concentration. If a solution containing

Figure E5.5

Lineweaver-Burk plot for noncompetitive inhibition.

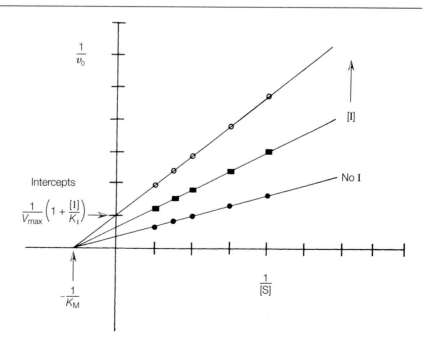

enzyme converts 10 μmoles of substrate to product in 5 minutes, the solution contains 2 enzyme units. A new unit of activity, the katal, has been recommended. One **katal** of enzyme activity represents the conversion of 1 mole of substrate to product in 1 second. One international unit is equivalent to 1/60 μkatal or 0.0167 μkatal. One katal, therefore, is equivalent to 6×10^7 international units. It is convenient to represent small amounts of enzyme in millikatals (mkatals), microkatals (μkatals), or nanokatals (nkatals). The enzyme activity in the above example was 2 units which can be converted to katals as follows: Since one katal is 6×10^7 units, 2 units are equivalent to $2/6 \times 10^7$ or 33 nkatals (0.033 μkatals). If the enzyme is an impure preparation in solution, the activity is most often expressed as units/mL or katals/mL.

Another useful quantitative definition of enzyme efficiency is **specific activity.** The specific activity of an enzyme is the number of enzyme units or katals per milligram of protein. This is a measure of the purity of an enzyme. If a solution contains 20 mg of protein that express 2 units of activity (33 nkatals), the specific activity of the enzyme is 2 units/20 mg = 0.1 units/mg or 33 nkatals/20 mg = 1.65 nkatals/mg. As an enzyme is purified, its specific activity increases. That is, during purification, the enzyme concentration increases relative to the total protein concentration until a limit is reached. The maximum specific activity is attained when the enzyme is homogeneous or in a pure form.

The activity of an enzyme depends on several factors, including substrate concentration, cofactor concentration, pH, temperature, and ionic

Figure E5.6

Direct linear plot for competitive inhibition.

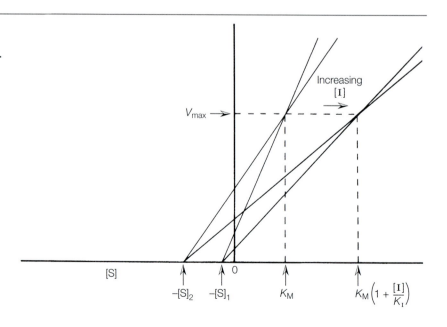

V_{max} Increasing [I]

[S] $-[S]_2$ $-[S]_1$ 0 K_M $K_M \left(1 + \dfrac{[I]}{K_I}\right)$

strength. The conditions under which enzyme activity is measured are critical and must be specified when activities are reported.

Design of an Enzyme Assay

Whether an enzyme is obtained commercially or prepared in a multistep procedure, an experimental method must be developed to detect and quantify the specific enzyme activity. During isolation and purification of an enzyme, the assay is necessary to determine the amount and purity of the enzyme and to evaluate its kinetic properties. An assay is also essential for a further study of the mechanism of the catalyzed reaction.

The design of an assay requires certain knowledge of the reaction:

1. The complete stoichiometry.
2. What substances are required (substrate, metal ions, cofactors, etc.) and their kinetic dependence.
3. Effect of pH, temperature, and ionic strength.

The most straightforward approach to the design of an enzyme assay is to measure the change in substrate or product concentration during the reaction.

If an enzyme assay involves continuous monitoring of substrate or product concentration, the assay is said to be **kinetic.** If a single measurement of substrate or product concentration is made after a specified reaction time, a **fixed-time assay** results. The kinetic assay is more desirable because the time course of the reaction is directly observed and any discrepancy from linearity can be immediately detected.

Figure E5.7 displays the kinetic progress curve of a typical enzyme-catalyzed reaction and illustrates the advantage of a kinetic assay. The rate of product formation decreases with time. This may be due to any combination of factors such as decrease in substrate concentration, denaturation of the enzyme, and product inhibition of the reaction. The solid line in Figure E5.7 represents the continuously measured time course of a reaction (kinetic assay). The true rate of the reaction is determined from the slope of the dashed line drawn tangent to the experimental result. From the data given, the rate is 5 μmoles of product formed per minute. Data from a fixed-time assay are also shown on Figure E5.7. If it is assumed that no product is present at the start of the reaction, then only a single measurement after a fixed period is necessary. This is shown by a circle on the experimental rate curve. The measured rate is now 16 μmoles of product formed every 5 minutes or about 3 μmoles/minute, considerably lower than the rate derived from the continuous, kinetic assay. Which rate measurement is correct? Obviously, the kinetic assay gives the true rate because it corrects for the decline in rate with time. The fixed-time assay can be improved by changing the time of the measurement, in this example, to 2 minutes of reaction time, when the experimental rate is still linear. It is possible to obtain

Kinetic progress of an enzyme-catalyzed reaction. See text for details.

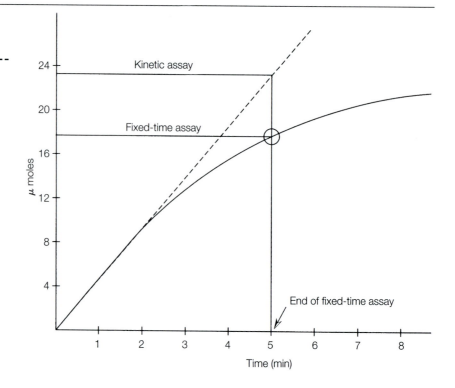

true rates with the fixed-time assay, but one must choose the time period very carefully. In the laboratory, this is done by removing aliquots of a reaction mixture at various times and measuring the concentration of product formed in each aliquot. Figure E5.7 reinforces an assumption used in the derivation of the Michaelis-Menten equation. Only measurements of initial velocities lead to true reaction rates. This avoids the complications of enzyme denaturation, decrease of [S], product inhibition, and reversion of the product to substrate.

Several factors must be considered when the experimental assay conditions are developed. The reaction rate depends on the concentrations of substrate, enzyme, and necessary cofactors. In addition, the reaction rate is under the influence of environmental factors such as pH, temperature, and ionic strength. Enzyme activity increases with increasing temperature until the enzyme becomes denatured. The enzyme activity then decreases until all enzyme molecules are inactivated by denaturation. During kinetic measurement, it is essential that the temperature of all reaction mixtures be maintained constant.

Ionic strength and pH should also be monitored carefully. Although some enzymes show little or no change in activity over a broad pH range, most enzymes display maximum activity in a narrow pH range. Any assay developed to evaluate the kinetic characteristics of an enzyme must be performed in a buffered solution, preferably at the optimal pH.

Applications of an Enzyme Assay

The conditions used in an enzyme assay depend on what is to be accomplished by the assay. There are two primary applications of an enzyme assay procedure. First, it may be used to measure the concentration of active enzyme in a preparation. In this circumstance, the measured rate of the enzyme-catalyzed reaction must be proportional to the concentration of enzyme; stated in more kinetic terms, there must be a linear relationship between initial rate and enzyme concentration (the reaction is first order in enzyme concentration). To achieve this, certain conditions must be met: (1) the concentrations of substrate(s), cofactors, and other requirements must be in excess; (2) the reaction mixture must not contain inhibitors of the enzyme; and (3) all environmental factors such as pH, temperature, and ionic strength should be controlled. Under these conditions, a plot of enzyme activity (μmole product formed/minute) vs. enzyme concentration is a straight line and can be used to estimate the concentration of active enzyme in solution.

Second, an enzyme assay may be used to measure the kinetic properties of an enzyme such as K_M, V_{max}, and inhibition characteristics. In this situation, different experimental conditions must be used. If K_M for a substrate is desired, the assay conditions must be such that the measured initial rate is first order in substrate. To determine K_M of a substrate, constant amounts of enzyme are incubated with varying amounts of substrate. A Lineweaver-Burk plot ($1/v_0$ vs. $1/[S]$) or direct linear plot may be used to determine K_M and V_{max}. If a reaction involves two or more substrates, each must be evaluated separately. The concentration of only one substrate is varied while the other is held constant at a saturating level. The same procedure holds for determining the kinetic dependence on cofactors. Substrate(s) and enzyme are held constant and the concentration of the cofactor is varied.

Inhibition kinetics are included in the second category of assay applications. An earlier discussion outlined the kinetic differentiation between competitive and noncompetitive inhibition. The same experimental conditions that pertain to evaluation of K_M and V_{max} hold for K_I estimation. A constant level of inhibitor is added to each assay, but the substrate concentration is varied as for K_M determination. In summary, a study of enzyme kinetics is approached by measuring initial reaction velocities under conditions where only one factor (substrate, enzyme, cofactor) is varied and all others are held constant.

Properties of Tyrosinase

Tyrosinase, also commonly called polyphenol oxidase, has two catalytic activities; *o*-hydroxylation of monophenols and aerobic oxidation of *o*-diphenols (Equations E5.7 and E5.8).

>> $$\text{monophenol} + O_2 \longrightarrow \text{catechol} + H_2O$$ *Equation E5. 7*

>> $$2 \text{ catechol} + O_2 \longrightarrow 2 \text{ } o\text{-quinone} + 2 \text{ } H_2O$$ *Equation E5. 8*

The official Enzyme Commission classification and number are monophenol, o-diphenol: O_2 oxidoreductase, EC 1.14.18.1. Tyrosinases are present in plant and animal cells. The biological function of tyrosinase in plant cells is unknown, but its presence is readily apparent. Upon injury, many plant products such as potatoes, apples, and bananas undergo a browning process. This is due to the formation and further oxidation of natural quinones as shown in Equation E5.8. In mammalian cells, tyrosinase catalyzes two steps in the biosynthesis of melanin pigments from tyrosine. The reaction sequence, shown in Figure E5.8, is localized in melanocytes or pigment cells and produces the melanin substances that impart color to skin, hair, and eyes. Tyrosinase located in skin is activated by the ultraviolet rays of the sun, causing greater melanin production and, ultimately, suntan.

The many tyrosinases in nature differ in cofactor requirement, metal ion content, substrate specificity, molecular weight, and oligomeric structure. The diverse properties of the tyrosinases will not be discussed; instead, the properties of a single enzyme, mushroom tyrosinase, will be outlined.

Figure E5.8

The oxidation of tyrosine and dopa as catalyzed by tyrosinase.

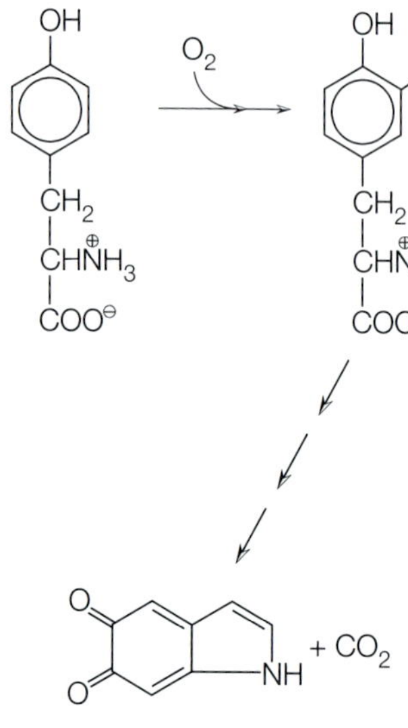

Dopachrome
λ_{max} = 475 nm

Mushroom tyrosinase is tetrameric, with a total molecular weight of 128,000. There are four atoms of Cu^+ associated with the active enzyme. Two types of substrate binding sites exist in the enzyme, one type for the phenolic substrate and one type for the dioxygen molecule. The copper atom is associated with the dioxygen binding site; hence, chemicals that form complexes with copper atoms are potent inhibitors of tyrosinase activity. The inhibitors azide, cyanide, phenylthiourea, and cysteine compete with oxygen binding. Nonphenolic, aromatic compounds compete with the binding and oxidation of phenols and catechols.

Overview of the Experiment

Several kinetic characteristics of mushroom tyrosinase will be examined in this experiment. A spectrophotometric assay of tyrosinase activity will be introduced and applied to the evaluation of substrate specificity, K_M of the natural substrate, 3,4-dihydroxyphenylalanine (L-dopa), and inhibition characteristics.

Many assays for tyrosinase activity have been developed. Procedures in the literature include use of the oxygen electrode, oxidation of tyrosine followed at 280 nm, and oxidation of dopa followed at 475 nm. The most convenient assay involves following the tyrosinase-catalyzed oxidation of dopa by monitoring the initial rate of formation of dopachrome at 475 nm (Figure E5.8).

Two substrates are required in the tyrosinase-catalyzed reaction, phenolic substrate (dopa) and dioxygen. The conditions described in the experiment are such that the reaction mixtures are saturated with dissolved dioxygen. Therefore, when measurements are made for K_M, only the concentration of dopa is limiting, so the rate of the reaction depends on dopa concentration. The dopachrome assay is extremely flexible, as it can be applied to a variety of studies of tyrosinase.

Time requirements for this experiment are:

Part A: Tyrosinase concentration–15 minutes.

Part B: Tyrosinase level for kinetic assay–1 hour.

Part C: K_M of dopa–1 hour.

Part D: Inhibition–1 hour.

It is recommended that part A be done as a group project so that time and enzyme are not wasted. If only one 3-hour laboratory period is allotted to this experiment, the following schedule is recommended:

Part A: Instructor or teaching assistant prepares the enzyme solution just before class and A_{280} is reported to students.

Part B: Do only three or four assays with D- or L-dopa.

Part C: Do three or four assays with the same dopa isomer as in part B.

Part D: Students choose one inhibitor and do four assays on the same dopa isomer as in parts B and C.

If 4–5 hours are available, students can do the following:

> Part A: Measure A_{280} as a group project.
>
> Part B: Do four or five assays with D- or L-dopa.
>
> Part C: Do four assays each with D- and L-dopa.
>
> Part D: Choose one inhibitor and do four assays with L- or D-dopa.

II.   MATERIALS AND SUPPLIES

- Sodium phosphate buffer, 0.05 M, pH 7. 0
- Mushroom tyrosinase, 100 units/mL

  1 unit represents the amount of enzyme that causes an A_{280} change of 0.001 using tyrosine as substrate. Prepare solution in phosphate buffer. A_{280} should be about 0.20. Store in ice bath during laboratory period.

- L-Dihydroxyphenylalanine (L-dopa), 1.5 mg/mL in sodium phosphate buffer
- D-Dihydroxyphenylalanine (D-dopa), 1.5 mg/mL in sodium phosphate buffer
- Cinnamic acid, 50 mg/100 mL in sodium phosphate buffer
- Thiourea, 30 mg/100 mL in sodium phosphate buffer
- Cuvettes, glass and quartz
- Constant-temperature bath at 25–30°C
- UV-VIS spectrophotometer with recorder and constant-temperature circulating bath or colorimeter

III.   EXPERIMENTAL PROCEDURE

A. Measurement of Tyrosinase Concentration

In order to determine the specific activity of the enzyme, the exact concentration of the enzyme must be known. The concentration of the solution of tyrosinase may be determined as a class project by the following procedure. Turn on the spectrophotometer and the UV lamp. Adjust the wavelength to 280 nm. Allow the instrument and lamp to warm up for 15 to 20 minutes. Transfer 1.0 or 3.0 mL of the phosphate buffer to a 1- or 3-mL quartz cuvette. Place it in the sample position of the spectrophotometer and adjust the balance to zero absorbance. Discard the buffer, and clean and dry the cuvette. Transfer 1.0 or 3.0 mL of the tyrosinase solution into the quartz cuvette. Place in the sample position and record the absorbance at 280 nm. Calculate the tyrosinase concentration as described in the Analysis of Results section.

B. Determination of the Tyrosinase Level for Kinetic Assays

All solutions listed under Materials and Supplies, except tyrosinase, should be stored in a constant-temperature bath. The tyrosinase solution should be stored in an ice bath. If there is no constant-temperature circulating bath on the spectrophotometer, maintain all solutions except tyrosinase at room temperature. Otherwise, set both the constant-temperature bath and the circulating bath at the same temperature. The recommended temperature range is 25 to 30°C. Turn on the spectrophotometer and recorder for warm-up. Adjust the wavelength to 475 nm.

Before kinetic constants can be evaluated, it is critical to find the correct concentration of enzyme to use for the assays. If too little enzyme is used, the overall absorbance change for a reaction time period will be so small that it is difficult to detect differences due to substrate concentration changes or inhibitor action. On the other hand, too much enzyme will allow the reaction to proceed too rapidly, and the leveling off of the time course curve as shown in Figure E5.7 will occur very early because of the rapid disappearance of substrate. A rate that is intermediate between these two extremes is best. For the dopachrome assay, it is desirable to use the level of tyrosinase that gives a linear absorbance change at 475 nm for 2 minutes.

The general procedure for the assay is as follows. Pipet phosphate buffer and L-dopa into a glass cuvette according to the directions in Table E5. 1. Mix by inversion (cover the cuvette with hydrocarbon foil), and place the cuvette in the spectrometer sample position. If a recorder is available, adjust it to a full-scale absorbance of 1 and set the pen to 0 or the 0.10 absorbance line with the recorder zero adjust dial. Set the recorder speed to 1 or 2 cm/min and turn on the recorder. Observe whether there is a "blank" rate (no enzyme). Obtain a recorder trace for 2 minutes. Now, pipet the appropriate amount of enzyme into the cuvette. Mix by inversion of the cuvette after covering it with hydrocarbon foil **(Do not shake!),** and immediately place in the spectrophotometer. Turn on the recorder pen and observe the recorder trace for 2 minutes. Be sure you mark, on the recorder paper, the point of tyrosinase addition. With a straight-edge, draw a line tangent to the recorder trace as shown in Figure E5.7. Is the time course linear for the full 2 minutes? Determine the longest period of linearity and calculate the average ΔA/min over this time period. If you observed a "blank" rate with no enzyme present, calculate its ΔA/min and subtract it from the enzyme-catalyzed reaction rate.

If no recorder is available, you must take absorbance readings at 475 nm every 30 seconds. Proceed as follows. To the cuvette, add buffer and L-dopa according to Table E5.1. Mix well and place in the photometer. Immediately adjust the absorbance to 0. Begin timing with a stopwatch and record in your notebook an absorbance reading every 30 seconds. Continue for 2

Table E5.1

General Procedure for the Tyrosinase Assay[1]

Reagent	Assay 1	2	3	4	5
Phosphate buffer	1.45	1.40	1.30	1.20	1.10
L-Dopa	1.5	1.5	1.5	1.5	1.5
Tyrosinase	0.05	0.10	0.20	0.30	0.40

[1] Units are milliliters.

minutes. Now, add the appropriate level of tyrosinase, mix **(Do not shake)**, immediately place the cuvette in the photometer, and set the absorbance to 0. Start the stopwatch and record an absorbance reading every 30 seconds for 2 minutes. On graph paper, prepare a plot of A_{475} vs. time. Calculate ΔA/min over a period during which the assay is linear. On the same graph paper, plot A_{475} vs. time for the "blank." If there is a significant rate (greater than 5% of the enzyme rate), subtract it from the enzyme-catalyzed rate before preparing the graph.

Repeat the above procedures (with or without recorder) with a different concentration of tyrosinase. If the corrected rate (ΔA/min for enzyme minus ΔA/min for blank) was less than 0.03, double the amount of enzyme added. If the corrected rate was greater than 0.3/min, reduce the enzyme by one-half. Determine the rate for a total of five different enzyme levels. The enzyme levels chosen should cover a ΔA/min range from about 0.025/min to 0.25/min. The tyrosinase levels listed in Table E5.1 are only recommended; you may have to try higher or lower amounts of tyrosinase.

C. K_M of L-Dopa and D-Dopa

Now that the appropriate enzyme level has been determined, the kinetic constants may be evaluated. The K_M for L-dopa can be obtained by setting up the same assay as in part B, except that the factor to vary will be the concentration of L-dopa. The concentration of L-dopa in part B was sufficient to saturate all the tyrosinase active sites, so the rate depended only on the enzyme concentration. In part C, L-dopa levels will be varied over a range that is nonsaturating.

Choose the level of enzyme that will give a convenient rate, at saturating levels of L-dopa, of 0.10 to 0.15 ΔA/min. Prepare a table, for five assays, similar to Table E5.1. The total volume for each assay must be constant at 3.0 mL. Since the enzyme added to each assay remains constant, only the volume of buffer and L-dopa is varied. The recommended levels of L-dopa are 0.10, 0.20, 0.40, 0.80, 1.0 and 1.5 mL. The procedure to follow is identical to part B. The enzyme should be added last, after a "blank" rate is determined for each assay. If the recommended levels of L-dopa do not result in

Two inhibitors of
tyrosinase.

Thiourea   Cinnamic
                    acid

a saturating rate, increase the amount of L-dopa. From the slope of each
recorder trace or graph of absorbance vs. time, calculate the ΔA/min for
each level of L-dopa. In order to determine the K_M of D-dopa, repeat the
above procedure using D-dopa rather than L-dopa. Calculate the ΔA/min
for each concentration level of D-dopa.

D. Inhibition of Tyrosinase Activity

Whether an inhibitor acts in a competitive or noncompetitive manner is de-
duced from a Lineweaver-Burk or direct linear plot using varying concentrations of inhibitor and substrate. In separate assays, two substances will be
added to the dopa-tyrosinase reaction mixture, and the effect on enzyme activity will be quantified. The structures of the potential inhibitors, cinnamic
acid and thiourea, are shown in Figure E5.9. The inhibition assays must be
done immediately following the K_M studies. To measure inhibition, reaction
rates both with and without inhibitor must be used and the tyrosinase activity must not be significantly different. If it is necessary to do the inhibition
studies later, the K_M assay for L-dopa must be repeated with freshly prepared
tyrosinase solution.

To set up the inhibition assay, prepare a table similar to Table E5. 1. Inhibitor should appear in the list of reagents before tyrosinase. Use the same
level of tyrosinase and the same dopa stereoisomer as in part C. Vary the
amount of dopa as in part C. A constant amount of inhibitor (cinnamic
acid or thiourea) should be added to each cuvette. You will have to determine this level of inhibitor by trial and error. The desired inhibition rate
with saturating substrate is about 50% of the uninhibited rate. Add all
reagents except tyrosinase, mix well, and determine the blank rate, if any.
Add tyrosinase, mix, and immediately record A_{475} for 2 minutes. From
recorder traces or graphs of A_{475} vs. time, calculate ΔA/min for each assay.

IV. ANALYSIS OF RESULTS

A. Measurement of Tyrosinase Concentration

If the tyrosinase preparation is homogeneous, the concentration of enzyme
may be determined by absorbance measurement at 280 nm. The absorption

coefficient, $E_{280}^{1\%}$, for tyrosinase is 24.9. That is, the absorbance of a 1% (w/v) solution of pure tyrosinase in a 1-cm cell at 280 nm is 24.9. Beer's law states that

$$A = Elc$$

where

> A = absorbance
>
> E = absorption coefficient
>
> c = concentration of solution
>
> l = cell path length (usually 1 cm)

For a 1% solution of tyrosinase, A is equal to E if l is 1 cm,

$$A = E_{280}^{1\%}$$

Therefore, the concentration of tyrosinase in solution is calculated from a simple ratio,

$$\frac{c_1}{A_1} = \frac{c_2}{A_2}$$

where

> c_1 = concentration of a standard solution of tyrosinase = 1%
>
> A_1 = absorption of the standard solution of tyrosinase at 280 nm = 24.9
>
> c_2 = concentration of unknown solution of tyrosinase in % (w/v)
>
> A_2 = absorption of unknown solution of tyrosinase at 280 nm

Convert your concentration (now in w/v%) to mg of tyrosinase/mL.

B. Determination of the Tyrosinase Level for Kinetic Assays

Prepare a table of rate (ΔA/min) vs. enzyme concentration (mg per assay). ΔA/min units are converted to a more useful form, amount of product formed/time, usually μmole/minute. The conversion is straightforward when Beer's law is used in the form of Equation E5.9.

>> $$c = \frac{A}{El}$$ ***Equation E5.9***

where A refers to the absorbance change that occurs per minute, l is the light path length through the cuvette (1 cm), E is the molar absorption coefficient for the product dopachrome ($3600\ M^{-1}\ cm^{-1}$), and c is the concentration of product. According to this relationship, the absorbance change during 1 minute for production of 1 molar dopachrome is 3600:

$$1\ molar\ =\ \frac{A}{3600\ M^{-1}\ cm^{-1} \times 1\ cm}$$

$$A = 3600$$

To convert the raw data to moles of product formed per minute per liter, each $\Delta A/min$ is divided by 3600. For example, a $\Delta A/min$ of 0.1 is converted to μmoles of product formed (Δc) in the following way:

$$\Delta c = \frac{\Delta A/min}{El} = \frac{0.1/min}{3600\ M^{-1}\ cm^{-1}\ cm} = 2.78 \times 10^{-5}\ \frac{M}{min}$$

Now convert to moles/min:

$$2.78 \times 10^{-5}\ \frac{moles}{liter - min} \times 0.003\ liters\ = 8.33 \times 10^{-8}\ \frac{moles}{min}$$

Now convert to μmoles/min:

$$\Delta c = 8.33 \times 10^{-2}\ \frac{\mu moles}{min}$$

Add another column to your table, label it μmoles product formed per minute, and calculate the appropriate rate for each enzyme concentration. Prepare a graph of rate (μmole/min on the ordinate, y axis) vs. enzyme concentration in each assay (mg, on the abscissa, x axis). Connect as many of the points as possible with a straight line passing through the origin. If most of the points are on this line, the assay and the standard curve can be used to quantify an unknown level of tyrosinase. The standard curve also provides the experimenter with a choice of enzyme levels to use for further kinetic studies.

C. K_M of L-Dopa and D-Dopa

Kinetic analysis of tyrosinase and calculation of constants will be described using graphical analysis by the Michaelis-Menten equation, Lineweaver-Burk equation, or the direct linear curve. Procedures for preparing these graphs are described below. Alternatively, students may use available computer software to graph data and calculate kinetic constants. Recommended enzyme kinetic computer software packages include "Enzyme

Kinetics" (from ChemSW, Inc. or Trinity Software) and "ENZYPLOT" (*Biochem. Educ.* **23,** 35–37, 1995).

The treatment of results will be described for L-dopa. The procedure for D-dopa is identical. Prepare a table of L-dopa concentration per assay (mmolar) vs. ΔA/min. Convert all ΔA/min units to μmoles/min as described in part B. Prepare a Michaelis-Menten curve (μmoles/min vs. [S]) as in Figure E5.1 and a Lineweaver-Burk plot (1/μmole/min vs. 1/[S]) as in Figure E5.2. Alternatively, you may wish to use the direct linear plot. Estimate K_M and V_{max} from each graph. The intercept on the rate axis of the Lineweaver-Burk plot is equal to $1/V_{max}$. For example, if the line intersects the axis at 0.02, then $V_{max} = 1/0.02$ or 50 μmoles of product formed per minute. The line intersects the 1/[S] axis at a point equivalent to $-1/K_M$. If the intersection point on the 1/[S] axis is -0.67, then $K_M = -1/-0.067 = 15$ μmolar. Repeat this procedure for the data obtained for D-dopa. Compare the K_M and V_{max} values and explain any differences.

Some enzymes are able to differentiate between D and L isomers of substrates. Dopa, with a chiral center, exists in enantiomeric pairs, D and L. The naturally occurring dopa molecule has the L configuration, so it might be expected that the enzyme-catalyzed oxidation of L-dopa is more facile than that of D-dopa. The relative reactivity can be determined in a quantitative way by comparing K_M or V_{max} values. Does tyrosinase exhibit stereoselectivity?

D. Inhibition of Tyrosinase Activity

A Lineweaver-Burk plot is prepared for each chemical inhibitor tested. Each plot of $1/v_0$ vs. 1/[S] consists of two lines, one representing the absence of inhibitor, the other representing a specific concentration of inhibitor.

Convert all ΔA/min data into μmole/min and calculate the reciprocal of each data point. Plot against the reciprocal of the corresponding substrate concentration (1/mmolar). The computer programs listed above may also be used for graphing inhibition. One line on each graph is obtained from the set of data acquired from the L-dopa K_M experiment in part C. Draw the two separate lines with a straight-edge, making sure the appropriate data points are used. Compare the two graphs with Figures E5.4 and E5.5 to determine the type of inhibition. Explain the inhibitory effect of the substance.

Study Problems

▶ *1.* One of the many straight-line modifications of the Michaelis-Menten equation is the Eadie-Hofstee equation:

$$v_0 = -K_M \frac{v_0}{[S]} + V_{max}$$

Beginning with the Lineweaver-Burk equation, derive the Eadie-Hofstee equation. Explain how it may be used to plot a straight line from experimental rate data. How are V_{max} and K_M calculated?

▶ 2. The table below gives initial rates of an enzyme-catalyzed reaction
       along with the corresponding substrate concentration. Use any graphi-
       cal method to determine K_M and V_{max}.

v_0 (μmole/min)	[S] (mole/liter)
130	65×10^{-5}
116	23×10^{-5}
87	7.9×10^{-5}
63	3.9×10^{-5}
30	1.3×10^{-5}
10	0.37×10^{-5}

▶ 3. The enzyme concentration used in each assay in Problem 2 is 5×10^{-7}
       mole/liter. Calculate k_3, the turnover number, in units of $\sec^{-1}$.

   4. Using the data in Problems 2 and 3, calculate the specific activity of the
      enzyme in units/mg and katal/mg. Assume the enzyme has a molecular
      weight of 55,000 and the reaction mixture for each assay is contained
      in a total volume of 1.00 mL.

▶ 5. Each of the compounds listed below is known to inhibit tyrosinase
       activity.

   sodium azide, NaN_3 sodium cyanide, $NaCN$

   L-phenylalanine   8-hydroxyquinoline

   tryptophan   diethyldithiocarbamate

   cysteine   thiourea

   4-chlororesorcinol   phenylacetate

   (a) Study the structure of each and try to predict whether the sub-
       stance is a competitive or noncompetitive inhibitor of tyrosinase
       activity as measured by the dopachrome assay. Assume that the in-
       hibition data were evaluated using a Lineweaver-Burk plot of $1/v_0$
       vs. 1/[dopa].

   (b) An alternative method for measuring tyrosinase activity is the use
       of an oxygen electrode to measure the rate of dioxygen utilization
       during phenol oxidation (see Equations E5.7 and E5.8). Predict
       whether each substance is a competitive or noncompetitive in-
       hibitor of tyrosinase as measured by an oxygen electrode. Assume
       that a Lineweaver-Burk plot of $1/v_0$ vs. $1/[O_2]$ was used to evaluate
       the rate data.

▶ 6. Compound X was tested as an inhibitor of the enzyme in Problem 2.
       Use the rate data in Problem 2 and the following inhibition data to
       evaluate compound X. Is it a competitive or noncompetitive inhibitor?
       Calculate K_I for the inhibitor.

$[X] = 3.7 \times 10^{-4}M$		$[X] = 1.58 \times 10^{-3}M$
v_0 (μmole/min)	[S] (mole/liter)	v_0 (μmole/min)
125	6.5×10^{-4}	110
102	2.3×10^{-4}	80
70	7.9×10^{-5}	40
45	3.9×10^{-5}	20
20	1.3×10^{-5}	—
5	0.37×10^{-5}	—

▶ 7. Predict what each of the following experimental changes will do to the initial rate of dopa oxidation catalyzed by tyrosinase. Answer each with "increase," "decrease," or "no change."

(a) Increase the temperature of the reaction mixture from 25°C to 37°C.

(b) Change the concentration of dopa from a value equal to 1 K_M to a value equal to $2K_M$.

(c) Add a few drops of a $1 \times 10^{-3}M$ cysteine solution.

(d) Increase the reaction temperature from 37°C to 100°C.

(e) Decrease the pH of the reaction mixture from pH = 7.0 to pH = 1.0.

(f) Change the concentration of tyrosinase from 1 μg/assay to 2 μg/assay.

▶ 8. Show the mathematical steps required to derive the Lineweaver-Burk equation beginning with the Michaelis-Menten equation.

▶ 9. Study Figure E5.7, which displays the kinetic progress of an enzyme-catalyzed reaction. What time limit must be imposed on rate measurements taken using the fixed-time assay? Why?

▶ 10. What is the concentration (in mg/mL and μg/mL) of tyrosinase in a solution that has an A_{280} of 0.25?

Further Reading

I. Behbahani, S. Miller, and D. O'Keeffe, *Microchem. J.* **47**, 251–260 (1993). "A Comparison of Mushroom Tyrosinase Dopaquinone and Dopachrome Assays."

R. Boyer, *Concepts in Biochemistry* (1999), Brooks/Cole (Pacific Grove, CA), pp. 142–173. An introduction to enzymes and kinetics.

J. Busch, *Biochem. Educ.* **27**, 171–173 (1999). "Enzymic Browning in Potatoes."

R. Dean, *The American Biology Teacher,* **61**(7), 523–527 (1999). "A Commentary on Experiments with Tyrosinase."

H. Duckworth and J. Coleman, *J. Biol. Chem.* **245**, 1613–1625 (1970). Kinetic properties of mushroom tyrosinase.

R. Garrett and C. Grisham, *Biochemistry,* 2nd ed. (1999), W. B. Saunders (Orlando, FL), pp. 426–459. Introduction to enzyme kinetics.

A. Lehninger, D. Nelson, and M. Cox, *Principles of Biochemistry,* 3rd ed. (1999), Worth Publishers (New York), pp. 243–292. Introduction to enzymes.

C. Matthews, K. van Holde, and K. Ahern, *Biochemistry,* 3rd ed. (2000), Benjamin/Cummings (San Francisco), pp. 360–413. Introduction to enzyme kinetics.

L. Stryer, *Biochemistry,* 4th ed. (1995), W. H. Freeman (New York), pp. 181–204. Introduction to enzymes and kinetics.

D. Voet and J. Voet, *Biochemistry,* 2nd ed. (1995), John Wiley & Sons (New York), pp. 332-410. An introduction to enzymes.

D. Voet, J. Voet, and C. Pratt, *Fundamentals of Biochemistry,* (1999), John Wiley & Sons (New York), pp. 281–347. Enzymes and kinetics.

C. Whiteley, *Biochem. Educ.* **25,** 144–146 (1997). Enzyme kinetics.

W. Wood et al., *Biochemistry, A Problems Approach,* 2nd ed. (1981), Benjamin/Cummings (San Francisco), pp. 144–172. Enzyme kinetics with an introduction to the direct linear plot.

C. Worthington, Editor, *Worthington Enzyme Manual* (1988), Worthington Biochemical Corporation (Freehold, NJ), pp. 288–291. Properties and analysis of tyrosinase. http://www.worthingtonbiochem.com/manual/P/TY.html

Enzyme Kinetics on the Web

http://2sg-www.mit.edu:8001/esgbio/
   A review of enzyme kinetics.

http://www.chem.qmw.ac.uk/iubmb/enzyme/
   The Enzyme List maintained by the International Union of Biochemistry and Molecular Biology.

6

PURIFICATION AND CHARACTERIZATION OF TRIACYLGLYCEROLS IN NATURAL OILS

- **Recommended Reading**

Chapter 3, Sections A, B, C, D.

- **Synopsis**

Lipids, relatively nonpolar chemical substances found in plant, bacterial, and animal cells, are among the most ubiquitous of biomolecules. In this experiment, a lipid extract of ground nutmeg will be purified by chromatography on a silica gel column. Analysis of the lipid extract by thin-layer chromatography will provide the classification of the components in the extract. The unknown lipids will be further characterized by saponification and analysis of the fatty acid content by gas chromatography. For an abbreviated experiment, students may be provided samples of natural oils and fats that can be analyzed by saponification and gas chromatography.

I. INTRODUCTION AND THEORY

Lipids are low-molecular-weight biomolecules that can be extracted from plant, microbial, and animal tissues by organic solvents. The major classes of compounds represented in the lipid group are triacylglycerols, glycerophosphatides, sphingolipids, glycolipids, and sterols (Figure E6.1). Lipids are found in most cells and tissue, but rarely are they present in a "free" or "uncombined" state. They are usually associated with proteins or carbohydrates, which complicates their extraction and structural identification. Lipids not only have a wide variety of molecular structures but also are involved in a whole array of biological functions, including membrane structural integrity, vitamin activity, cell recognition, and cellular energy.

In this experiment we will approach the world of lipids with an introduction to analytical procedures currently applied to lipid extraction, purification, characterization, and structure elucidation.

Lipid Extraction

The extraction procedure used to isolate lipids from biological tissue depends on the class of lipid desired and the nature of the biological source (animal tissue, plant leaf, plant seed, bacteria, cell membranes, etc.). Because lipids are generally less polar than other cell constituents, they may be selectively extracted with the use of organic solvents. Early studies of lipids used ether, acetone, hexane, and other organic solvents for extraction; however, these solvents extract only lipids bound in a nonpolar or hydrophobic manner. In the 1950s, Folch's group reported the use of chloroform and methanol (2:1) in

Figure E6.1

Structures of the major lipid classes.

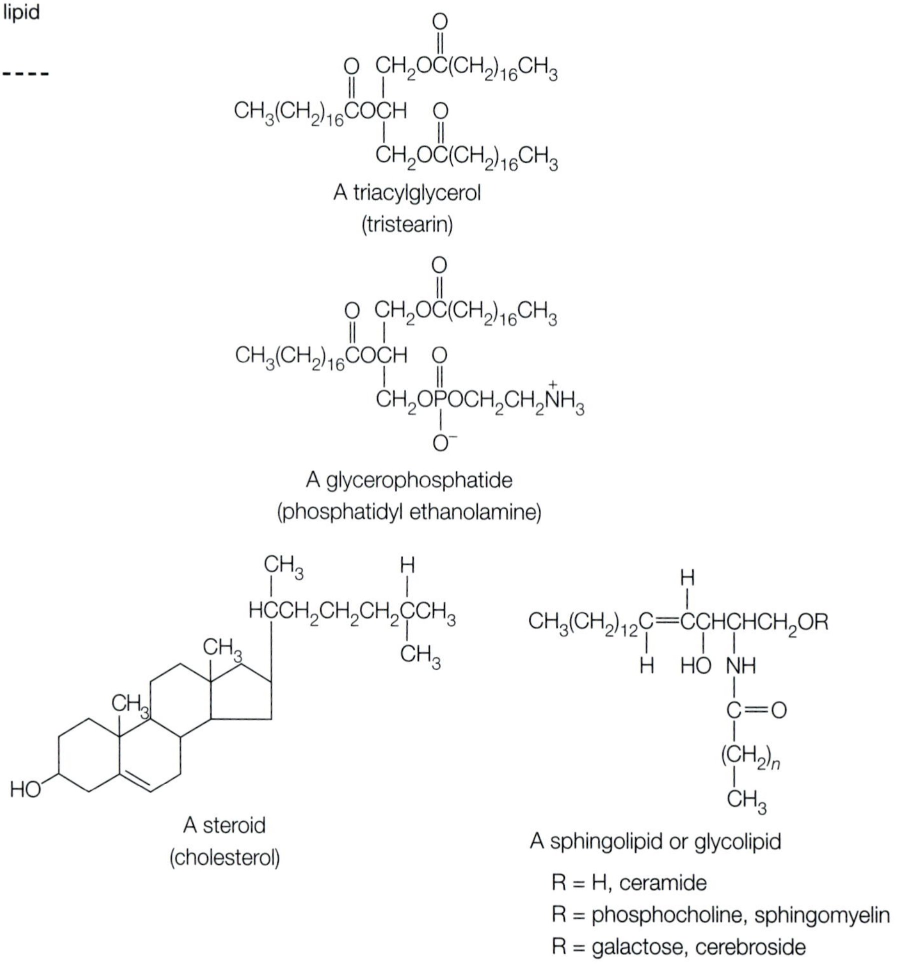

A triacylglycerol
(tristearin)

A glycerophosphatide
(phosphatidyl ethanolamine)

A steroid
(cholesterol)

A sphingolipid or glycolipid

R = H, ceramide
R = phosphocholine, sphingomyelin
R = galactose, cerebroside

lipid extractions (Folch, 1957). This became the solvent system of choice for total lipid extraction because it was especially useful in dissociating lipid-protein complexes of plasma membranes. The only apparent disadvantage of this solvent system was its tendency to dissolve some nonlipids, including proteins. Increasing awareness of safety in the laboratory (both methanol and chloroform are toxic) led to a search for other solvent systems. A more desirable lipid solvent is hexane and isopropanol (3:2) (Radin, 1981). In addition to lower toxicity, it has the advantage over chloroform-methanol that it dissolves very little nonlipid material. Whatever solvent system is used, chloroform-methanol or hexane-isopropanol, a total lipid extraction process yields a complex mixture of organic compounds in which all components have one common characteristic, a relatively nonpolar structure.

Lipid Purification

Separation of crude lipid extracts into individual lipid classes is difficult and time-consuming. In some cases a crude separation of lipids can be attained by selective solvent extraction. For more extensive purification of lipids, the researcher must turn to chromatography. Chromatographic methods can both resolve a complex lipid mixture into the various classes of lipids and separate the individual components of a single class of lipids.

Adsorption chromatography (Chapter 3) on columns of silica gel is the best preparative method for the separation and purification of neutral lipids. Silica gel (sometimes called silicic acid) is chemically represented as $SiO_2 \cdot xH_2O$ and is a relatively polar adsorbent. Neutral or nonpolar lipids are not readily adsorbed by the support and can be separated from the more polar lipids, which are preferentially adsorbed. In practice, a glass column is packed with a slurry of silica gel in a hydrocarbon solvent (petroleum ether, pentane, or hexane). The lipid fraction to be purified is dissolved in the same hydrocarbon solvent and applied to the top of the column. To elute the various lipid classes, solvents of increasing polarity are allowed to percolate through the silica gel column and are collected at the bottom of the column. Table E6.1 outlines the order of elution of lipids from silica gel columns. The polarity of the eluting solvent is increased by adding a gradually increasing proportion of ethyl ether. For example, 1% ether in petroleum ether (fraction I) elutes only the least polar lipids—paraffins and fatty acid esters; 4% ethyl ether in petroleum ether has sufficient polarity to elute triacylglycerols, and 25 to 75% ethyl ether in petroleum ether elutes diacylglycerols and monoacylglycerols. For elution of phospholipids and glycolipids, solvent systems of increasing polarity such as chloroform-methanol must be used.

Once several silica gel column fractions have been collected, the eluted fractions must be analyzed for the presence of lipids. Thin-layer chromatography on silica gel plates will be introduced in this experiment for the detection and identification of neutral lipids.

As discussed in Chapter 3, thin-layer chromatography utilizes a thin coating of silica gel on a glass or plastic plate to separate components of a mixture. The mixture of chemical substances to be analyzed is applied near one edge

Table E6.1

Order of Elution of Lipids from Silica Gel Columns

Fraction Number	Solvent	Lipid Component
Neutral		
I	1% Ethyl ether/petroleum ether	Paraffins
II	1% Ethyl ether/petroleum ether	Squalene, waxes, fatty acid esters
III	1% Ethyl ether/petroleum ether	Cholesterol esters
IV	4% Ethyl ether/petroleum ether	Triacylglycerols, fatty acids
V	8% Ethyl ether/petroleum ether	Steroids (cholesterol)
VI	25% Ethyl ether/petroleum ether	Diacylglycerols
VII	100% Ethyl ether	Monoacylglycerols
Polar		
VIII	5% Methanol/chloroform	Ceramides, phosphatidic acid, cardiolipins
IX	20% Methanol/chloroform	Phosphatidylethanolamine, phosphatidylserine phosphatidylinositol
X	30% Methanol/chloroform	Phosphatidylcholine
XI	50% Methanol/chloroform	Sphingomyelin
XII	70% Methanol/chloroform	Lysophosphatidylcholine

of the plate, and development of the plate with a solvent system allows the applied samples to move a distance that is related to the polarity of the sample. A solvent system consisting of hexane, diethyl ether, and acetic acid will separate triacylglycerols, cholesterol esters, fatty acid esters, and fatty acids (see Figure E6.2). More polar lipids, such as glycerophosphatides, sphingolipids, and glycolipids, remain at the origin. Pure standard samples of lipids are evaluated on the same TLC plate in order to help identify unknowns. The lipids on the solvent-developed plate are not colored but will react with iodine vapor to produce dark red-brown spots on a yellow background.

Triacylglycerol Structure

Triacylglycerols, the neutral, saponifiable lipids found in most organisms, serve as biochemical energy reserves in the cell. Neutral lipids may be isolated from natural sources by extraction with nonpolar solvents (ethyl ether, chloroform, hexane-isopropanol) as in the isolation of the unknown lipid from nutmeg. Chemically, the triacylglycerols are fatty acid esters of the trihydroxy alcohol glycerol (Figure E6.3). The figure also illustrates the recommended stereospecific numbering (sn) scheme for glycerol derivatives. According to IUPAC, "The carbon atom that appears on top in that Fischer projection that shows a vertical carbon chain with the secondary hydroxyl group to the left is designated as C-1." To differentiate such numbering from conventional numbering conveying no steric information, the prefix "sn" (stereospecific numbering) is used; therefore, the structure in Figure E6.3 is designated 1,2,3-triacyl-*sn*-glycerol.

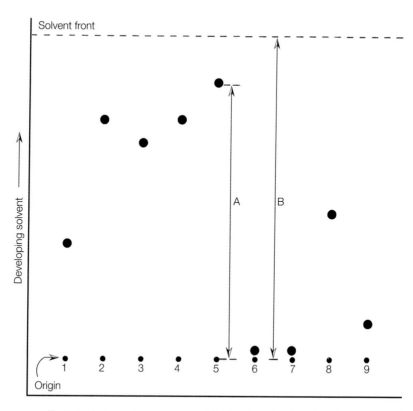

Figure E6.2

Typical thin-layer chromatogram of lipids. The solvent system, hexane–diethyl ether–acetic acid (90:10:1). (1) cholesterol, (2) fatty acid, (3) triacylglycerol, (4) fatty acid methyl ester, (5) cholesterol ester, (6) glycerophosphatide, (7) sphingolipid, (8) 1,3- or 1,2-diacylglycerol, (9) 1- or 2-monoacylglycerol. The R_f for (5), cholesterol ester, is calculated as follows:

$$R_f = \frac{\text{distance of migration of cholesterol ester}}{\text{distance of migration of solvent}} = \frac{A}{B} = \frac{72\text{ mm}}{85\text{ mm}} = 0.85$$

Figure E6.3

General structure for a triacylglycerol. R represents the alkyl chains of fatty acids.

A variety of saturated and unsaturated fatty acids is present in triacylglycerols. Among the most abundant fatty acids are myristic (14:0), palmitic (16:0), stearic (18:0), and oleic (18:1$^{\Delta 9}$). To completely characterize the neutral lipids, one must know the identity of the fatty acids *and* the position that each fatty acid occupies on the glycerol skeleton (carbon number 1, 2, or 3). The biosynthetic distribution of fatty acids in positions 1, 2, or 3 of glycerol was once considered to be random; however, more recent studies indicate some selectivity. For example, the proportion of unsaturated to saturated fatty acids in position 2 is quite often greater than at positions 1 and 3. Analytical techniques are available to determine the identity and position of each fatty acid in a triacylglycerol sample. Since glycerol is a **prochiral** molecule, it should be recognized that the 1 and 3 positions of glycerol are *not* equivalent.

Characterization of Triacylglycerols

The development of procedures in gas chromatography now makes it possible to do structural analysis of triacylglycerols. Fatty acid content (identity and percentage composition) of triacylglycerols is most easily determined by complete saponification (NaOH/methanol) followed by esterification of the released fatty acids (Equation E6.1). Gas chromatographic analysis of the fatty acid methyl esters (FAMEs), in conjunction with analysis of the appropriate standard FAMEs, conveniently provides both qualitative and quantitative fatty acid analysis. Polar or nonpolar columns may be used to analyze FAMEs. Polar columns, which have liquid phases containing diethyleneglycol succinate (DEGS) or ethyleneglycol adipate (EGA), separate FAMEs on the basis of carbon chain length and degree of unsaturation. The most popular nonpolar columns have either saturated hydrocarbon (Apiezon) or silicone (OV-101, SE-30) liquid phases. Capillary (open tabular) columns have the advantages that they require smaller sample injection volumes and offer excellent resolution of FAMEs.

Equation E6.1

Overview of the Experiment

In this experiment you will isolate an unknown lipid mixture from its natural source by solvent extraction. The source of the unknown lipid is *Myristica fragrans* seed (nutmeg). This particular seed is unusual in that it contains an appreciable quantity of a single type of lipid. In fact, a single compound makes up 70 to 80% of the total lipid extract. The lipid is of sufficient homogeneity to allow its isolation as a solid with a relatively sharp melting point. The crude isolated material will be purified by silica gel column chromatography or recrystallization and classified by thin-layer chromatography with standard lipids. For a flowchart of the procedure, see Figure E6.4.

In the second part of this experiment you will characterize the purified lipids (triacylglycerols) isolated from nutmeg. The fatty acids in the triacylglycerols are released by saponification and their identities determined by gas chromatography. Alternatively, students may be provided various fat and oil samples for analysis. For example, the fatty acid content of triacylglycerols in lard, butter, vegetable oils, and other natural fats and oils may be used in place of the nutmeg lipid. The times required for each part of the experiment are as follows:

A. Isolation and purification of the unknown lipid: $1\frac{1}{2}$–2 hours

B. Characterization by thin-layer chromatography: $1\frac{1}{2}$ hours

C, D. Saponification and gas chromatography: 2–3 hours

If only one period (3 hours) is available for this experiment, parts C and D may be completed. Students should be provided up to three oil samples for analysis. The oils are saponified as directed and the fatty acid percent composition determined by gas chromatography.

II.   MATERIALS AND SUPPLIES

A. Isolation and Purification of the Unknown Lipids

− Nutmeg, ground or whole

− Hexane-isopropanol, 3:2

− Acetone

Figure E6.4

Flowchart for isolation of lipid from nutmeg.

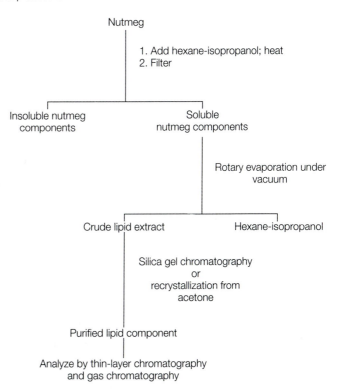

Flowchart of procedure

Nutmeg

1. Add hexane-isopropanol; heat
2. Filter

Insoluble nutmeg components   Soluble nutmeg components

Rotary evaporation under vacuum

Crude lipid extract   Hexane-isopropanol

Silica gel chromatography
or
recrystallization from acetone

Purified lipid component

Analyze by thin-layer chromatography and gas chromatography

- Diethyl ether
- Hexane
- Silica gel, 100–200 mesh
- Glass column (1 × 25 cm)
- Glass wool or cotton
- Small funnel and qualitative grade filter paper
- Silica gel thin-layer plates (20 × 20 cm and 5 × 10 cm)
- Chromatography solvent system, hexane–diethyl ether–acetic acid (80:20:1) in a chromatography jar
- Lipid standards (2% solutions in chloroform): a triacylglycerol (triolein), cholesterol ester (cholesterol linoleate), fatty acid (palmitoleic, oleic, etc.), fatty acid methyl ester (linolenic acid, methyl ester), a glycerophosphatide (phosphatidylcholine, phosphatidylethanolamine, etc.), a diacylglycerol (diolein), and a monoacylglycerol (monoolein).
- Microcapillaries or tapered capillaries
- Iodine chamber. Place a few crystals of iodine in an empty chromatography jar and cover.
- Rotary evaporator

B, C, D. Characterization of the Unknown Lipids

- Several fat and oil samples (lard, butter, vegetable oil, etc.)
- Unknown lipid from nutmeg
- Methanolic sodium hydroxide, 0.5 M
- BF_3 solution (14% in methanol)
- Hexane
- Saturated NaCl
- Anhydrous $MgSO_4$
- Fatty acid methyl ester standards (12:0, 14:0, 16:0, 18:0, 16:1$^{\Delta 9}$, 18:1$^{\Delta 9}$, 18:2$^{\Delta 9,12}$, 18:3$^{\Delta 9,12,15}$, solutions in hexane; 2 mg/mL). Standard samples containing mixtures of FAMEs are commercially available. Rapeseed oil is an appropriate standard for this experiment.
- Ethyl ether, peroxide free
- Methanol containing 1% acetic acid
- Silica gel thin-layer plates (2.5 × 10 cm)
- Chromatography solvent system, hexane–ethyl ether–acetic acid (80:20:1) in chromatography jar
- Centrifuge tubes, 10-mL conical

- Iodine chamber. Place a few crystals of iodine in an empty chromatography jar and cover.
- Capillary tubes
- Constant-temperature water bath at 37°C
- Steam bath
- Gas chromatograph

Conditions:

gas flow: 50–70 mL/min

$T_{injector}$: 200°C

T_{column}: 185°C

$T_{detector}$: 200°C

Attenuation: 2 or 4

Injection sample size: 5–10 μL

Column: 6 ft, 10% diethylene glycol succinate (DEGS) on Chrom W, 60/80 mesh

> **CAUTION**
>
> All procedures in parts A, B, and C requiring the heating or evaporating of hexane, ethyl ether, and methanol should be carried out in a well-ventilated hood. There should be no flames of any kind in the laboratory.

III.   EXPERIMENTAL PROCEDURE

A. Isolation and Purification of the Unknown Lipids

> **CAUTION**
>
> Acetone, ether, isopropanol, and hexane cause eye and skin irritation. Vapor or mist of these solvents is irritating to the eyes, mucous membranes, and upper respiratory tract. Prolonged exposure should be avoided. Harmful if swallowed.
>
> The solvents, also, are volatile and extremely flammable. Always measure quantities in a hood. No lighted burners or flames should be in the vicinity of your work. In case of fire, use a CO_2 extinguisher.
>
> First aid: eyes–flush with copious amounts of water for at least 15 minutes; skin–wash with soap and copious amounts of water; inhalation–remove to fresh air.
>
> Dispose of organic solvents in the organic waste container.

Place 5 g of finely crushed or ground nutmeg into a 125-mL Erlenmeyer flask. Add 50 mL of hexane-isopropanol (3:2) and warm on a steam bath or hot plate in a hood for 15 minutes. Mix well while heating. Filter the extraction mixture rapidly through fluted qualitative grade filter paper and pour an additional 20 mL of warm hexane-isopropanol through the solid residue in the filter. Remove the solvent from the extract under vacuum on a rotary evaporator to obtain a yellow oil or off-white solid. Alternatively, the solvent may be removed on a steam bath **(Hood!)** while flushing the inside of the flask with N_2 gas. Weigh the crude product and retain a 1- to 2-mg portion for chromatographic analysis.

Partial and rapid purification of the lipid is possible by recrystallization from acetone. If more time is available, and a purer sample of the lipid is desired, a silica gel column should be run.

Recrystallization

To recrystallize the unknown lipid, dissolve the crude residue in a minimum amount (about 20 mL) of warm acetone on a steam bath **(Hood!).** After slow cooling of the acetone solution to room temperature and then in ice to allow crystals to form, filter by suction, and wash the solid product with ice-cold acetone. A typical yield is about 0.5 g of purified lipid. Save the purified lipid for characterization.

Silica Gel Chromatography

To prepare the silica gel column, clamp the glass column to a ring stand. Prepare a slurry of 6 g of silica gel in 20 mL of hexane. Put a small piece of glass wool or cotton into the bottom of the column and pour about 5 mL of hexane into the column. Be sure the stopcock is closed! With a long glass rod, tap the glass wool or cotton to drive out air bubbles and push it tightly into the column. Place a small funnel on top of the column and begin to pour the well-stirred slurry of silica gel into the column. As the silica gel begins to pack into a column, open the stopcock to allow release of solvent. Continue to add the slurry of silica gel until a tightly packed column of 10 to 12 cm is attained. *Do not allow the column to run dry during packing.* Allow 1 to 2 cm of solvent to remain above the level of the silica gel.

Prepare the lipid for chromatography by dissolving 50 to 75 mg of crude lipid in a minimum of hexane (5 to 10 mL). Open the stopcock and allow excess solvent to drain from the column until the level of solvent just reaches the top of the silica gel column. Very carefully add the solution of crude lipid to the top of the column. This should be done by using a Pasteur pipet and allowing drops of the solution to run down the inside of the glass column to the top of the silica gel bed. It is important not to disturb the top of the silica gel column. After addition of the lipid solution, begin to collect a fraction from the bottom of the column into a 25-mL Erlenmeyer flask. Set the flow rate to about 2 drops per second. When the level of solution in the column reaches the top of the silica gel, turn off the stopcock and carefully rinse the inside of the glass column with 1 mL of hexane. Open the stopcock until the solvent just enters the column. This solvent washes the lipid material that may have adhered to the inside glass walls. Now fill the column with hexane and continue to collect the eluting solvent at 2 drops per second. Collect a total of 20 mL of solvent in the first fraction. During this collection, continue to add solvent to the top of the column. Now change the eluting solvent to 90% hexane and 10% diethyl ether. Pour this into the column and collect four 20 mL fractions. Be sure to number the collection flasks. Add solvent to the column as necessary. Finally, elute the column with 80% hexane and 20% diethyl ether and collect two more 20-mL fractions.

Each column fraction is analyzed for lipid material by spotting on a 5 × 10 cm silica gel thin-layer plate and exposing it to iodine vapor as follows. Prepare seven tapered capillary tubes and use these to place a spot of each solution on the TLC plate. Put at least 10 capillary applications from a single fraction on a spot. Your final plate should then have seven spots, one for each fraction. Set the plate in an iodine chamber and allow it to remain for about 15 minutes or until some spots are yellow or red-brown. The presence of lipid in a fraction is indicated by the red-brown color. Retain all the column fractions that appear to have lipid. Each of these fractions will be analyzed in part B by thin-layer chromatography.

B. Characterization of the Unknown Lipid

> **⟩⟩ CAUTION**
>
> Chloroform is harmful if inhaled, swallowed, or absorbed through the skin. Vapor or mist is irritating to the eyes, mucous membranes, and upper respiratory tract. Chloroform is suspected to be a cancer-causing agent.
>
> First aid: eyes—prolonged irrigation with copious amounts of water; skin—wash extensively with soap and water; inhalation—remove to fresh air.
>
> Dispose in the organic waste container.

Thin-Layer Chromatography

Analysis of the purity of each silica gel column fraction and classification of the unknown lipid can be accomplished by thin-layer chromatography on silica gel plates. On a single plate will be spotted (1) a solution of the crude lipid extract from part A, (2) aliquots from each lipid fraction of the column (or recrystallized lipid), and (3) solutions of standard lipids (listed in the Materials section). On a 20 × 20 cm silica gel plate there is room for nine different analyses. Prepare a 1% solution of the crude lipid from part A in chloroform (10 mg/1 mL). If recrystallized lipid is to be analyzed rather than column-purified lipid, prepare a 1% solution in chloroform as for the crude lipid. Apply each sample in a row across one side of the silica gel plate, 2 cm from the edge (see Figure E6.2). Use small, tapered capillary tubes or microcapillaries. To apply each solution, dip the capillary into the solution, and gently touch the end of the filled capillary on the TLC plate where you desire the spot. Allow the solution from the capillary to spread onto the TLC plate to form a circle no wider than 0.5 cm diameter. Lift the capillary from the plate and allow the solvent on the plate to dry. Again touch the end of the capillary to the same spot and reapply solution. Repeat this application process about 20 times for each sample. Of course, each sample to be analyzed is placed on a different location along the edge of the TLC plate.

Develop the plate in hexane–diethyl ether–acetic acid (80:20:1) by placing the TLC plate in a chamber containing the solvent system, making sure the edge with the applied samples is down. The solvent level in the chamber must not be above the application spots on the plate. (Why?) Leave the chromatogram in the chromatography jar until the solvent front rises to about 1 cm from the top of the plate (45 to 60 min). Remove the plate and make a small scratch at the solvent level. Allow the chromatogram to dry **(Hood!)** and then place it in an iodine chamber for several minutes. Remove the plate and lightly trace, with a pencil or other sharp object, around each red-brown spot. This should be done promptly, as the colors will fade with time. Calculate the mobility of each standard and unknown lipid relative to the solvent front (R_f):

$$R_f = \frac{\text{distance from the origin migrated by a compound}}{\text{distance from origin migrated by solvent}}$$

C. Saponification of Triacylglycerol for Fatty Acid Composition

Select a triacylglycerol sample for analysis. This may be the unknown lipid that was isolated and purified nutmeg or a fat or oil supplied by your instructor. Weigh 25 to 50 mg of the sample into a small conical test tube. Add 3 mL of 0.5 M methanolic sodium hydroxide. Heat the mixture over a steam bath **(Hood!)** until a homogeneous solution is obtained. Add 5 mL of BF_3-methanol to the saponification reaction mixture and boil 2 to 3 minutes. Cool and transfer the solution into a separatory funnel containing 25 mL of hexane and 20 mL of saturated NaCl solution. Shake well, but gently, and allow the layers to separate. Vigorous shaking will give rise to an emulsion. The hexane layer, containing the fatty acid methyl esters, is dried with about 1 g of anhydrous $MgSO_4$ and filtered or decanted into a small vial. Concentrate the solution on the steam bath to a volume of about 0.5 mL **(Hood!).** This solution of fatty acid methyl esters should be capped to prevent air oxidation and stored in a refrigerator until it is to be analyzed by gas chromatography.

D. Gas Chromatography of Fatty Acid Methyl Esters

Study the section on gas chromatography in Chapter 3. If you are not familiar with the use of the instrument, have your instructor assist you. Be sure to record all instrumental conditions in your notebook (temperatures of the column, oven, and detector; rate of gas flow (mL/min); attenuation setting; sample size; and recorder chart speed). It is recommended that you begin with the FAME standards, using 5- to 10-μL injection samples (1 μL for a capillary column) with an attenuator setting of 2. Two peaks should appear, one for the hexane solvent and a second representing the FAME. The hexane peak will be off scale; if the FAME peak is also off the recorder paper, reduce the injection sample or increase the attenuation setting to 8 or 16. If the recorder peaks for the FAMEs are too small, decrease attenuation setting (1 or 2). When you are analyzing the standard FAMEs, you may inject a second sample as soon as the FAME from the first sample has been eluted from the column. This occurs when the recorder pen returns to the original baseline.

Use the same instrumental conditions to analyze the samples from part A. The samples will contain a mixture of fatty acid methyl esters, so several recorder peaks will be obtained. Do not inject a second sample until you are sure all the fatty acids have been eluted from the first sample. Determine the retention time for each standard FAME and for each FAME in the unknown samples. Retention time is the time interval between injection of sample and maximum response of the recorder.

IV.   ANALYSIS OF RESULTS

If you completed only parts A, C, and D, it is possible to identify all the fatty acids present and to calculate the composition in weight percentage.

Prepare a table listing the retention time for each standard FAME. Use this table to identify the fatty acids present in each triacylglycerol you analyzed. Alternatively, plot the log of the retention time against the chain length of each saturated FAME (see Figure E6.5). Unknown saturated fatty acids can be identified from experimental retention times using this plot. Unsaturated fatty acids cannot be identified from the plot of saturated FAMEs. A separate plot of log retention time vs. the chain length must be prepared for each level of saturation (saturated, monounsaturation, diunsaturation, etc.).

The amount of each FAME present is proportional to the area under its peak on the recorder chart. If the peaks are symmetrical, the peak area may be calculated by triangulation:

Peak area = peak height $\times$ peak width at half height

or

Peak area = $HW_{1/2}$

A plot of log retention time vs. carbon number for gas chromatography of a series of saturated fatty acid methyl esters.

- -

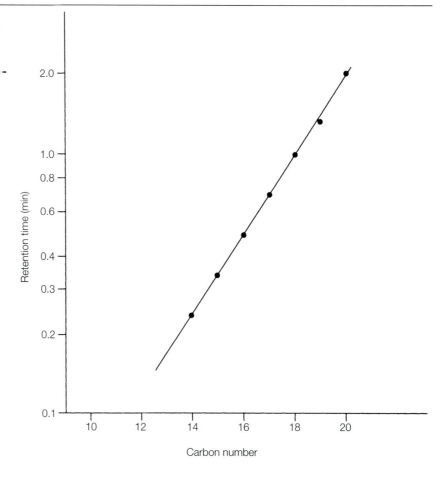

Thermal conductivity detectors used in gas chromatographs do not respond equally to all FAMEs. To correct for varying detector sensitivity, peak area for each FAME should be multiplied by the proper **response correction factor (RCF)** (Table E6.2). If extra time is available, you may want to calculate your own response correction factors for the fatty acid methyl esters. The factors are experimentally determined on a gas chromatograph by comparing the area under a GC peak due to a known amount of compound to the area under a GC peak represented by a reference compound.

$$RCF_1 = \frac{\text{amount of compound 1}}{\text{peak area of 1}} \times \frac{\text{peak area of reference}}{\text{amount of reference}}$$

The response correction factors for compound 1 and other compounds are calculated relative to the response correction factor for the reference, which is assumed to be 1.00.

Table E6.2

Fatty Acid Methyl Ester Response Factors Measured with a Thermal Conductivity Detector, DEGS Column at 190°C

FAME	Response Correction Factor (RCF)
Myristic (14:0)	0.908
Palmitic (16:0)	0.954
Stearic (18:0)	1.010
Linoleic (18:2$^{\Delta 9,12}$)	1.071
Linolenic (18:3$^{\Delta 9,12,15}$)	1.172

The mole percentage of each FAME is calculated according to the following sequence of equations.

1. Correction of peak area of $FAME_1$ for varying detector sensitivity.

$$\text{Peak area}_1 \times RCF_1 = \text{corrected peak area}_1$$

2. Conversion of corrected peak area$_1$ (now in weight %) to moles.

$$\frac{\text{Corrected peak area}_1}{\text{Molecular weight}_1} = \text{mole-corrected peak area}_1$$

3. Conversion of mole-corrected peak area$_1$ to mole % of $FAME_1$.

>>

$$\text{Mole \%}_1 = \frac{\text{mole-corrected peak area}_1}{\text{total of mole-corrected peak area for all FAMES}} \times 100 \quad \textbf{\textit{Equation E6.2}}$$

The denominator for Equation E6.2 is calculated by performing steps 1 and 2 for all of the FAMEs present in a single triacylglycerol sample. The mole-corrected peak areas are added together to obtain the total of all mole-corrected areas. If we assume that the extent of conversion of free fatty acids to FAMEs is essentially quantitative, or at least equal for all fatty acids in our experiment, the above calculation leads directly to the combined mole percent composition of fatty acids in all positions of the triacylglycerol.

For the lipid sample that you analyzed, report the fatty acid composition in mole %. Compare your results with those of other students.

Calculate the approximate percentage lipid composition of the nutmeg (assume the lipid recovery is quantitative). Prepare a table listing the R_f values obtained for each standard lipid and for the crude and purified unknown. By comparing R_f values, identify the general class of lipid to which your purified extract belongs. How effective was your purification step? If impurities are present in the crude or purified lipid, can you identify them? If you recrystallized the crude lipid, calculate the recovery yield.

$$\text{Recovery yield (\%)} = \frac{\text{g purified lipid}}{\text{g crude lipid}} \times 100$$

Study Problems

1. If you wished to determine the complete structure of the isolated triacylglycerol, what experimental approach would you follow?

2. Why is acetic acid added as a minor component (1%) of the chromatography solvent system?

3. Try to explain the relative order of R_f values you obtained for the standard lipids.

4. How does iodine react to produce the red-brown spots on a developed chromatogram? Why do some lipids give darker spots than others?

5. Calculate the R_f for standard lipid number 2 (a fatty acid) in Figure E6.2.

6. Why should whole nutmeg be ground before the solvent extraction?

7. Why are FAMEs rather than free fatty acids used for gas chromatographic analysis?

8. What is the identity of the lipid that was extracted from nutmeg seed?

9. What is the function of BF_3 in the formation of FAMEs?

10. Using data from your experiment, write the order of gas chromatographic elution (first to last) for each set of fatty acid methyl esters below. Explain your answers.

    (a) 14:0, 20:0, 12:0, 16:0, 18:0

    (b) $16:1^{\Delta 9}$, 16:0, $16:2^{\Delta 9, 12}$

    (c) 16:0, 18:0, $16:1^{\Delta 9}$, $18:1^{\Delta 9}$

Further Reading

R. Ackman, in *Methods in Enzymology,* Vol. XIV, J. Lowenstein, Editor (1969), Academic Press (New York), pp. 329–381. Gas-liquid chromatography of fatty acid esters.

R. Boyer, *Concepts in Biochemistry* (1999), Brooks/Cole (Pacific Grove, CA), pp. 238–256. Introduction to lipid structure and function.

J. Folch et al., *J. Biol. Chem.* **226,** 497–509 (1957). Use of chloroform-methanol as a lipid solvent.

R. Garrett and C. Grisham, *Biochemistry,* 2nd ed. (1999), Saunders (Orlando, FL), pp. 238–258. Lipid structure and function.

M. Gurr and J. Harwood, *Lipid Biochemistry,* 4th ed. (1991), Chapman & Hall (New York). An introduction to lipid structure and function.

D. Holme and H. Peck, *Analytical Biochemistry,* 3rd ed. (1998), Addison Wesley Longman (New York), pp. 406–442. Structure, function, and analysis of lipids.

C. Matthews, K. van Holde, and K. Ahern, *Biochemistry,* 3rd ed. (2000), Benjamin/Cummings (San Francisco), pp. 315–357. Lipid structure and function.

G. Patton, S. Cann, H. Brunengraber, and J. Lowenstein, in *Methods in Enzymology,* Vol. 72, J. Lowenstein, Editor (1981), Academic Press (New York), pp. 8–20. Separation of fatty acid methyl esters by gas chromatography on capillary columns.

N. Radin, in *Methods in Enzymology,* Vol. 72, J. M. Lowenstein, Editor (1981), Academic Press (New York), pp. 5–7. Extraction of lipids with hexane-isopropanol.

L. Stryer, *Biochemistry,* 4th ed. (1995), W. H. Freeman (New York), pp. 263–270; 603–606. Lipid structure and function.

D. Voet and J. Voet, *Biochemistry,* 2nd ed. (1995), John Wiley & Sons (New York), pp. 277–284. Lipid structure and function.

D. Voet, J. Voet, and C. Pratt, *Fundamentals of Biochemistry* (1999), John Wiley & Sons (New York), pp. 219–233. Lipid structure and function.

Lipids on the Web

http://birdsong.cudenver.edu/~davidp/chromato.htm
Principles of gas chromatography.

http://dir.yahoo.com/Science/Chemistry/Chromatography/
An introduction to chromatography.

http://www.nidlink.com/~jfromm/chem301/chem302p.htm
Review of lipid structure and function.

http://www.compusmart.ab.ca/plambeck/che/p265/p06211.htm
Lipid structure and function.

7

IDENTIFICATION OF SERUM GLYCOPROTEINS BY SDS-PAGE AND WESTERN BLOTTING

- **Recommended Reading**

Chapter 4, Sections A, B, C; *Experiment 4.*

- **Synopsis**

Western blotting has become an important, modern technique for analysis and characterization of proteins. The procedure consists of, first, the electrophoretic transfer (blotting) of proteins from polyacrylamide gels to synthetic membranes. The transferred blots are then probed using immunological detection methods to identify proteins of specific structure and/or function. In this experiment, bovine serum will be fractionated by SDS-PAGE and the proteins blotted onto a nitrocellulose membrane. Serum glycoproteins will be identified by their specific interaction with the lectin concanavalin A.

I. INTRODUCTION AND THEORY

In Experiment 4, your sample of α-lactalbumin extracted from bovine milk was subjected along with other proteins to sodium dodecyl sulfate–polyacrylamide gel electrophoresis (SDS-PAGE). After staining with the dye Coomassie Blue, deeply colored bands appeared on the gel wherever there was a protein. You suspected that some of the blue bands on the gel were due to α-lactalbumin. If molecular weight standards were included on the slab gel, you were able to estimate the molecular weight for α-lactalbumin and other proteins. SDS-PAGE is indeed a very effective analytical tool to achieve fractionation of protein mixtures, to analyze purity, and to estimate molecular weight, but it provides no experimental data to prove the identity of any of the dyed protein bands. A Coomassie Blue stain simply indicates the presence and location of each and every protein on the gel. It is often possible to identify proteins by treating gel bands directly with chemical reagents that react with a specific protein. For example, the identity and location of an enzyme may be noted by treating the gel with a substrate that is converted to a colored product by the enzyme. However, proteins are deeply embedded in the polyacrylamide gel matrix and are not readily accessible to most analytical reagents. This hinders specific analysis of the protein bands in order to identify individual proteins. Proteins separated by PAGE may be transferred (or blotted) from the gel to a thin support matrix, usually a nitrocellulose membrane, which strongly binds and immobilizes proteins. The protein blots on the membrane surface are more accessible to chemical or biochemical reagents for further analysis. When the transfer process is coupled with protein identification using highly specific and sensitive immunological detection techniques, the procedure is called **Western blotting** (see page 136). Western blotting or immunoblotting assays of proteins have many advantages including the need for only small reagent volumes, short processing times, relatively inexpensive equipment, and ease of performance.

The Western Blot

To begin the Western blot procedure, a protein mixture for analysis and further characterization is fractionated by PAGE. Since denaturing, SDS-PAGE results in better resolution than PAGE performed under native conditions, SDS-PAGE is usually preferred; however, the detection method used at the conclusion of the blotting experiment must be able to recognize denatured protein subunits. The next step involves selection of the membrane matrix for transfer. Three types of support matrices are available for use: nitrocellulose, nylon, and polyvinyl-difluoride (PVDF). Nitrocellulose membranes, currently the most widely used supports, have a satisfactory protein binding capacity ($100 \ \mu g/cm^2$), but they display weak binding of proteins of molecular weights smaller than 14,000 and they are subject to tearing. Binding of proteins to nitrocellulose membranes is noncovalent, most likely hydrophobic. Nylon membranes are stronger than nitrocellulose and some have a binding capacity up to $450 \ \mu g/cm^2$. However, since they are cationic, they only weakly bind basic proteins. During detection procedures, nylon membranes often display high background colors, so it is difficult to visualize proteins of interest. PVDF membranes bind proteins strongly ($125 \ \mu g/cm^2$) and, because of their hydrophobic nature, give light background color after analysis. For overall general use in protein transfer and immunoblotting, nitrocellulose membranes are the most common choice.

The actual blotting process may be accomplished by one of two methods: **passive (or capillary) transfer** and **electroblotting.** In passive transfer, the membrane is placed in direct contact with the polyacrylamide gel and organized in a sandwich-like arrangement consisting of (from bottom to top) filter paper soaked with transfer buffer, gel, membrane, and more filter paper. The sandwich is compressed by a heavy weight. Buffer passes by capillary action from the bottom filter paper through the gel, transferring the protein molecules to the membrane, where the macromolecules are immobilized. Passive transfer is very time consuming, sometimes requiring 1–2 days for complete protein transfer. Faster and more efficient transfer is afforded by the use of an electroblotter. Here a sandwich of filter paper, gel, membrane, and more filter paper is prepared in a cassette, which is placed between platinum electrodes. An electric current is passed through the gel, causing the proteins to electrophorese out of the gel and onto the membrane.

Detection of Blotted Proteins

The Western blot procedure is concluded by probing the blotted protein bands and detecting a specific protein or group of proteins among the blots. In other words, visualization of specific protein blots is now possible. The most specific identification techniques are based on immunology (antigen-antibody interactions; see Chapter 4, page 132). A general procedure for immunoblotting is outlined in Figure E7.1. Before the protein detection process can begin, it is necessary to block protein binding sites on the membrane that are not occupied by blotted proteins. This is essential because antibodies used to detect blotted proteins are also proteins and will bind to the membrane and interfere with detection procedures. Protein binding sites still remaining on blotted membranes may be blocked by treatment with solutions of casein (major protein in milk), gelatin, or bovine serum albumin.

The blotted membrane, with all protein binding sites occupied, can now be treated with analytical reagents for detection of specific proteins. Typically, the blotted membrane is incubated with an antibody specific for the protein of interest. This is called the **primary antibody,** which is a protein of the immunoglobulin G (IgG) class (see Figure E7.1). The primary antibody binds to the desired protein, forming an antigen-antibody complex. The interaction between the protein and its antibody does not usually result in a visible signal. The blot is then incubated with a **secondary antibody,** which is directed against the general class of primary antibody. For example, if the primary antibody was produced in rabbit serum, then the second antibody would be anti-rabbit IgG, usually from a goat or horse. The second antibody is labeled (conjugated) so that the interaction of the second antibody with the primary antibody produces some visual signal. For most detection procedures, the secondary antibody may be tagged with an enzyme, usually horseradish peroxidase (HRP) or alkaline phosphatase (AP). When the treated blot is incubated in a substrate solution, the conjugated enzyme catalyzes the conversion of the substrate into a visible product that precipitates at the blot site. The presence of a colored band indicates the position of the protein of interest (see Figure E7.2). This general procedure is also the basis for the widely used enzyme-linked immunosorptive assay (ELISA). The reactions catalyzed by conjugated enzymes are shown below:

$$\text{4-chloro-1-naphthol}_{\text{reduced}} + H_2O_2 \xrightleftharpoons{\text{HRP}} \text{4-chloro-1-naphthol}_{\text{oxidized}} + H_2O + \tfrac{1}{2}O_2 \qquad \textit{Equation E7.1}$$

Figure E7.1

Specific detection of nitro-cellulose membrane–bound proteins using a conjugated enzyme. (1) Proteins are transferred from electrophoresis gel to nitrocellulose membrane. Blocker proteins bind to unoccupied sites on the membrane. (2) The membrane is incubated with a primary antibody directed against the protein of interest. (3) A secondary antibody is directed against the primary antibody. (4) The second antibody is conjugated with an enzyme to provide a detection mechanism. Substrate solution is added to the blot. The conjugated enzyme (HRP or AP) catalyzes the conversion of substrate (S) to product (P) to form a colored precipitate at the site of the protein-antibody complex.

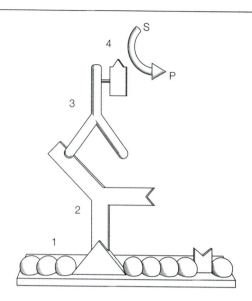

$$5\text{-bromo-4-chloro-3-indolylphosphate} + \text{nitroblue tetrazolium}_{\text{oxidized}} \overset{\text{AP}}{\rightleftharpoons}$$

>>

Equation E7.2

$$5\text{-bromo-4-chloro-3-indole} + P_i + \text{nitroblue tetrazolium}_{\text{reduced}}$$

A modification of these coloring systems has recently been developed that leads to more sensitive detection. Chemiluminescent substrates have been designed that are converted by the enzymes to products that generate a light signal that can be captured on photographic film. This increases the level of sensitivity about 1000-fold over standard color detection methods.

Even though primary and secondary antibodies are widely used in Western blotting detection systems, they do have some disadvantages. For proteins to be detected, specific antibodies must be available. It is often very time-consuming and expensive for a research laboratory to generate the proper antibodies if they are not available commercially. Even if antibodies are commercially available, they are very expensive.

In the experiment described here, an alternative and more economic method of specific detection will be introduced. A specific class of proteins, glycoproteins, will be identified by the carbohydrate binding ability of a lectin. **Lectins** are proteins found in plants and bacteria that bind to carbohydrates in glycoproteins. The lectin used here, concanavalin A (Con A) from jack bean, has a high binding affinity for α-D-mannosyl residues and related carbohydrates in proteins. Similar to the secondary antibody in traditional Western blotting, the Con A can be conjugated to horseradish peroxidase, allowing the application of the 4-chloro-1-naphthol/H_2O_2 color development system (Equation E7.1). To summarize, protein mixtures con-

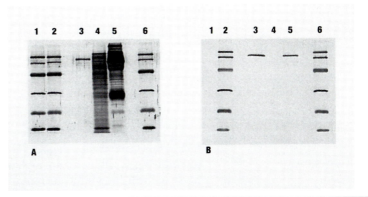

Figure E7.2

Results from a Western blot. **A** SDS-PAGE gels, 12%, were run and transferred to nitrocellulose. Lane 1, MW standards; lane 2, biotinylated standards; lane 3, human transferrin; lane 4, *E. coli* lysate; lane 5, total human serum; lane 6, biotinylated standards. Gel A was stained with a protein dye. Blot **B** was assayed using rabbit anti-human transferrin as the first antibody. The second antibody solution contained anti-rabbit HRP conjugates. Only the transferrin bands and the prestained biotinylated standards were detected by the antibodies and the avidin-HRP treatment.

---

taining glycoproteins that bind Con A may be fractionated by SDS-PAGE and glycoproteins can be visualized on membrane blots by treatment with Con A–HRP conjugate followed by incubation with substrate solution.

The Western blot has gained widespread use in biochemical and clinical investigations. It is one of the best methods for identifying the presence of specific proteins in complex biological mixtures. The Western blot procedure has been modified to develop a diagnostic assay that detects the presence in serum of antibodies to the AIDS virus. The presence of AIDS antibodies in a patient is an indication of viral infection. Western blotting techniques are also the basis for many home diagnostic tests such as pregnancy tests.

Overview of the Experiment

The techniques introduced in this experiment may be applied to the analysis of complex protein mixtures such as human or other animal serum. The Con A–HRP probe will recognize and visualize blots containing several serum glycoproteins including IgG heavy chains, transferrin, and α_1-antitrypsin. Other major serum proteins including albumin are not visualized by the reagents. The mannose-containing glycoprotein egg albumin (ovalbumin), often present in standard protein markers, reacts positively with the Con A–HRP detection system. In this experiment, bovine serum will be fractionated by SDS-PAGE and the proteins blotted onto a nitrocellulose membrane. The blotted membrane will be blocked with nonspecific protein and incubated with Con A–HRP conjugate. Treatment of the membrane with substrate solution will produce colored bands localizing glycoproteins on the blot.

This experiment requires two 3-hour lab periods. Assuming the use of preformed gels, the following schedule may be completed:

Period 1: Fractionation of protein mixture by SDS-PAGE; begin passive blot procedure or electroblotting.

Period 2: Membrane blocking and incubation of blot with Con A–HRP and substrate solution.

It is recommended that students be provided precast gels for this experiment, especially if they prepared their own in Experiment 4. Availability of preformed gels will ensure completion of the experiment in two 3-hour periods and will allow students to learn new procedures and techniques.

II. MATERIALS AND SUPPLIES

A. SDS-Polyacrylamide Gel Electrophoresis

- Reagents and equipment for SDS-PAGE (see Experiment 4)
- Precast polyacrylamide-SDS gels: 12% acrylamide, 0.75 mm thickness, 8 × 10 cm. A gradient gel also provides excellent resolution.
- Sample application buffer: Tris, glycerol, bromophenol blue, pH 6.8
- Protein solution buffer: Tris, pH 6.8
- Bovine serum samples
- Electrophoresis buffer: Tris, glycine, SDS, pH 8.2
- 2-Mercaptoethanol (STENCH!)
- Standard protein mixture; molecular weight range 10,000–70,000, prestained
- Boiling-water bath
- Coomassie Blue staining solution
- Destaining solution

B. Blotting and Detection

- Nitrocellulose membrane, 5 × 7.5 cm, 0.45 μm
- Transfer buffer: Tris, glycine, methanol, pH 8.2
- Filter paper, Whatman 3MM
- Blocking solution: BSA (10%) in Tris, NaCl, Tween, pH 8.0
- Electrophoretic blotter
- TBST buffer: Tris, NaCl, Tween, pH 8.0
- TBS buffer: Tris, NaCl, pH 8.0
- Con A–HRP conjugate, diluted in 10% BSA solution
- Substrate solution: 4-chloro-1-naphthol in methanol
- H_2O_2, 30% in water

– Knife to cut gel

– Trays for staining, blotting, and incubation

III. EXPERIMENTAL PROCEDURE

> ▶ **CAUTION**
>
> Acrylamide in the unpolymerized form is a skin irritant and a potential neurotoxin. Wear gloves and a mask while weighing the dry powder. Do not breathe the dust. Prepare all acrylamide solutions in the hood. Do not mouth pipet any solutions used for gel formation, staining, blotting, or detection.
>
> Do not touch the electrophoresis chamber or wires while the electrophoretic operation is in progress. Voltages may be as high as 300 V and shocks may be fatal.

A. Procedure for SDS-PAGE

Obtain a precast SDS-polyacrylamide slab gel or prepare one according to instructions in Experiment 4. The recommended gel is 12% acrylamide with a thickness of 0.75 mm. Protein samples are prepared as follows: Purified proteins (transferrin, bovine serum albumin, α_1-antitrypsin, α-lactalbumin from Experiment 4, and molecular weight standards) are supplied in Tris buffer, pH 6.8 solutions at a concentration of 1 mg/mL. Sera samples have been diluted and are ready for use. Prepare protein samples for electrophoresis in 0.5-mL microcentrifuge tubes with attached caps. Label the tubes from 1 to 5 as below or per your Instructor's directions.

Tube No.	Protein Sample
1	Standard mixture for molecular weight
2	Transferrin
3	α-lactalbumin
4	α_1-antitrypsin or bovine serum albumin
5	Bovine serum

Before loading on the gel, the protein samples must be incubated with denaturing reagents as follows:

Sample	Amount
Purified proteins or mixtures	50 μL
10% SDS	10 μL
2-Mercaptoethanol (STENCH!)	3 μL
Sample application buffer, pH 6.8	40 μL

Each tube will have the following components: single protein or protein mixture, SDS, 2-mercaptoethanol, and buffer. Add all reagents as indicated and cap each tube tightly. Mix well with gentle shaking and place in a boiling-water bath for 3 minutes. Cool to room temperature and load 10 μL of a sample onto the top of the appropriate sample well (lane) as follows:

Lane No.	Tube No.
1, 6	1 (standard mixture)
2, 7	2 (transferrin)
3, 8	3 (α-lactalbumin)
4, 9	4 (α_1-antitrypsin or bovine serum albumin)
5, 10	5 (bovine serum)

Run gels at constant voltage, 25–30 mA per slab gel, until the bromophenol blue marker dye has reached the bottom of the gel (about 1 hour). When completed, *turn off and unplug the electrophoresis apparatus* and remove the gel. Carefully cut the gel between lanes 5 and 6 so that you have two identical halves. Cut a notch in the corner of each gel half to indicate proper orientation. The half containing lanes 1–5 will be stained with Coomassie Blue to detect the location of all proteins. The other half (lanes 6–10) will be blotted onto a nitrocellulose membrane to analyze for glycoproteins.

Stain lanes 1–5 by placing the gel in a small dish of Coomassie Blue staining solution for 30 minutes. Remove background dye by incubating the gel in destaining solution that is changed every 4–6 hours. A time period of 24–48 hours may be required for complete destaining. Alternatively, the gel may be placed in a Pyrex dish containing water and heated to near boiling in a microwave. The gel in its final state should have a clear, colorless background and intense blue-colored bands in each lane.

B. Procedure for Blotting and Detection

Prepare the gel lanes 6–10 and the nitrocellulose membrane for blotting by incubating in separate baths of transfer buffer for 10–15 minutes. For passive transfer, obtain a glass bowl and cut two pieces of 3MM filter paper to the exact size of the nitrocellulose membrane (5 × 7.5 cm). In the glass bowl add transfer buffer and prepare a stack consisting of a glass plate (to lift gel above buffer), filter paper, gel, nitrocellulose membrane, more filter paper, a 1- to 2-inch stack of paper towels, and finally a heavy object such as a book. Two ends of the bottom filter paper should dip into the transfer buffer in order to act as a wick to allow buffer movement by capillary action. Cover the entire stack and buffer with plastic wrap to avoid buffer evaporation. Passive transfer should be allowed to continue for about 2 days. For electroblotting, follow instructions for the particular apparatus you are using. Prepare a sandwich of filter paper, gel, nitrocellulose membrane, and more filter paper and place in the apparatus. Applying 15–30 volts for about 12 hours should be sufficient for complete transfer.

After blotting, the nitrocellulose membrane must be prepared for the detection process. These steps may be performed in a petri dish holding the 5 × 7.5 cm membrane and about 20 mL of each solution. Block the unoccupied protein binding sites on the blotted membrane by incubating in a solution of bovine serum albumin (10 mg/mL in TBST) for about 1 hour at room temperature. Wash the membrane twice (5 minutes each) with TBST solution. Now incubate the membrane in 20 mL of Con A–HRP solution for a 2-hour period. This solution should be gently mixed on a shaker or agitated every 15 minutes. Wash the membrane 2 × 5 minutes in TBST and once for 5 minutes in Tris-NaCl buffer. For color formation, incubate the membrane in 20 mL of substrate solution. Prepare the substrate solution immediately before use by adding 10 μL of 30% H_2O_2 to 20 mL of 4-chloro-1-naphthol–methanol solution. Incubate the membrane in the substrate solution until blue-purple bands appear (about 10–15 minutes). To remove excess substrate and stop color development, rinse the membrane with distilled water. Air dry the blotted membrane on a paper towel and store for future use. Obtain a photocopy of the blotted membrane for a permanent record.

IV. ANALYSIS OF RESULTS

Draw a picture of your Coomassie Blue–stained gel in your notebook. Use standard proteins on the gel to prepare a table of molecular weights and distances migrated during electrophoresis. You may also prepare a graph of distance migrated (x axis) versus log MW. Use the standard proteins to identify as many bands in the bovine serum as possible. Draw a picture of the gel blot in your notebook or insert a photocopy. What proteins on the gel appear to be glycoproteins? Known glycoproteins in bovine serum include transferrin, IgG, and α_1-antitrypsin. Are there other blue-purple bands present on the blot besides these? Can you identify the proteins producing these bands? If bovine α-lactalbumin from Experiment 4 was one of the samples you ran, determine whether or not it is a glycoprotein.

Study Problems

▶ 1. The name "Western blot" is derived from other blotting procedures called the Southern blot (developed by Earl Southern) and the Northern blot. What are the differences among these three types of blotting techniques and what is the purpose of each?

▶ 2. Is the detection system used in this experiment, treatment with a single identifying protein, Con A–HRP, as specific as a traditional Western blot using two proteins (a primary antibody and a secondary antibody)? Why or why not?

▶ *3.* An economical way to "block" a blotted membrane is to incubate it in a 10% solution of nonfat milk powder. How does this solution function as a blocking reagent?

▶ *4.* If a protein you wish to analyze by Western blotting is acidic (anionic under blotting conditions), what type of membrane would be best to ensure the tightest binding?

▶ *5.* Design a diagnostic test based on the Western blot that would give an indication of infection by the AIDS virus. Assume that a blood serum sample is available from the patient.

▶ *6.* A protein mixture can be fractionated either by native PAGE or by denaturing, SDS-PAGE, before Western blotting. What factors would determine your choice of electrophoresis method?

▶ *7.* Some proteins in standard molecular weight mixtures will give a positive test in the Con A–HRP assay. Did this happen in your experiment? Why did this happen and are you able to identify the glycoprotein in the standard mixture?

▶ *8.* In this experiment, you used bromophenol blue dye as a marker to tell you when to stop the electrophoresis process. What assumptions must you make about the relative movement of the dye versus sample proteins during electrophoresis?

9. Define each of the following items in terms of their use in Western blotting.

(a) SDS-PAGE

(b) Nylon membranes

(c) Electroblotting

(d) Primary antibody

(e) Secondary antibody

(f) Lectins

(g) Protein molecular weight standard mixture

(h) Conjugated enzyme

▶ *10.* In your biochemistry research project, you have isolated a new protein from spinach leaves. You wish to do a Western blot experiment with the protein to help determine its chemical structure and/or biological function. An amino acid analysis of the protein showed a great abundance of phenylalanine, leucine, and valine. What would be your choice of membrane for the blotting experiments?

Further Reading

G. Anderson and L. McNellis, *J. Chem. Educ.* **75,** 1275–1277 (1998). "Enzyme-Linked Antibodies: A Laboratory Introduction to the ELISA."

W. Burnette, *Anal. Biochem.* **112,** 195–203 (1981). "Western Blotting."

P. Cummings, *Biochem. Educ.* **25,** 39–40 (1997). "Simulated Western Blot for the Science Curriculum."

B. Dunbar, *Protein Blotting, A Practical Approach* (1994), IRL Press (Oxford). Complete and up-to-date book on blotting.

S. Farrell and L. Farrell, *J. Chem. Educ.* **73,** 740–742 (1995). "A Fast and Inexpensive Western Blot Experiment for the Undergraduate Laboratory."

B. Hames and D. Rickwood, Editors, *Gel Electrophoresis of Proteins: A Practical Approach,* 2nd ed. (1990), IRL Press (Oxford). An excellent book for starters.

C. Mathews, K. van Holde, and K. Ahern, *Biochemisty,* 3rd ed. (2000), Benjamin/Cummings (San Francisco), pp. 253–254. An introduction to Western blotting.

A. Paladichuk, *The Scientist,* March 15, pp. 18–21 (1999). "A Profile of Tools and Kits Available for Western Blotting."

R. Sanchez et al., *Rev. Cubana Med. Trop.* **50**(3), 221–2 (1998). Western blotting for assay of HIV antibodies.

E. Snyder and R. Fall, *Biochem. Educ.* **18,** 147–148 (1990). "Western Blotting with a Concanavalin A–Horseradish Peroxidase Conjugate."

L. Stryer, *Biochemistry,* 4th ed. (1995), W. H. Freeman (New York), p. 62. An introduction to Western blotting.

Western Blotting on the Web

http://www.tropix.com/westbak.htm
    Overview of Western blotting.

http://www.med.unc.edu/wrkunits/3ctrpgm/pmbb/mbt/WESTERN.htm
    For introduction and procedures read Unit 1: Western Blotting.

http://www.biology.arizona.edu/immunology/activities/western_blot/w_main.html
    Review of procedures in Western blotting.

http://www.uct.ac.za/microbiology/western.html
    Review of the technique of Western blotting.

http://www.uct.ac.za/microbiology/sdspage.html
    Review of SDS-PAGE.

http://www.msichicago.org/ed/AIDS/hivts3.htm
    The Western Blot test and HIV.

8

ISOLATION AND CHARACTERIZATION OF PLANT PIGMENTS

- **Recommended Reading**

Chapter 3, Section B; *Chapter 5*, Section A; *Experiment 9.*

- **Synopsis**

Many of the colors associated with higher plants (green leaves in the spring and summer, yellow or red leaves in the fall, the orange color of carrots, some colors in flower petals) are due to the presence of pigment molecules such as chlorophylls and carotenoids. In this experiment a mixture of these pigments will be isolated by solvent extraction of plant tissue, separated by chromatography, and the components identified by visible spectrophotometry.

I. INTRODUCTION AND THEORY

Chlorophylls and Carotenoids

The major photosynthetic pigments of higher plants can be divided into two groups, the *chlorophylls* and the *carotenoids*. Both types of pigments are present in the subcellular organelles called chloroplasts, where they are bound to proteins in the thylakoids, the photochemically active photosynthetic biomembranes (see Experiment 9). Intact pigment-protein complexes, which are held together by weak, noncovalent bonds, can be isolated from chloroplasts and characterized by polyacrylamide gel electrophoresis and isoelectric focusing. The pigments are released in a protein-free form by grinding plant tissue in a solvent such as acetone or methanol. Since chlorophyll and the carotenoids are readily soluble in organic solvents, they are biochemically classified as lipids.

333

The most abundant plant pigments are **chlorophyll** *a* and **chlorophyll** *b,* which occur in a ratio *(a:b)* of approximately 3:1. As shown in Figure E8.1, the chlorophylls have a porphyrin ring with a coordinated magnesium atom at its center, a fused five-membered ring, and a C_{20} phytyl side chain. This nonpolar hydrocarbon side chain enhances the solubility of chlorophyll in nonpolar solvents. The only difference between chlorophylls *a* and *b* is the substituent on position 3 (ring 3); chlorophyll *a* has a methyl group, whereas *b* has an aldehyde functional group. The chlorophylls are biosynthesized by condensation of δ-aminolevulinic acid in a process similar to the synthesis of heme in mammals. In the living plant cell, the chlorophylls are the primary photosynthetic pigments. They absorb light in the blue region (450 nm) and the red region (650–700 nm). Absorption of light, which excites an electron in the chlorophyll molecule, provides energy for initiation of the photosynthetic process producing NADPH and ATP to be used for carbohydrate biosynthesis by the plant (see Experiment 9).

The second group of plant pigments, the **carotenoids,** can be divided into two different types: (1) the **carotenes,** which contain only carbon and hydrogen, and (2) the **xanthophylls,** which contain carbon, hydrogen, and oxygen atoms in the form of hydroxyl or epoxide functional groups. The structures of several major carotenoids are shown in Figure E8.2. Note that they all contain 40 carbon atoms and, because of the extensive conjugation, are highly colored. Most carotenoids are yellow, red, or orange; however, some are green, pink, and even black. Many of the bright colors of flower petals are due to the presence of carotenoids. The yellow colors of fall result from the preferential destruction of the green chlorophylls, revealing the carotenoid color (Goodwin, 1958).

The percentage composition of carotenoids in plants depends on growing conditions. The average weight percent ranges are β-carotene, 25–40%; lutein, 40–60%; violaxanthin, 10–20%; and neoxanthin, 5–13%. Carotenoids are biosynthesized by condensation of acetyl CoA through the mevalonic acid pathway and the Porter-Lincoln pathway.

The physiological function of the carotenoids is not completely understood; however, two probable functions have been described. First, the ca-

---

Figure E8.1

Structures of chlorophyll *a* and chlorophyll *b.*

Chlorophyll *a*   Chlorophyll *b*

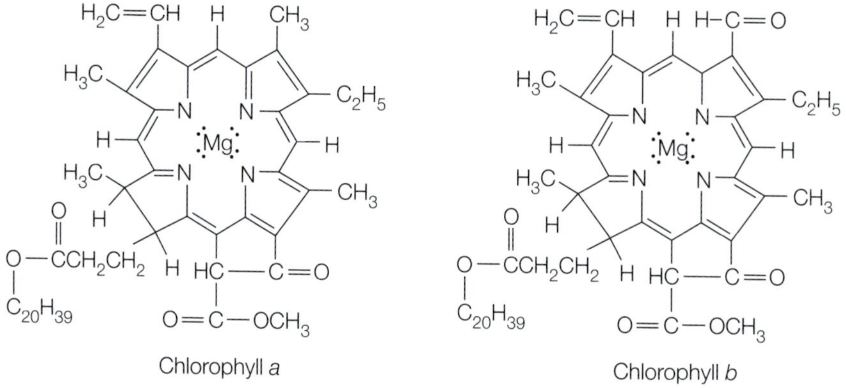

Figure E8.2

Structures of major
carotenoids: β-carotene, lutein,
violaxanthin, and neoxanthin.

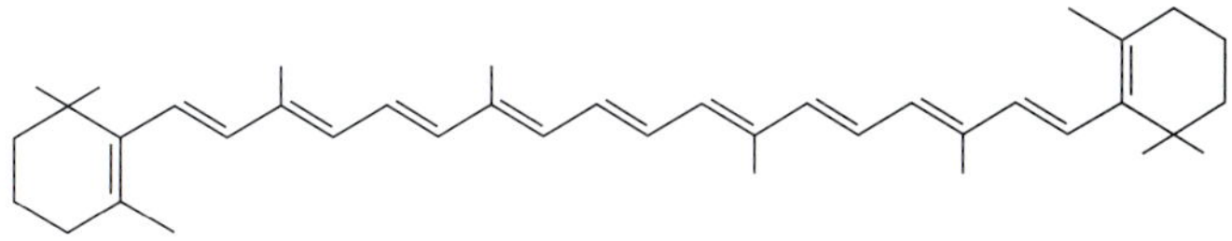

β-Carotene

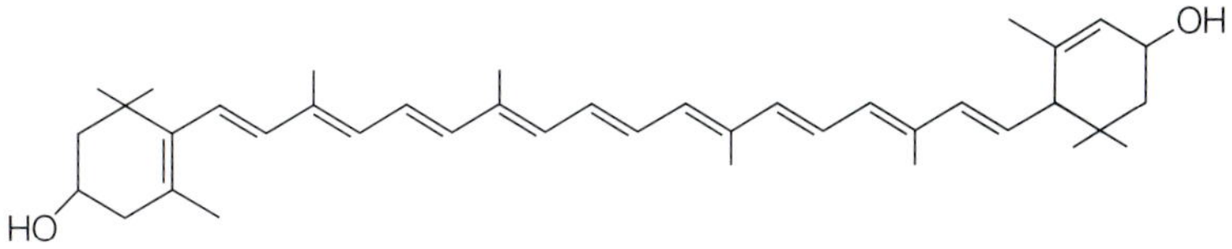

Lutein

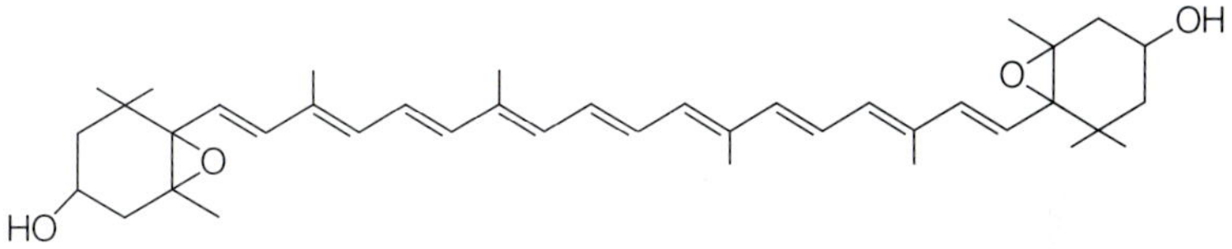

Violaxanthin

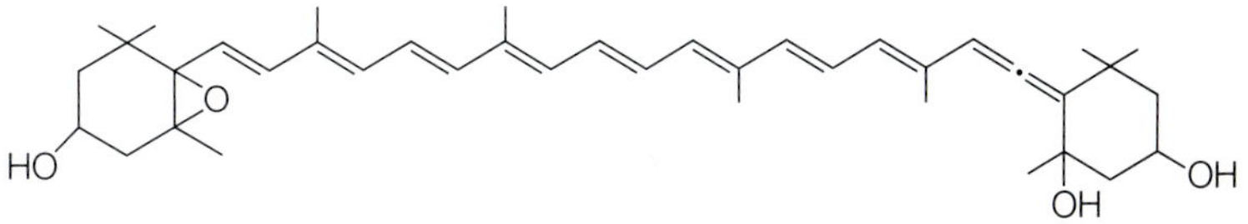

Neoxanthin

rotenoids, particularly the xanthophylls, may participate in light absorption
for photosynthesis. It has been shown that illuminated xanthophylls can
transfer excitation energy directly to chlorophyll *a*. Second, the carotenes
may inhibit the photo-oxidative destruction of chlorophyll, a process that
consumes oxygen.

General information on photosynthetic pigments can be obtained from
standard biochemistry texts and other reference books listed at the end of
the experiment.

Spectral Characteristics of Plant Pigments

The intense colors of the chlorophylls and carotenoids make them ideal
candidates for absorption spectroscopy studies (Tan and Soderstrom, 1989).
In fact, each plant pigment studied in this experiment has a unique visible
spectrum that can provide a positive identification. As shown in Figure
E8.3, chlorophylls *a* and *b* have absorption maxima in the 600–675 and

Visible absorption spectra of **A** chlorophyll *a* and **B** chlorophyll *b*.

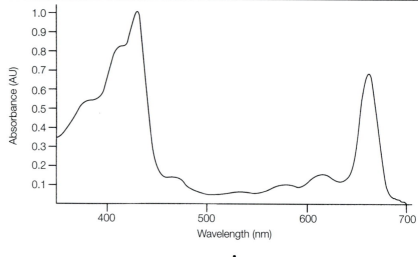

A

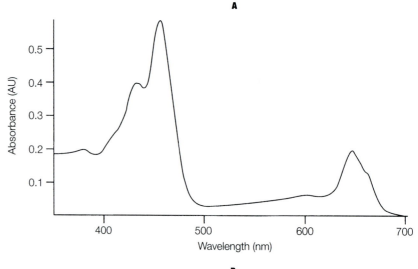

B

400–475 nm ranges. The absorption maximum of each peak depends on solvent polarity. For example, the two most intense absorption peaks for chlorophyll *a* in diethyl ether are 660 and 428 nm, whereas in more polar methanol the peaks are shifted to 665 and 432 nm.

In contrast to the chlorophylls, which absorb light in two regions of the visible spectrum, the carotenoids exhibit intense absorption in just one, 350–500 nm. Figure E8.4 compares the absorption spectra of four common carotenoids. As with the chlorophylls, the absorption maxima of the carotenoids vary with polarity of the solvent. β-Carotene in diethyl ether has a λ_{max} of 449.8 nm, but in the more polar acetone, the λ_{max} is 454 nm.

In addition to their use in pigment identification, spectrophotometric data can be used to determine the concentration of pigments in a plant extract. Data from spectral analysis were used to calculate absorption coefficients for each pigment in order to derive the following equations (Lichtenthaler, 1987):

$$C_a = 11.24A_{661.6} - 2.04A_{644.8}$$

$$C_b = 20.13A_{644.8} - 4.19A_{661.6}$$

$$C_{a+b} = 7.05A_{661.6} + 18.09A_{644.8}$$

$$C_{x+c} = \frac{1000A_{470} - 1.90C_a - 63.14C_b}{214}$$

where

C_a = concentration of chlorophyll *a* in units of micrograms per milliliter of plant extract solution (μg/mL)

C_b = concentration of chlorophyll *b* in μg/mL

C_{a+b} = concentration of total chlorophyll in μg/mL

C_{x+c} = concentration of total carotenoids (xanthophylls and carotenes) in μg/mL

Figure E8.4

Visible absorption spectra for four carotenoids: β-carotene (—), lutein (- -), violaxanthin (._.), and neoxanthin (. . .).

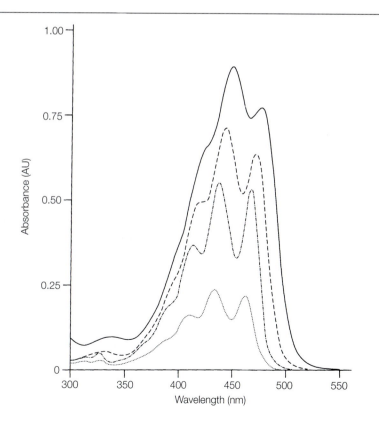

$$A_z = \text{absorbance measurement at wavelength } z$$

These equations are designed for the use of 100% acetone as solvent.

Practical Aspects of the Experiment

Plant pigments will be isolated by careful extraction of plant tissue (green or dried leaves) with 100% acetone. The extraction can be carried out with either a mortar and pestle or a glass homogenizer. Addition of an abrasive agent such as quartz sand enhances the extraction process. Both types of plant pigments, chlorophylls and carotenoids, are extremely light sensitive and are readily destroyed by photobleaching; therefore, the extraction process should be performed rapidly and in subdued lighting. If quantitative determinations are to be made for each pigment, the absorption data must be obtained immediately after extraction.

A wide range of chromatographic techniques, including column, paper, thin-layer, and high-performance liquid chromatography, have been applied to the separation of photosynthetic pigments. Theoretical and practical information on these techniques may be found in Chapter 3. There is great historical interest here, because Tswett, in the first demonstration of chromatography in 1906, separated a plant pigment extract on $CaCO_3$ and MgO columns. In this experiment, you will use either paper or thin-layer chromatography or both to separate the pigments. Either technique provides excellent resolution of the two chlorophylls and the major carotenoids. It is instructive for students to use both techniques. After solvent elution of the pigments from the paper or silica gel, the visible absorption spectrum of each is determined.

Overview of the Experiment

This experiment provides students with the opportunity to isolate a biomolecule from its natural source, followed by its purification and identification. In addition, students will follow a procedure that is typical of the general extraction and characterization of lipids. However, unlike most lipids, the plant pigments are highly colored and may be characterized and quantified by visible spectrophotometry. Several types of plant tissue may be used. Some recommendations are fresh leaves (tree, plant, grass, spinach), green algae, or mosses. For variety, students may be asked to bring their own samples for analysis.

The approximate time requirements for the experiment are:

A. Extraction of the pigments–30 minutes
B. Determination of pigment concentration–15 minutes
C. Chromatography and elution of pigments–1 hour
D. Measurement of visible absorption spectrum–1 hour

Since the pigment extract deteriorates rapidly by photo-oxidation, it is critical that students work rapidly, but safely and efficiently. Part B must be completed immediately after the extraction. Parts C and D may be completed at a later time, but, in the meantime, the plant extracts must be refrigerated in a concentrated form and under nitrogen gas.

II.   MATERIALS AND SUPPLIES

- Plant tissues (leaves, moss, algae, etc.)
- Acetone, 100%
- Mortar and pestle or glass homogenizer
- Quartz sand
- Small funnel and qualitative grade filter paper
- Test tubes, 100×12 mm
- Microcapillaries or tapered capillaries
- Anhydrous sodium sulfate, Na_2SO_4
- UV-VIS spectrophotometer
- Glass cuvettes, 3 or 1 mL
- Materials for paper chromatography:
  - Whatman 3MM paper, 13.5×7 cm
  - Solvent system, petroleum ether–acetone (9:1 v/v)
  - Chromatography chamber, 250-mL beaker with watch glass
  - Stapler
  - Scissors
- Materials for thin-layer chromatography:
  - Silica gel plates, 10×10 cm
  - Solvent system, petroleum ether–dioxane–acetone (70:30:10, v/v/v)
  - Chromatography chamber
  - Metal spatula or clean razor blade

III.   EXPERIMENTAL PROCEDURE

A. Extraction of the Pigments

▶ **CAUTION**

Exposure to acetone by inhalation, ingestion, or skin absorption is harmful. Contact with the eyes will cause severe irritation.

First aid: eyes–flush with copious amounts of water for at least 15 minutes; skin–flush with water for a few minutes; inhalation–remove to fresh air.

Acetone is volatile and extremely flammable. No open flames should be in the vicinity of your work. In case of fire, use a CO_2 extinguisher.

Dispose of acetone extract and solutions in the organic waste container.

This experiment should be completed in subdued lighting! Obtain five tree leaves or equivalent amount of grass or other plant tissue, tear into small pieces, and place in a mortar or homogenizer. Measure 15 mL of acetone and pour into the container. Add approximately 0.5 g of pure sand and thoroughly grind the mixture until the solution becomes a deep green. The plant cell walls of cellulose are tough, so you need to be persistent. After extraction is complete, add 1–2 g of anhydrous sodium sulfate to remove water from the extract. (What is the source of the water?) Rapidly filter the extract by gravity through fluted filter paper. Measure the volume of the filtered extract and rapidly proceed to part B.

B. Determination of Pigment Concentration

In order to determine the chlorophyll and carotenoid content of your extract, you must measure the absorbance at several wavelengths, 661.6, 644.8, and 470 nm. If you are using a single-beam spectrophotometer, use a cuvette of acetone to zero the instrument at each wavelength. A double-beam instrument should have a cuvette of acetone in the reference beam and the acetone extract solution in the sample beam. If desired, it is instructive to obtain a complete spectrum of the extract in the range 400–700 nm. This can be compared to the spectrum obtained for each of the individual pigments in part D.

C. Chromatography and Elution of Pigments

The general procedures for paper and thin-layer chromatography are similar. Obtain a piece of chromatography paper (13.5 × 7 cm) or a thin-layer plate (10 × 10 cm). With a pencil, draw a very light line 1.5 cm from and parallel to the long edge of the paper or TLC plate. Do not scrape silica off the plate. Using a Pasteur pipet or tapered capillary, evenly apply the acetone extract along the line. Begin on the penciled line approximately 1 cm from the edge and stop approximately 1 cm from the opposite edge. The narrower the line, the better your separation will be. Allow the first coating of extract to dry and repeat this procedure with another application of extract. Repeat a third time if the line is not dark green. Allow the paper to dry completely, and roll it into a circle with the pencil line and extract outside and on the circular edge. Use two or three staples to hold the circle in place. (Do not overlap the edges.) Place the paper (extract line down) into a chamber containing no more than 1 cm of appropriate solvent: petroleum ether–acetone (9:1). A chamber may be prepared by pouring 7 mL of solvent into a 250-mL beaker. Cover with a watch glass. If a thin-layer plate was used, place directly in a chamber containing petroleum ether–dioxane–acetone (70:30:10) and cover. When the solvent has risen to within 0.5 cm from the top of the paper or TLC plate, remove from the chamber and mark the position of the solvent front. Allow the chromatogram to dry. Measure the position of each pigment band and the solvent front and draw a full-scale picture of the completed chromatogram in your notebook. Since each chromatogram is complete in a few minutes, it is possible to experiment with the amount of extract applied to the paper or plate. If good separation is not attained and pigment bands are very dark, try a chromatogram with less extract. On the other hand, if the bands are light and difficult to distinguish, prepare and develop a chromatogram with more extract.

Each pigment is eluted from the paper or TLC plate using acetone. With the paper chromatogram, use a scissors to cut each colored band from the paper. Cut each paper strip into smaller pieces and place pieces from each band into separate, labeled test tubes containing 4 mL of acetone. Cover the tubes with a cork and allow to sit for 5–10 minutes with gentle mixing at 1- to 2-minute intervals. Proceed directly to part D. To extract the pigments from the TLC plate, use a metal spatula or razor blade to scrape the silica gel at each colored band into labeled test tubes containing 4 mL of acetone. Cover with a stopper and allow to sit for 5–10 minutes with mixing at intervals. Centrifuge or filter each solution using fluted filter paper. Proceed directly to part D.

D. Measurement of Visible Absorption Spectrum

Using glass cuvettes, measure the absorption spectrum of each fraction from 400 to 700 nm. A cuvette containing acetone should be used as reference. A Pasteur pipet may be used to transfer solutions to and from the cuvette. After all spectra have been obtained, dispose of the acetone solutions in the waste organic container in the laboratory.

IV.   ANALYSIS OF RESULTS

A. Extraction of the Pigments

Describe the color of the extract. Explain the action of acetone in the extraction process. What is the purpose of the sand?

B. Determination of Pigment Concentration

Using the absorbance data obtained on the crude extraction mixture and the equations provided earlier in the experiment, calculate the concentrations of chlorophyll *a*, chlorophyll *b*, and total carotenoids.

C. Chromatography and Elution of Pigments

Calculate the mobility of each pigment relative to the solvent front (R_f):

$$R_f = \frac{\text{distance from origin migrated by a compound}}{\text{distance from origin migrated by solvent}}$$

Explain the relative order of the R_f values. Can you make any general statements about the polarity of each pigment? In your notebook, describe the color of each eluted pigment fraction.

D. Measurement of Visible Absorption Spectrum

Prepare a table of spectral data (wavelengths of major peaks and absorbance) for each pigment. Using these data and the standard spectra shown earlier in this experiment, try to assign the identity of each pigment isolated. Now that each compound is identified, is there a correlation between the experimental R_f value and polarity as determined from the structure? Which pigments are more polar—those near the top or those near the bottom of the chromatogram?

Study Problems

1. What other solvents in addition to acetone could be used for extracting plant photosynthetic pigments?

2. Study the structures of the chlorophylls and carotenoids and predict the relative polarity of each set of pigments below. List each set in order of increasing polarity.

   (a) Chlorophyll *a*, chlorophyll *b*

   (b) Lutein, neoxanthin, violaxanthin, β-carotene

3. Several pigments were isolated from blueberry leaves using the procedures described in this experiment. Each purified pigment was characterized by spectrophotometry. Use Figures E8.3 and E8.4 to identify each unknown below.

   (a) Unknown 1 in diethyl ether: absorption maxima at 410, 428, and 660 nm

   (b) Unknown 2 in diethyl ether: absorption maxima at 450 and 480 nm

   (c) Unknown 3 in diethyl ether: absorption maxima at 430, 455, and 650 nm

4. What extraction solvent system would you use if you wanted to isolate the intact protein-pigment complexes? Describe experimental techniques that could be used to determine the molecular weight of each protein-pigment complex.

5. What are the structural features that make chlorophyll and the carotenoids so highly colored?

6. Explain the following experimental observation. If aqueous acetone (80% acetone, 20% H_2O) is used as the extraction solvent instead of 100% acetone, the extracted yields of chlorophyll *a* and β-carotene are lowered. Why?

7. How would you modify this experiment so that larger quantities of the pigments (mg or g) could be isolated?

8. If you wished to extract large amounts of chlorophylls *a* and *b*, would it be better to pick leaves in the summer or fall?

➤ *9.* Study the synthesis of the chlorophylls in your biochemistry text-book. What two small molecules are used as the starting point for the chlorophylls?

➤ *10.* Polyunsaturated fatty acids (PUFAs) have several carbon-carbon double bonds as present in β-carotene. Why are PUFAs not colored like β-carotene?

Further Reading

R. Boyer, *Concepts in Biochemistry* (1999), Brooks/Cole (Pacific Grove, CA), pp. 528–541. Introduction to photosynthesis and plant pigments.

R. Boyer, *Biochem. Educ.* **18,** 203–206 (1990), "Isolation and Spectrophotometric Characterization of Photosynthetic Pigments."

H. Dailey, Editor, *The Biosynthesis of Heme and Chlorophylls* (1990), McGraw-Hill (New York). An excellent reference on the chemistry and biochemistry of chlorophyll.

R. Garrett and C. Grisham, *Biochemistry,* 2nd ed. (1999), W. B. Saunders (Orlando, FL), pp. 709–741. Introduction to photosynthesis and pigments.

T. Goodwin, *Biochem. J.* **68,** 503–511 (1958), "Studies in Carotenogenesis 24. The Changes in Carotenoid and Chlorophyll Pigments in the Leaves of Deciduous Trees During Autumn Necrosis."

H. Lichtenthaler, in *Methods in Enzymology,* Vol. 148, L. Packer and R. Douce, Editors (1987), Academic Press (New York), pp. 350–382, "Chlorophylls and Carotenoids: Pigments of Photosynthetic Biomembranes."

C. Mathews, K. van Holde, and K. Ahern, *Biochemistry,* 3rd ed. (2000), Benjamin/Cummings (San Francisco), pp. 594–626. An introduction to photosynthesis and photosynthetic pigments.

L. Stryer, *Biochemistry,* 4th ed. (1995), W. H. Freeman (New York), pp. 655–661. An introduction to photosynthesis.

B. Tan and D. Soderstrom, *J. Chem. Educ.* **66,** 258–260 (1989), "Qualitative Aspects of UV-VIS Spectrophotometry of β-Carotene and Lycopene."

D. Voet and J. Voet, *Biochemistry,* 2nd ed. (1995), John Wiley & Sons (New York), pp. 626–661. An introduction to the pigments of photosynthesis.

D. Voet, J. Voet, and C. Pratt, *Fundamentals of Biochemistry* (1999), John Wiley & Sons (New York), pp. 529–549. An introduction to photosynthesis.

Photosynthesis on the Web

http://gened.emc.maricopa.edu/bio/bio181/BIBOBK/BioBooksPS.html
     Pigments in photosynthesis.

PHOTOINDUCED PROTON TRANSPORT
THROUGH CHLOROPLAST MEMBRANES

- **Recommended Reading**

Chapter 2, Section A; *Chapter 5*, Section A; *Chapter 7*, Sections A, B, C; *Experiment 8*.

- **Synopsis**

Photosynthesis occurs in the plant cell organelle called the chloroplast. During the process of photosynthesis, electrons are transferred from H_2O to $NADP^+$ via an electron carrier system. The energy released by electron transport is converted into the form of a proton gradient and coupled to ADP phosphorylation. In this experiment a method is introduced to demonstrate the formation of the proton gradient across the chloroplast membranes.

I.   INTRODUCTION AND THEORY

The Photosynthetic Process

Photosynthesis occurs only in plants, algae, and some bacteria, but all forms of life are dependent on its products. In photosynthesis, electromagnetic energy from the sun is used as the driving force for a thermodynamically unfavorable chemical reaction, the synthesis of carbohydrates from CO_2 and H_2O (Equation E9.1).

>> $$CO_2 + H_2O \xrightarrow{\ hv\ } (CH_2O) + O_2$$ *Equation E9.1*

Here (CH$_2$O) represents a carbohydrate molecule. The simplicity of Equation E9.1 is deceiving, because the process of photosynthesis is a complex one; the details are still under investigation. The fundamental photosynthetic event is initiated by the interaction of a photon with a chlorophyll molecule, causing excitation of an electron in chlorophyll. The activated electron is passed through a chain of electron carriers aligned in order of decreasing reduction potential. The electron transport chain is similar to that in mitochondrial respiration. Energy released during electron transport is coupled to the synthesis of ATP from ADP and phosphate, P$_i$. This process is outlined in Figure E9.1. Two photosystems, I and II, are at work during photosynthesis in plants. Each photosystem contains reaction centers (chlorophylls P680 and P700) that collect the light. The two systems are linked by the electron transport chain, allowing electron flow from H$_2$O (to produce O$_2$) ultimately to NADP$^+$. In other words, H$_2$O is oxidized and the electrons are carried to NADP$^+$, forming the reduced cofactor, NADPH. The overall reaction is shown in Equation E9.2. Note that

---

Figure E9.1

The photosynthetic process, showing coupling of electron transport and ADP phosphorylation. The dashed line shows electron flow in cyclic photophosphorylation. See text for details.

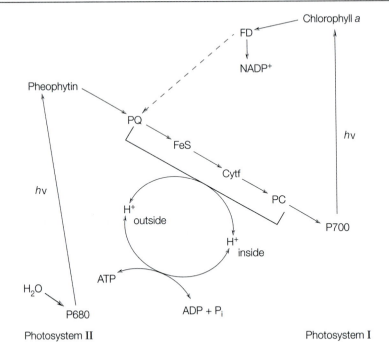

PQ = Plastoquinone
FeS = Iron sulfur protein
Cytf = Cytochrome f
PC = Plastocyanin
FD = Ferredoxin

this is the reverse of the respiratory process occurring in mitochondria, where NADH is oxidized.

>>
$$H_2O + NADP^+ + nADP + nP_i \xrightarrow{hv} NADPH + H^+ + nATP + \tfrac{1}{2}O_2 \qquad \textit{Equation E9.2}$$

When electrons flow from photosystem I to photosystem II, protons are transported across the chloroplast membranes as indicated in Figure E9.1. This aspect of photosynthesis will be discussed in a later section.

The light reactions illustrated in Figure E9.1 are accompanied by a sequence of "dark reactions" leading to the formation of carbon intermediates and sugars. This sequence of reactions, called the Calvin cycle, incorporates CO_2 into carbohydrate structures.

Figure E9.1 illustrates the photosynthetic process as it occurs in higher plants. This is called **noncyclic photophosphorylation** to distinguish it from **cyclic photophosphorylation** in photosynthetic bacteria. Cyclic photophosphorylation requires only photosystem I and a second series of electron carriers to return electrons to the electron-deficient chlorophyll. The dashed line in Figure E9.1 indicates the flow of electrons in cyclic photophosphorylation. ATP is produced during the cyclic process just as in the noncyclic process, but NADPH is not.

The photosynthetic process in green plants occurs in subcellular organelles called **chloroplasts.** These organelles resemble mitochondria; they have two outer membranes and a folded inner membrane called the **thylakoid.** The apparatus for photosynthesis, including the chlorophyll reaction centers and electron carriers, is in the thylakoid membrane. The chemical reactions of the Calvin cycle take place in the **stroma,** the region around the thylakoid membrane.

Mechanism of Photophosphorylation

One very active area of biochemical investigation is the study of energy transfer in mitochondria and chloroplasts. In both oxidative phosphorylation and photophosphorylation, electron flow through a carrier chain is linked to the generation of ATP. Most research is directed toward determining how the energy provided by electron transport is coupled to the phosphorylation of ADP. Three theories of energy coupling have received attention: chemical coupling, conformational coupling, and chemiosmotic coupling. The **chemical coupling hypothesis** proposes the formation of high-energy chemical intermediates; the **conformational coupling hypothesis** proposes a high-energy molecular conformation: and **chemiosmotic coupling** involves a charge separation or concentration gradient across a membrane.

The chemiosmotic coupling hypothesis, proposed by P. Mitchell, is the most attractive explanation, and many experimental observations now support this idea. Simply stated, Mitchell's hypothesis suggests that electron transfer is accompanied by transport of protons across the membrane, resulting in a proton concentration gradient. When the proton gradient (a high-energy condition) collapses, the released energy is coupled to the phosphorylation of ADP. In mitochondria, the proton transport is through the inner membrane, from the inside (matrix side) to the outside (cytoplasmic side), whereas in chloroplasts protons are transported across the thylakoid membrane from the outside to the inside.

Experimental results supporting the Mitchell hypothesis were first reported from studies using chloroplasts. Two critical experiments providing unequivocal evidence for linkage between proton transport and ADP phosphorylation were reported. Neumann and Jagendorf (1964) generated a high-energy proton gradient by illuminating chloroplasts in the absence of ADP and phosphate. As one would predict from the Mitchell hypothesis, Jagendorf observed a rise in the pH of the medium indicating proton uptake by the chloroplasts. In later experiments, Jagendorf and Uribe (1966) induced an artificial proton gradient across the thylakoid membrane by incubating chloroplasts under acidic conditions (pH 4) and then rapidly transferring them to pH 8 buffer. When ADP and P_i were added to the chloroplast suspension, the synthesis of ATP was observed, and the previously induced pH gradient disappeared. Similar experiments with mitochondria also confirm the transport of protons in oxidative phosphorylation, but in the opposite direction from photophosphorylation. Continued study of light-induced proton uptake by chloroplasts has led to the observation that proton uptake is accompanied by Mg^{2+} and K^+ extrusion from the chloroplasts.

Practical Aspects of the Experiment

Chloroplasts will be isolated by careful extraction of spinach leaves, using tricine buffer containing sucrose. The crude extract contains both whole and fragmented chloroplasts, but both contain all the necessary photosynthetic components and are capable of photophosphorylation. The preparation described in this experiment retains almost all of the chlorophyll in the chloroplasts. The total chlorophyll content of the chloroplasts will be determined by extracting the pigment with aqueous acetone and measuring the absorption at $\lambda = 652$ nm. The chlorophyll concentration is calculated according to Equation E9.3 (Arnon, 1949),

$$C = \frac{A}{El} = \frac{A_{652}}{34.5 \times l}$$

Equation E9.3

where

C = concentration of chlorophyll in mg/mL
A_{652} = absorption of chlorophyll at 652 nm
E = absorption coefficient for chlorophyll, 34.5 mL/(mg × cm)
l = path length of cuvette

The average range of chlorophyll concentration for the described chloroplast preparation is 0.3 to 1.0 mg/mL. For an alternative method of measuring chlorophyll and other pigment concentrations in plant extracts, see Experiment 8.

The light-induced pH changes are measured by monitoring the pH of a suspension of illuminated chloroplasts. If protons are taken up into the chloroplasts, the pH of the suspension medium should increase. Figure E9.2 shows typical results of such an experiment. The proton transport process is reversed by eliminating the light source (indicated by the dashed line in Figure E9.2). Figure E9.2 shows the result for illuminated chloroplasts alone. It has been found that the presence of a light-sensitive redox dye stimulates the proton transport process. In this experiment phenazine methosulfate (PMS) will be used, although others, including methyl viologen and trimethylhydroquinone, may be used as an electron transport cofactor.

Overview of the Experiment

Chloroplasts will be isolated from spinach leaves and used to explore some intricacies of proton transport and photophosphorylation. The chloroplast preparation should be used as soon as possible (within 2 to 3 hours) since biological activity will decrease with time. The experimental apparatus for

Figure E9.2

Light-induced pH changes in chloroplasts. Light is turned on and off at indicated times.

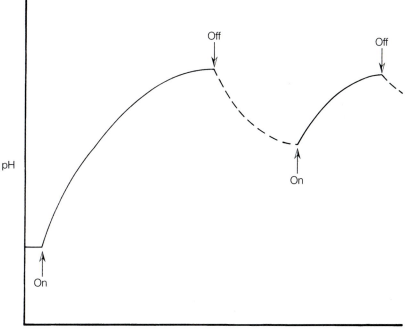

measurement of proton uptake should be assembled and all reagents prepared before the chloroplast extract is begun. The time required for the various parts of the experiment is as follows.

A. Preparation of spinach chloroplasts—30 minutes.

B. Determination of chlorophyll content—15 minutes.

C. Measurement of proton uptake—1 to 2 hours, depending on the choice of experiments.

Other optional experiments may be completed if time allows. For example, the effectiveness of various redox dyes may be analyzed. In addition to those listed in the text, FMN, ferricyanide, and dichlorophenolindophenol may be tested (Neumann and Jagendorf, 1964). It has been shown that NH_4Cl and amines stimulate proton uptake. If a potassium ion-specific electrode is available, the light-induced efflux of K^+ from spinach chloroplasts may be studied (Dilley, 1972).

II.   MATERIALS AND SUPPLIES

A. Preparation of Spinach Chloroplasts

- Fresh spinach leaves, about 50 g. These should be washed well with cold water, placed in plastic bags, and kept cold for several hours before use.
- Homogenizing buffer, 0.02 M tricine, pH 8, containing 0.01 M NaCl and 0.4 M sucrose. **Keep cold.**
- Sharp knife
- Blender
- Cheesecloth
- Refrigerated centrifuge
- Chloroplast suspension solution, 0.4 M sucrose containing 0.01 M NaCl

B. Determination of Chlorophyll Content

- Chloroplast preparation from part A
- 80% acetone solution in water
- Conical centrifuge tube
- Glass cuvette, one pair
- Spectrophotometer for measurement at 652 nm
- Clinical or table top centrifuge

C. Measurement of Proton Uptake

- pH meter, with recorder if possible. This should have a full-scale pH change of 0.5 pH unit.

- 500-watt tungsten lamp, with a Corning 1-69 filter or a solution containing a trace of $CuSO_4$ to remove infrared radiation
- Magnetic stirrer and small stir bar
- Phenazine methosulfate (PMS)
- Timer
- Large test tube, 2 × 20 cm
- Phosphate solution, 0.01 M K_2HPO_4
- Adenosine diphosphate, 0.02 M

III. EXPERIMENTAL PROCEDURE

A. Preparation of Spinach Chloroplasts

Chloroplasts are fragile and unstable; therefore, this part of the experiment should be done as rapidly as possible and in subdued lighting. Maintain all solutions at 4°C or below.

Cut and discard the midribs from 50 g of spinach leaves and tear the leaves into small pieces. Immediately place the leaves into a precooled blender containing 100 mL of ice-cold homogenizing buffer. Grind the leaves at top speed no longer than 5 seconds. Strain the homogenate through eight layers of cheesecloth. Squeeze all the liquid from the cheesecloth. Centrifuge the filtrate in precooled tubes at 1000 × g for 1 minute to remove whole cells. Transfer the supernatant to precooled centrifuge tubes and spin at 6000 × g for about 15 seconds. Decant the supernatant and suspend the chloroplasts (pellet) in 50 mL of tricine homogenizing buffer. Centrifuge the suspension at 6000 × g for about 10 seconds. (Note the color of the supernatant; the deeper the green, the more chlorophyll was lost from the chloroplasts). Pour off and discard the supernatant. Resuspend the isolated chloroplasts by stirring into 50 mL of suspension solution (sucrose containing NaCl). Store the chloroplast preparation in ice until use. It will remain stable for 2 to 4 hours.

B. Determination of Chlorophyll Content

Extract the chlorophyll from the chloroplasts by mixing, in a conical centrifuge tube, 0.05 mL of well-mixed chloroplast suspension with 9.9 mL of 80% acetone in water. Spin in a tabletop centrifuge for 10 minutes. Transfer the supernatant to a glass cuvette and read the absorbance at 652 nm using 80% acetone in water as reference. Calculate the concentration of chlorophyll in the chloroplast suspension using Equation E9.3.

C. Measurement of Proton Uptake

> ⮞ **CAUTION**
>
> Phenazine methosulfate is a skin irritant, so it should be handled with care. Avoid contact with skin and do not breathe dust.

The experimental arrangement is shown in Figure E9.3. Maintain the temperature of the water bath at 10°C by adding ice. Add a trace of $CuSO_4$ solution to the water bath to eliminate infrared radiation from the lamp. Obtain a test tube that will hold the pH electrode and 10 to 15 mL of solution. Dilute the chloroplast solution with suspension buffer to a chlorophyll concentration of about 300 $\mu g/mL$. Adjust the pH meter with standard pH 7 buffer. The following three experimental conditions are recommended. Each condition should be tested individually and done in duplicate.

1. Chloroplasts with no redox cofactor

    10 mL of chloroplast suspension (about 3 mg chlorophyll)
    1 mL of suspension buffer

2. Chloroplasts with redox cofactor

    10 mL of chloroplast suspension
    5 mg of phenazine methosulfate or other redox dye
    1 mL of suspension buffer

3. Chloroplasts with redox cofactor, ADP, and phosphate

    10 mL of chloroplast suspension
    5 mg of phenazine methosulfate or other redox dye

Figure E9.3

Experimental arrangement for measuring proton uptake by illuminated chloroplasts. See text for details.

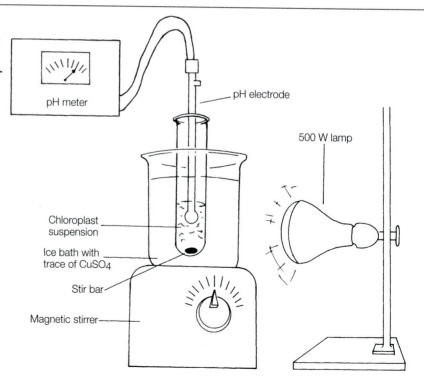

0.5 mL of ADP solution

0.5 mL of phosphate buffer

To complete each experiment, prepare the reaction mixture as described above. Mix well and transfer to the reaction tube. Insert a small magnetic stir bar and place the pH electrode into the reaction mixture. Turn on the magnetic stirrer to a slow rate and check to see whether the stirrer affects the pH meter. Adjust the pH of the reaction mixture to a pH reading in the range of 6 to 6.2 with acid or base. Turn on the pH meter and recorder, if available, and then turn on the lamp to illuminate the chloroplasts. Mark the recorder sheet at the time the lamp is turned on. Allow the recorder to trace the pH reading with time for 60 seconds, and then turn off the light but continue to record the pH. Turn the lamp on after a dark interval of 30 seconds. Continue to record the pH for 60 seconds, again turn the lamp off, and continuously record the pH. Repeat the experiment with a fresh portion of chloroplasts.

If no pH recorder is available, pH readings can be taken from the meter or digital readout and a graph of pH vs. time prepared. Record a pH reading every 10 seconds.

Experimental conditions 2 and 3, with their various additions, are completed in an identical fashion. If chloroplasts still remain stable and time permits, other experiments outlined in the overview may be completed.

IV.   ANALYSIS OF RESULTS

A. Preparation of Spinach Chloroplasts

Write all observations regarding the preparation in your notebook. Record the color of each supernatant and pellet.

B. Determination of Chlorophyll Content

Use Equation E9.3 to calculate the concentration of chlorophyll in the chloroplast suspension. What is the total yield of chlorophyll in mg?

C. Measurement of Proton Uptake

If no recorder tracing is available, prepare a graph of pH vs. time for each experiment. Prepare and complete a table with the following headings:

Addition	Rate of pH Change (sec/0.5 pH)	Relative Rate
None		
+ PMS		
+ PMS, ADP, P_i		

Compare the relative rates of proton uptake for each experimental condition and explain any differences. Did the addition of light-sensitive redox cofactor affect the rate of pH shift? What is the effect of the addition of

ADP and phosphate on the rate of pH shift? Explain the effect of alternating light and dark intervals.

Study Problems

 1. Design a flowchart outlining the chloroplast preparation procedure and describe the purpose of each step.

▶ 2. Explain the chlorophyll extraction procedure in chemical terms. Why is acetone effective in the extraction? How could the experiment be modified to increase the yield of chlorophyll extraction?

▶ 3. Describe other methods for measuring the rate of ATP generation during photophosphorylation.

 4. Why is the amount of chlorophyll kept constant in all experiments described here?

▶ 5. Describe methods that could be used to measure extrusion of K^+, Mg^{2+}, or Ca^{2+} from chloroplasts.

 6. Explain the effect, if any, of the redox cofactor on the rate of proton uptake.

 7. Why does the pH of the suspension solution decrease during dark intervals?

▶ 8. A student in the lab completes part B of this experiment to measure the chlorophyll content of the chloroplast preparation. The final A_{652} of the 80% acetone extract was 0.13. What is the concentration of chlorophyll in mg/mL in the chloroplast preparation?

▶ 9. Would you expect the rate of proton uptake in this experiment to be dependent on the wavelength of light used? What wavelength(s) is most effective? Explain.

▶ 10. Examine Equation E9.2 and answer the following questions.

   (a) What is the oxidizing agent (acceptor of electrons)?

   (b) What is the reducing agent (donor of electrons)?

   (c) What is the source of energy for phosphorylation of ADP?

Further Reading

D. Arnon, *Plant Physiol.* **24,** 1–15 (1949). Determination of chlorophyll content of chloroplasts.

D. Arnon, *Trends Biochem. Sci.* **9,** 258–262 (1984). "The Discovery of Photosynthetic Phosphorylation."

R. Boyer, *Concepts in Biochemistry* (1999), Brooks/Cole (Pacific Grove, CA), pp. 528–549. Review of photosynthesis.

S. Brown, *Biochem. Educ.* **26,** 164–167 (1998). Photosynthesis and respiration in leaf slices.

R. Dilley, in *Methods in Enzymology*, Vol. 24B, A. San Pietro, Editor (1972), Academic Press (New York), pp. 68–74. Transport of H^+, K^+, and Mg^{2+} in isolated chloroplasts.

R. Garrett and C. Grisham, *Biochemistry*, 2nd ed. (1999), W. B. Saunders (Orlando, FL), pp. 709–741. An introduction to photosynthesis.

Govindjee and W. Coleman, *Sci. Am.* **262**(2), 50–58 (1990). How plants make oxygen.

J. Ho and E. Po, *Biochem. Educ.* **24,** 179–180 (1996). Light-induced proton transfer through chloroplast membrane.

A. Jagendorf and E. Uribe, *Proc. Natl. Acad. Sci. USA,* **55,** 170–177 (1966). "ATP Formation Caused by Acid-Base Transition of Spinach Chloroplasts."

C. Mathews, K. van Holde, and K. Ahern, *Biochemistry*, 3rd ed. (2000), Benjamin/Cummings (San Francisco), pp. 594–626. An introduction to photosynthesis.

J. Neumann and A. Jagendorf, *Arch. Biochem. Biophys.* **107,** 109–119 (1964). "Light-Induced pH Changes Related to Phosphorylation by Chloroplasts."

W. Nitschke and A. Rutherford, *Trends Biochem. Sci.* **16,** 241–245 (1991). "Photosynthetic Reaction Centres: Variations on a Common Structural Theme."

L. Stryer, *Biochemistry*, 4th ed. (1995), W. H. Freeman (New York), pp. 653–680. An introduction to photosynthesis.

D. Voet and J. Voet, *Biochemistry*, 2nd ed. (1995), John Wiley & Sons (New York), pp. 626–661. An introduction to photosynthesis.

D. Voet, J. Voet, and C. Pratt, *Biochemistry*, (1999), John Wiley & Sons (New York), pp. 530–559. An introduction to photosynthesis.

D. Walker, in *Methods in Enzymology*, Vol. 69C, A. San Pietro, Editor (1980), Academic Press (New York), pp. 94–104. "Preparation of Higher Plant Chloroplasts."

Photosynthesis on the Web

http://www.life.uiuc.edu/govindjee/
    Educational material on photosynthesis including many Internet links.

http://gened.emc.maricopa.edu/biol/bio181/BIOBK/BioBooksPS.html
    A review of photosynthesis.

10

ISOLATION, SUBFRACTIONATION, AND ENZYMATIC ANALYSIS OF BEEF HEART MITOCHONDRIA

- **Recommended Reading**

Chapter 2, Section B; *Chapter 7,* Sections A, B, and C; *Experiments 4 and 5.*

- **Synopsis**

Many of the biochemical processes that generate chemical energy for the cell take place in the mitochondria. These organelles contain the biochemical equipment necessary for fatty acid oxidation, di- and tricarboxylic acid oxidation, amino acid oxidation, electron transport, and ATP generation. In this experiment, a mitochondrial fraction will be isolated from beef heart muscle. The mitochondria will be analyzed for protein content and fractionated into submitochondrial particles. Each fraction will be analyzed for malate dehydrogenase and monoamine oxidase activities.

I. INTRODUCTION AND THEORY

Mitochondrial Structure and Function

Mitochondria are intracellular centers for aerobic metabolism. They are cell organelles that are identified by well-defined structural and biochemical properties. In morphological terms, mitochondria are relatively large particles that are characterized by the presence of two membranes, a smooth outer membrane that is permeable to most important metabolites and an inner membrane that has unique transport properties. The inner membrane is highly folded, which serves to increase its surface area. Figure E10.1, which shows the structure of a typical mitochondrion, divides the organelle into four major components: inner membrane, outer membrane, intermembrane space, and the matrix. These regions are associated with different and

357

specific biological functions. Mitochondria are characterized by the presence of the proteins and enzymes of respiratory metabolism, many of which are present in the folded inner membrane. Hence, this is the location for the processes of electron transport and ATP formation for energy production. The enzymes of the tricarboxylic acid cycle, fatty acid oxidation, and amino acid metabolism are present in the matrix region. The outer membrane contains many enzymes, including monoamine oxidase for chemical reactions on neuroactive aromatic amines. The intermembrane space contains enzymes such as adenylate kinase and nucleoside diphosphokinase, which function in nucleotide balance.

Isolation of Intact Mitochondria

Because of the biological significance of these intracellular particles, biochemists have devised several methods for isolating mitochondria. Most isolation methods take advantage of the relatively large size of mitochondria. In animal cells only the nucleus is larger. The standard method for preparing a mitochondrial fraction is differential centrifugation of homogenized tissue (Figure 7.11, Chapter 7). It should be pointed out that only rarely does one isolate and characterize a preparation of pure mitochondria.

Figure E10.1

Structure of the mitochondrion with associated functions. From *Concepts in Biochemistry*, p. 482, by Rodney Boyer. © 1999 by Brooks/Cole. Reprinted by permission of Brooks/Cole, Pacific Grove, CA 93950.

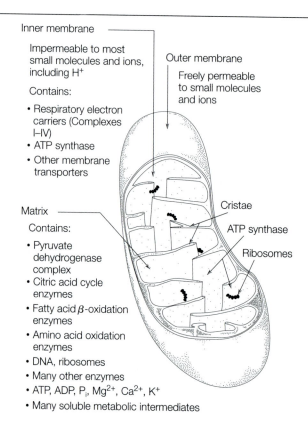

Inner membrane

Impermeable to most small molecules and ions, including H^+

Contains:

- Respiratory electron carriers (Complexes I–IV)
- ATP synthase
- Other membrane transporters

Outer membrane

Freely permeable to small molecules and ions

Matrix

Contains:

- Pyruvate dehydrogenase complex
- Citric acid cycle enzymes
- Fatty acid β-oxidation enzymes
- Amino acid oxidation enzymes
- DNA, ribosomes
- Many other enzymes
- ATP, ADP, P_i, Mg^{2+}, Ca^{2+}, K^+
- Many soluble metabolic intermediates

Cristae

ATP synthase

Ribosomes

Rather, the isolated subcellular fraction is a **mitochondrial fraction,** which indicates that its major components are mitochondria. Other cellular components that may be present are lysosomes, cell fragments, nuclear fragments, and microbodies (peroxisomes). The purity of the fraction depends on the source of the extract and the method chosen for isolation.

The general procedure for preparation of a mitochondrial fraction consists of the following steps:

1. Select and procure a suitable biological material. Mitochondrial fractions have been prepared from many animal organs including liver, heart, muscle, and brain, many plant tissues including spinach and avocado, and yeast.

2. The tissue is minced or ground and suspended in sucrose buffer. The choice of buffer conditions is critical in order to avoid loss of protein and other components that may leak from the mitochondria. Cytochrome c is one of the easiest mitochondrial proteins to dislodge, and some is almost always lost. The buffer should be isotonic or hypotonic with a low ionic strength. Sucrose is an ideal buffer component because mitochondria are especially stable and remain intact under these conditions.

3. Homogenize the tissue. This is carried out with a glass-Teflon homogenizer or a blender. This procedure gently breaks open the cells and allows the release of the subcellular organelles into the buffer. The mechanical disruption of cells should be gentle so the fragile mitochondria are not damaged.

4. Centrifuge the homogenate at low speeds. The heavier particles such as whole and fragmented cells, nuclei and nuclear fragments, and membrane fragments are sedimented during this step.

5. Centrifuge the supernatant from step 4 at a higher speed. This step sediments lighter particles including mitochondria. Two distinct layers usually are observed in the pellet: (a) a loosely packed, fluffy upper layer, which consists of damaged mitochondria and is sometimes called **light mitochondria,** and (b) a dark brown layer that consists of **heavy mitochondria.**

6. The heavy mitochondrial fraction is suspended in buffer and homogenized for a brief period in a glass-Teflon homogenizer to complete cell lysis.

7. The mitochondrial fraction is then obtained by centrifugation at relatively high speeds (15,000 to 20,000 rpm; $25,000 \times g$).

8. The pellet from step 7 now consists primarily of heavy mitochondria suspended in sucrose buffer. Typical yields are approximately 1 mg of protein per gram of starting minced tissue.

Several variations of this procedure may be found in the literature. Often, homogenization is facilitated by adding a proteolytic enzyme to the original suspension in step 2, but the proteolysis must be carefully controlled to avoid degradation of mitochondrial enzymes. The various methods lead to mitochondrial fractions of different quality and biochemical characteristics.

This experiment describes the preparation of a mitochondrial fraction from beef heart muscle. Heart muscle is an excellent choice of tissue because isolated mitochondria are stable and most enzyme activities remain high for a long period of time. Since heart muscle is more fibrous than other tissues, some problems are encountered in homogenizing the tissue. The preparation described here is suitable for the study of characteristic enzymatic activity, electron transport, and ADP phosphorylation.

Subfractionation of Mitochondria

The discussion to this point has focused on the isolation of intact mitochondria. By various chemical and physical treatments, mitochondria may be separated into their four components. This allows biochemists to study the biological functions of each component. For example, by measuring enzyme activities in each fraction, one can assign the presence of a particular enzyme to a specific region of the mitochondria. Studies of mitochondrial subfractions have resulted in a distribution analysis of enzyme activities in the four locations (Table E10.1). This type of study is often referred to as an **enzyme profile** or **enzyme activity pattern** and the enzyme may be considered a **marker** enzyme. For example, cytochrome oxidase, which is involved in electron transport, is a marker enzyme for the inner membrane.

One of the most instructive fractionation procedures is the preparation of **submitochondrial particles** (SMPs). The particles are produced by sonication (see Experiment 4) and centrifugation. The pellet, which sediments between 12,000 and 100,000 $\times$ g after sonication, defines the submitochondrial particle fraction. Submitochondrial particles are actually chunks of inner membrane that have undergone circularization and inversion. In other words, the membrane has been turned "inside-out." Essentially all of the components for electron transport are still present; however, matrix enzymes are largely removed.

Table E10.1

Average Percent Distribution of Marker Enzyme Activities in Submitochondrial Fractions

| Fraction | Marker Enzyme[1] | | |
	Cytochrome c Oxidase	Monoamine Oxidase	Malate Dehydrogenase
Outer membrane	10	82	1
Inner membrane	90	11	15
Matrix	0	0	80
Intermembrane space	0	7	4

[1] The numbers represent the % total activity. For example, 82% of the total monoamine oxidase activity of mitochondria is found in the outer membrane.

Characterization of Mitochondria and SMPs

Mitochondrial fractions may be characterized by testing for the presence of known enzyme activities as previously discussed. The relative purity of each fraction can be estimated by measuring the specific activity of marker enzymes. Table E10.1 identifies marker enzymes for the matrix and membranes. Malate dehydrogenase, the tricarboxylic acid cycle enzyme that catalyzes the interconversion of malate and oxaloacetate (Equation E10.1), serves as a marker for the matrix enzymes.

>> malate + NAD$^+$ $\rightleftharpoons$ oxaloacetate + NADH + H$^+$ *Equation E10.1*

Malate dehydrogenase activity would be expected in intact mitochondria, but not in SMPs. The activity of this enzyme in mitochondrial fractions may be estimated by a spectrophotometric assay. Oxaloacetate and NADH are incubated, and the disappearance of NADH is monitored at 340 nm. NAD$^+$ does not have strong absorption at this wavelength. Note that the reverse reaction is studied because the reaction as shown above is very unfavorable in thermodynamic terms ($\Delta G^{o'} = +30$ kJ/mol).

Monoamine oxidase (MAO) serves as a marker enzyme for outer membrane. There is some MAO activity in the inner membrane and therefore also in SMPs; however, a high level of monoamine oxidase in the SMP preparation indicates a large contamination by outer membrane. Mitochondrial monoamine oxidase is an FAD-dependent enzyme that catalyzes the oxidation of amines to aldehydes (Equation E10.2). A convenient assay for this enzyme uses benzylamine as substrate and monitors the rate of benzaldehyde production at 250 nm.

>> RCH$_2$NH$_2$ + O$_2$ + H$_2$O $\rightleftharpoons$ RCHO + NH$_3$ + H$_2$O$_2$ *Equation E10.2*

Overview of the Experiment

Students will isolate intact mitochondria from beef heart and fractionate them to prepare submitochondrial particles. Each fraction will be characterized by protein estimation by the biuret method and measurement of malate dehydrogenase and monoamine oxidase activity.

The experiment has the following time requirements:

A. Isolation of the mitochondrial fraction–2 hours. It is recommended that the isolation procedure up to step 13 be done as a class project with each student group completing two or three sequential steps. The centrifugation step (step 13) may be done in several 50-mL tubes and each student group receives a centrifuge tube to continue through step 20 on their own.

B. Determination of protein–30 minutes.

C. Preparation of SMPs–1 to 2 hours beginning with intact mitochondria.

D. Measurement of enzyme activities–30 minutes per enzyme.

Completion of all parts will require 4–5 hours. If time is limited (3 hours), it is recommended that students complete parts A and B and then test only for the presence of malate dehydrogenase in the intact mitochondria (part D).

II. MATERIALS AND SUPPLIES

A. Preparation of the Mitochondrial Fraction

- Beef heart, from a local slaughterhouse. It should be obtained a few hours after the animal is slaughtered and transported in ice to the laboratory. A relatively fresh organ from a meat market is a suitable source for this experiment.
- Sharp knife
- Meat grinder, prechilled; plate holes: 4 to 5 mm
- Waring blender
- Sucrose-Tris homogenizing solution, 0.25 M sucrose and 0.01 M Tris-Cl buffer, pH 7.8. Keep ice-cold.
- 2 M Tris base solution, unneutralized, pH 10.8
- Cheesecloth
- Sucrose-Tris isolation solution, 0.25 M sucrose, 0.01 M Tris-Cl buffer, pH 7.8, 0.001 M succinic acid, and 0.2 mM EDTA. Keep ice-cold.
- Glass-Teflon homogenizer, manual or motorized
- Centrifuge, capable of 1200 $\times$ g and 26,000 $\times$ g

B. Determination of Protein

- Mitochondrial fraction from part A
- Sucrose-Tris isolation solution
- 10% sodium deoxycholate
- Bovine serum albumin solution, 10 mg/mL
- Biuret reagent
- Spectrophotometer

C. Preparation of SMPs

- Mitochondrial fraction from part A
- Sucrose-Tris isolation solution
- Sonicator
- Ice-salt bath
- Centrifuge (high-speed and ultracentrifuge)

D. Measurement of Enzyme Activities

Malate Dehydrogenase

- SMP fraction from part C
- Mitochondrial fraction from part A
- Phosphate buffer, 0.2 M, pH 7.4
- Oxaloacetic acid, 0.006 M in phosphate buffer, freshly prepared
- NADH, 0.00375 M in phosphate buffer, freshly prepared
- Spectrophotometer, suitable for measurements at 340 nm
- Cuvettes, 3-mL quartz

Monoamine Oxidase

- SMP fraction from part C
- Mitochondrial fraction from part A
- Phosphate buffer, 0.2 M, pH 7.4
- Benzylamine solution, 0.10 M in phosphate buffer
- Spectrophotometer, suitable for measurements at 250 nm
- Cuvettes, 3-mL quartz

III.   EXPERIMENTAL PROCEDURE

A. Preparation of the Mitochondrial Fraction

All of the following procedures must be carried out at 2–4°C.

1. With a sharp knife, trim all fat and connective tissue from the ice-cold beef heart.
2. Cut the tissue into cubes 4 to 5 cm wide.
3. Quickly weigh 200 to 300 g of the cubes and pass them through a prechilled meat grinder.
4. Suspend the minced tissue in 400 mL of ice-cold sucrose-Tris homogenizing solution.
5. Adjust the pH to 7.5 ± 0.1 by adding 2 M unneutralized Tris.
6. Pour the neutralized heart mince through two layers of cheesecloth and squeeze out the sucrose solution.
7. Suspend 200 g of the solid mince in 400 mL of ice-cold sucrose-Tris isolation solution.
8. (a) Homogenize 25- to 50-mL portions of the suspension in a glass homogenizing vessel with a motorized pestle. Homogenize each portion for about 10 seconds and then twice more for 5 seconds each. Combine all the homogenate solutions and adjust to pH 7.8 with 2 M Tris base. (b) Alternatively, the minced tissue may be homogenized in a blender.

For this procedure, suspend 200 g of mince in 400 mL of sucrose-Tris isolation solution and add 3 mL of 2 M Tris base. Turn on the blender at high speed for 15 seconds. Add 3 mL of 2 M Tris base and blend for another 5 seconds. Adjust the pH of the solution to 7.8 with 2 M Tris base.

9. Centrifuge the homogenate for 20 minutes at 1200 × g. The pellet consists of unfragmented cells and nuclei.

10. Carefully decant the supernatant without disturbing the loosely packed pellet.

11. Filter the supernatant through two layers of cheesecloth. This process removes lipid granules.

12. Adjust the pH of the filtrate to 7.8 with 2 M Tris base.

13. Centrifuge the pH-adjusted suspension for 15 minutes at 26,000 × g. At least two distinct layers will be observed in the pellet. The top layer consists of loosely packed light mitochondria (damaged mitochondria). The bottom layer of the pellet is brown and consists of heavy mitochondria.

14. Remove and discard the light mitochondria by pouring off about half of the supernatant and very gently shaking the centrifuge tube to release the top layer. Pour off and discard the suspension of light mitochondria.

15. Add 10 mL of sucrose-Tris isolation solution and stir the heavy mitochondria with a glass stirring rod.

16. Homogenize the mitochondrial suspension in a motorized or hand-held glass-Teflon homogenizer. Make two passes of the rotating pestle through the suspension, 5 seconds each.

17. Adjust the pH of the homogenate to 7.8 with 2 M Tris base and add sucrose-Tris isolation solution to a total volume of 180 mL. (Add sucrose-Tris isolation solution to a total volume of 20 mL if you began step 13 with a 50-mL centrifuge tube.)

18. Centrifuge the suspension at 26,000 × g for 15 minutes. The pellet this time consists primarily of a dark brown layer of heavy mitochondria. If an upper layer of light mitochondria is present, remove as in step 14.

19. Suspend the heavy mitochondria in about 60 mL of sucrose-Tris isolation solution and adjust the pH to 7.8 with 2 M Tris. (Suspend in only 6 mL if you began step 13 with a 50-mL centrifuge tube.)

20. Store the mitochondrial fraction on cracked ice and begin part B.

B. Determination of Protein

Before the mitochondrial fraction can be biochemically characterized, the protein content must be measured.

1. Obtain five small test tubes and set up the protein assay according to Table E10.2. Tubes 1 and 2 contain two different concentrations of mitochondrial protein. Tubes 3 and 4 are duplicates of a standard protein,

Reagents	Tube				
	1	2	3	4	5 (blank)
Sucrose-Tris isolation solution	0.5	0.5	0.5	0.5	0.5
Mitochondrial fraction	0.5	0.25	—	—	—
10% Sodium deoxycholate	0.2	0.2	0.2	0.2	0.2
Bovine serum albumin	—	—	0.1	0.1	—
H_2O	0.3	0.55	0.7	0.7	0.8
Biuret reagent	1.5	1.5	1.5	1.5	1.5

Table E10.2

Preparation of Tubes for the Biuret Protein Assay on Mitochondrial Fractions[1]

[1] Numbers are milliliters

bovine serum albumin (BSA). Tube 5 is used as a blank for the spectrophotometer. The purpose of sodium deoxycholate is to disrupt the mitochondria and release the protein material into solution. For a discussion of the biuret analysis of proteins, see Chapter 2.

2. Add the appropriate amount of each of the first four reagents (sucrose solution, mitochondria, sodium deoxycholate, and BSA) to each of the five test tubes. If the solutions are not clear, add more 10% sodium deoxycholate with a graduated pipet. Be sure you note the exact amount of deoxycholate added. Do not add more than a total of 0.4 mL.

3. Add sufficient water to each tube so that the total volume of all liquids is 1.5 mL. Table E10.2 is set up assuming that 0.2 mL of deoxycholate solution is sufficient. If, for example, tube 1 requires a total of 0.4 mL of deoxycholate, then only 0.1 mL of water should be added to the tube.

4. Add 1.5 mL of biuret reagent to each tube and mix well by inverting several times while holding a piece of hydrocarbon foil over the opening.

5. Incubate the tubes for 15 minutes at 37°C. Transfer the contents to cuvettes.

6. Measure and record the absorbance at 540 nm of tubes 1 through 4, using tube 5 to adjust the A_{540} to 0.0 absorbance.

7. Tube 1 should have approximately twice the A_{540} of tube 2. If the absorbance readings of tubes 1 and 2 are two to three times greater than the A_{540} of tubes 3 or 4, repeat the assay using less mitochondrial fraction in tubes 1 and 2. Be sure the new assays contain a total of 3.0 mL of liquid in each tube.

C. Preparation of SMPs

1. Dilute (with isolation solution) the appropriate amount of mitochondrial fraction from part A so that you have a 10-mL suspension at a concentration of 15 mg of protein per mL.

2. Transfer the suspension to a 10-mL stainless steel or glass beaker. Chill in an ice-salt bath. With continued cooling in the salt bath, sonicate the suspension at maximum energy for six 5-second bursts. The suspension will warm during sonication. Therefore, after each 5-second sonication, allow the suspension to cool for 30 seconds. The temperature of the suspension should not be allowed to rise above 10°C during sonication. The mitochondrial suspension will become less opaque during the sonication.

3. Add 10 mL of ice-cold isolation buffer and centrifuge for 10 minutes at 10,000 $\times$ g and 5°C.

4. Decant the supernatant into another tube and centrifuge at 100,000 $\times$ g for 20 minutes.

5. Repeat step 4 two more times.

6. Resuspend the pellet in 15 mL of isolation buffer.

7. Estimate the protein concentration using the biuret assay. If necessary, dilute to 1 mg protein/mL.

D. Measurement of Enzyme Activities

Malate Dehydrogenase

1. Turn on the spectrophotometer and UV lamp and allow to warm up for 15 minutes.

2. Obtain two quartz cuvettes and add the following reagents to each.

Cuvette 1	Reagent	Cuvette 2
1.3 mL	0.2 M phosphate, pH 7.4	1.3 mL
—	NADH	0.2 mL
0.1 mL	Oxaloacetic acid solution	0.1 mL
1.5 mL	H_2O	1.3 mL

3. If necessary, dilute the mitochondrial fraction to a protein concentration of 1 mg/mL. Add 0.1 mL (0.1 mg protein) of the diluted mitochondrial fraction from part B to cuvette 1. Cover the cuvette with hydrocarbon foil and gently invert two or three times.

4. Insert the cuvette into the sample beam of the spectrophotometer and adjust the absorbance at 340 nm to 0.00.

5. Add 0.1 mL (0.1 mg protein) of the diluted mitochondrial fraction to cuvette 2, mix as before, and immediately place it in the spectrophotometer.

6. If a recorder is available, monitor the change in absorbance at 340 nm for 5 minutes. If no recorder is available, read and record in your notebook the A_{340} at 30-second intervals for 5 minutes. If the reaction is too

slow or fast, repeat assay with more or less enzyme fraction. Final volume of assay must be 3.0 mL.

7. Prepare a plot of A_{340} (y axis) vs. time (x axis).

8. Repeat the above steps with the SMP fraction from part C.

Monoamine Oxidase

1. Adjust the spectrophotometer to 250 nm.

2. Add the following to a 3-mL quartz cuvette:

   2.75 mL 0.2 M phosphate buffer, pH 7.4

   0.1 mL 0.1 M benzylamine

   Mix well, place in the spectrometer, and record the baseline at 250 nm. Use a range of 0.5 absorbance units or less.

3. Initiate the enzyme-catalyzed reaction by adding 0.15 mL of the 1.0 mg/mL intact mitochondria or SMP suspension. Mix well and record the ΔA_{250} for 3–5 minutes. If the reaction rate is too slow or fast to measure, repeat using more or less protein. Be sure the total volume of each assay is 3.0 mL.

IV.   ANALYSIS OF RESULTS

A. Preparation of the Mitochondrial Fraction

Prepare a flowchart of the procedures followed in the isolation of mitochondria. Briefly explain the purpose of each step. Describe the appearance of the mitochondrial suspension.

B. Determination of Protein

Calculate the protein concentration in the mitochondrial fractions using Equation E10.3.

$$\frac{C_{std}}{A_{std}} = \frac{C_{unk}}{A_{unk}}$$

Equation E10.3

where

A_{std} = absorbance at 540 nm of the standard bovine albumin (tube 3 or 4)

A_{unk} = absorbance at 540 nm of the protein in the mitochondrial fractions (tube 1 or 2)

C_{std} = concentration of standard bovine serum albumin in mg/mL

C_{unk} = concentration of protein in the mitochondrial fractions in mg/mL

C. Preparation of SMPs

Describe the appearance of the SMPs. Use a diagram to show disruption of the mitochondria by sonication. Why are the SMPs washed several times (steps 4 and 5)? What proteins are present in the supernatant?

D. Measurement of Enzyme Activities

Malate Dehydrogenase

Use the plot of A_{340} vs. time to calculate ΔA/min over the linear portion of the curve. Convert the rate in absorbance terms to activity units. One enzyme unit is the amount of malate dehydrogenase that catalyzes the reduction of 1 micromole of oxaloacetate to L-malate in 1 minute under the described assay conditions. The reduction of 1 micromole of oxaloacetate leads to the oxidation of 1 micromole of NADH; therefore, Equation E10.4 may be used to calculate the specific activity of malate dehydrogenase.

$$\text{>>} \quad \text{Specific activity} = \frac{\Delta A_{340}/\text{min}}{6.2 \times \text{mg protein/mL reaction mixture}} \qquad \textit{Equation E10.4}$$

The millimolar absorption coefficient of NADH at 340 nm is 6.2 $\text{m}M^{-1}$ cm^{-1}.

Monoamine Oxidase

Convert the rate in ΔA_{250} units to activity units by using Equation E10.5.

$$\text{>>} \quad \text{Specific activity} = \frac{\Delta A_{250}/\text{min}}{13 \times \text{mg protein/mL reaction mixture}} \qquad \textit{Equation E10.5}$$

The millimolar absorption coefficient for benzaldehyde at 250 nm is 13 $\text{m}M^{-1}$ cm^{-1}.

Now calculate the proportion of total mitochondrial protein (malate dehydrogenase and monoamine oxidase) present in the submitochondrial fraction.

% of total mitochondrial protein in submitochondrial fraction =

$$100 \times \frac{\text{specific activity of marker enzyme in mitochondria}}{\text{specific activity of marker enzyme in submitochondrial fraction}}$$

This can be calculated separately for malate dehydrogenase and MAO activity. Do your results agree with those in Table E10.1? Explain.

Study Problems

▶ *1.* Why is sucrose used in the isolation buffer?

▶ *2.* Describe the function and mode of action of sodium deoxycholate in the protein assay in part B.

3. Explain the derivation of Equation E10.4 for malate dehydrogenase activity calculation.

▶ *4.* Briefly explain how the mitochondrial fraction prepared here could be used to study electron transport and ADP phosphorylation. Could the SMP fraction also be used for these studies?

5. Using structures, write the reaction for the monoamine oxidase–catalyzed oxidation of benzylamine.

▶ *6.* How would you prepare more highly purified mitochondria than described in this experiment?

7. Using a flowchart format, show how you would purify monoamine oxidase from beef heart mitochondria.

▶ *8.* What subfraction of mitochondria would be the best source of the pyruvate dehydrogenase complex?

▶ *9.* A student group completed the biuret assay on their mitochondrial preparation. Assume that the following absorbance values were obtained as described in Table E10.2.

Tube	A_{540}
1	0.25
2	0.12
3	0.08
4	0.09

(a) What is the protein concentration in the mitochondrial preparation in mg/mL?

(b) Describe how you would dilute the mitochondrial preparation in order to make a 1 mg/mL protein concentration.

▶ *10.* You are required to assay a solution of purified malate dehydrogenase (specific activity = 1.0). What ΔA_{340}/min will you observe if you use 1 mg of enzyme in a 3-mL assay as described in part D of this experiment?

Further Reading

M. Ator and P. Ortiz de Montellano, in *The Enzymes,* 3rd ed., Vol. XIX, D. Sigman and P. Boyer, Editors (1990), Academic Press (San Diego), pp. 234–240. Mechanistic discussion of monoamine oxidase.

R. Boyer, *Concepts in Biochemistry* (1999), Brooks/Cole (Pacific Grove, CA), pp. 480–495. An introduction to mitochondrial structure and function.

R. Capaldi, *Trends Biochem. Sci.* **25,** 212–214 (2000). "The Changing Face of Mitochondrial Research."

V. Darley-Usmar, D. Rickwood, and M. Wilson, Editors, *Mitochondria: A Practical Approach* (1987), IRL Press (Oxford). An excellent source for laboratory techniques.

R. Garrett and C. Grisham, *Biochemistry,* 2nd ed. (1999), W. B. Saunders (Orlando, FL), pp. 674–675. Mitochondrial structure and function.

C. Heisler, *Biochem. Educ.* **19,** 35–38 (1991). "Mitochondria from Rat Liver: Method for Rapid Preparation and Study."

C. Mathews, K. van Holde, and K. Ahern, *Biochemistry,* 3rd ed. (2000), Benjamin/Cummings (San Francisco), pp. 522–525. An introduction to mitochondrial structure and function.

A. Meyers, *J. Biol. Educ.* **24,** 123–127 (1990). Experiments using colorimetric methods to study mitochondria and respiration.

A. Smith, in *Methods in Enzymology,* Vol. X, R. Estabrook and M. Pullman, Editors (1967), Academic Press (New York), pp. 81–86. "Preparation, Properties and Conditions for Assay of Mitochondria: Slaughterhouse Material, Small-Scale."

L. Stryer, *Biochemistry,* 4th ed. (1995), W. H. Freeman (New York), pp. 529–533. Structure and function of mitochondria.

D. Voet and J. Voet, *Biochemistry,* 2nd ed. (1995), John Wiley & Sons (New York), pp. 564–568. Mitochondrial structure and function.

D. Voet, J. Voet, and C. Pratt, *Fundamentals of Biochemistry* (1999), John Wiley & Sons (New York), pp. 492–497. Mitochondrial structure and function.

C. Worthington, Editor, *Worthington Enzyme Manual* (1988), Worthington Biochemical Corporation (Freehold, NJ), pp. 224–228, 285–287.

Mitochondria on the Web

http://www.worthington-biochem.com/manual/M/MDH.html.
     Description of malate dehydrogenase.

http://www.worthington-biochem.com/manual/P/PAO.html.
     Description of amine oxidases.

http://www.msu.edu/~bchug/bch471/471cpeuc.htm
     Procedure for isolation of rat liver mitochondria.

11

M EASUREMENT OF C HOLESTEROL AND V ITAMIN C IN B IOLOGICAL S AMPLES

- **Recommended Reading**

Chapter 1, Section F; *Chapter 5,* Section A; *Experiment 5.*

- **Synopsis**

It is often possible to quantify the presence of natural molecules in biological samples without actually isolating the molecules. Two such analyses will be completed in this experiment: (1) serum cholesterol will be measured by coupling its enzyme-catalyzed oxidation to the peroxidase-catalyzed formation of a chromogen, and (2) the presence of vitamin C in dietary materials will be detected and quantified by redox titration with 2,6-dichlorophenol-indophenol.

I. INTRODUCTION AND THEORY

A. Measurement of Cholesterol

One of the most beneficial and interesting applications of biochemical methods is in the diagnosis of disease states. Most procedures in the clinical laboratory are used to measure the concentrations of various constituents in biological fluids and tissues. An abnormally high or low concentration of a biochemical (enzyme, metabolite, etc.) in a patient's blood or urine specimen is often a signal to the patient's physician that a pathologic condition may exist. Biochemical measurements aid clinicians in at least two ways: (1) they assist in the diagnosis (recognition and identification) of a diseased condition, or (2) they may be used for the confirmation of a suspected diseased condition.

Enzymes as Diagnostic Reagents

The availability of purified enzyme preparations and their unique substrate specificity make enzymes particularly attractive as diagnostic reagents. Since most enzymes act on a single type of reactant and the extent of an enzyme-catalyzed reaction depends on substrate concentration, it is possible to estimate the concentration of a single molecular species (the substrate). It is, of course, essential that the desired reaction be carried out under optimal conditions, that is, proper pH, temperature, and ionic strength, and with assured absence of interfering substances. Of equal importance are the concentrations of enzyme, cofactor, substrate to be measured, and other required reagents. **All reagents, except the substance to be measured, must be present in excess, so that the rate and extent of the enzyme-catalyzed reaction depend on only the concentration of the substance to be determined.** The initial velocity of an enzyme-catalyzed reaction is expressed by the Michaelis-Menten equation (Equation E11.1).

>>
$$v_0 = \frac{V_{max}[S]}{[S] + K_M}$$

Equation E11.1

The maximum rate of a reaction (V_{max}) is attained when all the enzyme active sites are saturated with substrate molecules. For the rate to approach V_{max}, the substrate concentration must be high; in fact, it must be much greater than K_M. When $[S] \gg K_M$, the Michaelis-Menten equation becomes

>>
$$[S] \gg K_M; \quad v_0 = \frac{V_{max}[S]}{[S]} = V_{max}$$

Equation E11.2

In words, under substrate-saturating conditions, the initial rate of the enzyme-catalyzed reaction (v_0) is independent of substrate concentration, [S] (Equation E11.2). However, when the substrate is present in much less than saturating amounts ($[S] \ll K_M$), the Michaelis-Menten equation becomes

>>
$$[S] \ll K_M; \quad v_0 = \frac{V_{max}[S]}{K_M}$$

Equation E11.3

The initial rate of the enzyme-catalyzed reaction is directly proportional to [S] (Equation E11.3). Most clinical assays using enzymes are performed under the conditions of Equation E11.3. From further study of this equation, you will note that v_0 also depends on enzyme concentration, since there is an enzyme concentration term hidden in V_{max}. (If you have forgotten this, review the derivation of the Michaelis-Menten equation in your biochemistry textbook.) This can be used to advantage, because if a reaction used for a clinical analysis is very slow (it probably will be, since [S] is low), extra enzyme can be used so that the reaction will proceed to completion in a reasonable period of time.

In general, the principle behind the clinical measurement of cholesterol and other biomolecules is the following. A biological fluid (serum, urine, etc.) containing the substance to be measured is incubated with an enzyme system that will interact only with that substance. The extent of the enzyme-catalyzed reaction is determined by monitoring some change associated with conversion of substrate to product. Ideally, there is a change in UV or VIS absorbance that can be measured at a convenient wavelength with a spectrophotometer. The spectrophotometric method might involve measurement of a product that appears in solution or the disappearance of substrate. The substance to be determined is incubated for a time interval sufficient to transform it completely into product or to attain equilibrium between substrate and product. Here you should recognize the importance of the enzyme concentration, a point raised earlier. Sufficient enzyme must be present so that an end point (complete conversion of substrate to product or attainment of equilibrium) is reached in the desired amount of time. Once the extent of reaction has been quantitatively measured, knowledge of the stoichiometry of the conversion of substrate to product allows one to calculate directly the initial concentration of the reactant.

The Cholesterol Assay

The measurement of serum cholesterol is one of the most common tests performed in the clinical laboratory. Hypercholesterolemia (high blood cholesterol levels) can be the result of a variety of medical conditions. Among the conditions implicated are diabetes mellitus, atherosclerosis, and diseases of the endocrine system, liver, or kidney. High blood cholesterol levels do not point to a specific disease; determination of cholesterol is used in conjunction with other clinical measurements mainly for confirmation of a particular diseased condition, rather than for diagnosis of a specific ailment.

Of current interest is the positive correlation between cholesterol levels and heart disease. Cholesterol in the blood is found associated with various lipoproteins. There are two major types of cholesterol-carrying lipoproteins, high-density lipoprotein (HDL) and low-density lipoprotein (LDL). High serum levels of cholesterol-bearing LDLs are positively correlated with the development of atherosclerosis. In contrast, high levels of HDL cholesterol are inversely related to a predisposition to coronary artery disease. In general, the higher the ratio of HDL-cholesterol to LDL-cholesterol, the lower the incidence of heart disease. In order to characterize the LDLs and HDLs and determine the amount of cholesterol associated with each, it is essential to separate them physically. They differ in density and size and may be separated by ultracentrifugation, but they can also be separated by electrophoresis and selective precipitation of LDLs with divalent cations and polyanions.

Clinical measurements of **total cholesterol** in serum or plasma detect cholesterol esters in addition to cholesterol. Between 60 and 70% of the cholesterol transported in blood is in an esterified form, where the β-3-OH group on the steroid skeleton is covalently linked to a naturally occurring fatty acid, shown as an R group in Figure E11.1. A sensitive and reproducible analysis of cholesterol and cholesterol esters is based on the three reactions shown below.

$$\text{Cholesterol esters} + H_2O \xrightarrow[\substack{\text{or} \\ \text{cholesterol esterase} \\ \text{(EC 3.1.1.13)}}]{H^+ \text{ or } OH^-} \text{cholesterol} + \text{fatty acids} \qquad \textit{Equation E11.4}$$

$$\text{Cholesterol} + O_2 \xrightarrow[\substack{\text{cholesterol} \\ \text{oxidase}}]{} \text{cholest-4-ene-3-one} + H_2O_2 \qquad \textit{Equation E11.5}$$

$$H_2O_2 + \text{phenol} + \text{4-aminoantipyrine} \xrightarrow[\text{peroxidase}]{} \underset{(\lambda_{max} = 510 \text{ nm})}{\text{quinoneimine chromogen}} \qquad \textit{Equation E11.6}$$

Equation E11.4, catalyzed by cholesterol esterase, shows the hydrolysis of cholesterol esters in the sample. In Equation E11.5, cholesterol is then oxidized to cholest-4-ene-3-one. Unfortunately, this reaction does not lead to a major absorbance change at an accessible wavelength, so the rate of the reaction cannot be directly measured. When it is not convenient or possible to monitor directly the progress of a reaction, it may be possible to "couple" it to another reaction that offers a measurable absorbance change. The oxidation of cholesterol in Equation E11.5 is accompanied by production of hydrogen peroxide. In a coupled reaction, H_2O_2 rapidly reacts with phenol and 4-aminoantipyrine in the presence of horseradish peroxidase to produce a quinoneimine chromogen, which has a maximum absorbance at 510 nm (Equation E11.6, Figure E11.2). For Equation E11.5 and E11.6 to be a properly coupled system, the coupled reaction must be fast enough to decompose H_2O_2 as rapidly as it is produced in Equation E11.5. This condition exists for the coupled reactions, so the absorbance change due to the production of quinoneimine is directly proportional to the concentration of total cholesterol in the sample analyzed. A more detailed kinetic analysis of coupled assays is beyond the scope of this chapter. Interested students should refer to specialized books on enzyme kinetics.

Figure E11.1

Structures of cholesterol and a cholesterol ester.

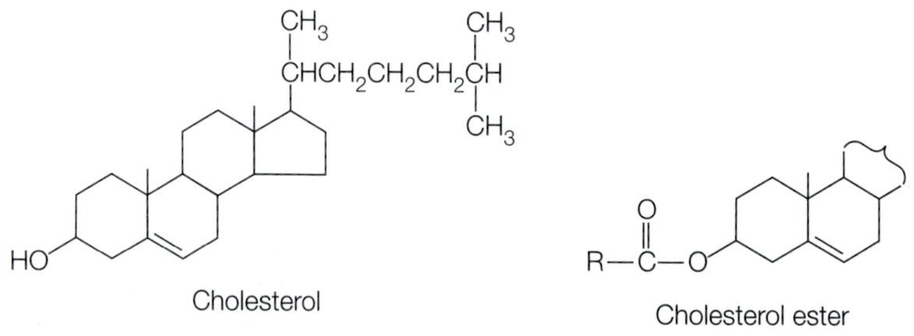

Cholesterol

Cholesterol ester

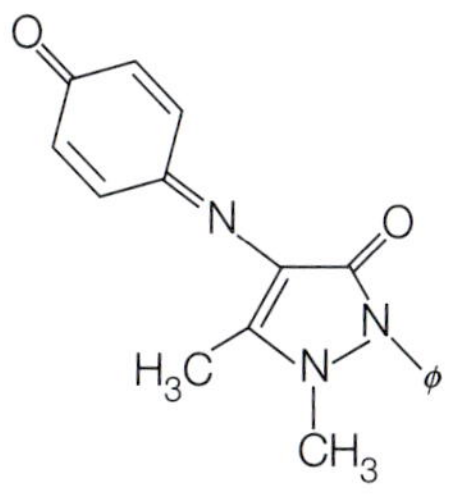

Quinoneimine
Chromogen

$\lambda_{max} = 510$ nm

Figure E11.2

Structure of quinoneimine chromogen formed in the assay of cholesterol.

In this experiment, two cholesterol measurements will be made: (1) total serum cholesterol and (2) HDL serum cholesterol, the amount of cholesterol associated with the HDL fraction. The following relationship leads to an estimate of LDL (Equation E11.7).

>> $$C_{LDL} = C_{serum} - C_{HDL}$$ **Equation E11.7**

where

C_{LDL} = concentration of cholesterol in the low-density lipoproteins and very low-density lipoproteins

C_{serum} = total concentration of serum cholesterol

C_{HDL} = concentration of cholesterol in the HDL fraction

C_{serum} will be determined by performing the described cholesterol assay directly on serum. C_{HDL} will be determined on a separated, soluble HDL fraction of serum. Very low-density lipoproteins and low-density lipoproteins are selectively removed from serum by precipitation with magnesium-phosphotungstate reagent.

The raw data collected in the experiment are in the form of absorbance measurements at 510 nm. These numbers are then converted to serum cholesterol concentration, which is reported in mg/100 mL serum.

>> Cholesterol concentration (mg/100 mL) $= \dfrac{A_{510(x)}}{A_{510(s)}} \times C_s$ **Equation E11.8**

where

$A_{510(x)}$ = absorbance at 510 nm obtained with unknown serum sample

$A_{510(s)}$ = absorbance at 510 nm obtained with standard cholesterol solution

C_s = concentration of cholesterol in standard (mg/100 mL)

Normal levels of serum cholesterol vary widely. For males, total cholesterol is in the range 130 to 320 mg/100 mL and HDL cholesterol ranges from 30 to 70 mg/100 mL. For females, the corresponding ranges are 130 to 295 and 35 to 80 mg/100 mL, respectively. Cholesterol levels depend on such factors as sex, age, diet, and emotional stress. Typical ranges at various age intervals are shown in Table E11.1.

B. Measurement of Vitamin C

Complete lack of vitamin C (ascorbic acid) in the diets of humans and other primates leads to a classic disease, scurvy. This nutritional disease, which was probably the first to be recognized, was widespread in Europe during the fifteenth and sixteenth centuries, but it is rare today.

Table E11.1

Average Total Human Serum Cholesterol Levels as a Function of Age

Age (years)	Total Cholesterol (mg/100 mL serum)
0–19	120–230
20–29	120–240
30–39	140–270
40–49	150–310
50–59	160–330

Ascorbic acid is widely distributed in nature, but it occurs in especially high concentration in citrus fruits and green plants such as green peppers and spinach. Ascorbic acid can be synthesized by all plants and animals with the exception of humans, other primates, and guinea pigs. Therefore, vitamin C must be present in our dietary substances.

The fundamental role of ascorbic acid in metabolic processes is not well understood. There is some evidence that it may be involved in metabolic hydroxylation reactions of tyrosine, proline, and some steroid hormones, and in the cleavage-oxidation of homogentisic acid. Its function in these metabolic processes appears to be related to the ability of vitamin C to act as a reducing agent.

Although there is much controversy about the exact requirement, the adult Recommended Daily Allowance of vitamin C is 70 mg per day. Some scientists and physicians have suggested doses up to 1 to 3 grams per day in order to help resist the common cold. Deficiency of vitamin C results in swollen joints, abnormal development and maintenance of tissue structures, and eventually scurvy.

Chemically, ascorbic acid is a water-soluble, slightly acidic carbohydrate that exists in an oxidized or reduced form (Equation E11.9).

>>   Ascorbic acid (reduced form)   Dehydroascorbic acid (oxidized form)   **Equation E11.9**

Both forms are biologically active. Ascorbic acid (the reduced form) is relatively stable to heat; however, dehydroascorbic acid (the oxidized form) is unstable. The lactone ring is easily hydrolyzed to diketogulonic acid, which has no antiscurvy activity. When fruits, vegetables, and other foods are heated, there is some loss of active vitamin C by conversion to diketogulonic acid.

Redox Titration of Vitamin C

Because of the clinical significance of vitamin C, it is essential to be able to detect and quantify its presence in various biological materials. Analytical methods have been developed to determine the amount of ascorbic acid in foods and in biological fluids such as blood and urine. Ascorbic acid may be assayed by titration with iodine, reaction with 2,4-dinitrophenylhydrazine, or titration with a redox indicator, 2,6-dichlorophenolindophenol (DCIP) in acid solution. The latter method will be used in this experiment because it is reasonably accurate, rapid, and convenient and can be applied to many different types of samples.

The reaction of DCIP with ascorbic acid is shown in Figure E11.3. Ascorbic acid reduces the indicator dye from an oxidized form (red in acid) to a reduced form (colorless in acid). The procedure is simple, beginning with dissolution of the sample to be tested in metaphosphoric acid. An aliquot of the sample is then titrated directly with a solution of DCIP. Although the original DCIP solution is blue, it becomes light red in the acid solution. Upon reaction with ascorbic acid in the sample, the dye becomes colorless. Titration is continued until there is a very slight excess of dye added (faint pink color remains in the acid solution).

Samples for analysis often contain traces of other compounds, in addition to ascorbic acid, that reduce DCIP. One way to minimize the interference of other substances is to analyze two identical aliquots of the sample. One aliquot is titrated directly and the total content of all reducing substances present is determined. The second aliquot is treated with ascorbic acid oxidase to destroy ascorbic acid and then titrated with DCIP. The

Figure E11.3

Reaction of the redox dye DCIP with ascorbic acid.

- - - - - - - - - - - - - - - - - - - -

Red in acid

Colorless in acid

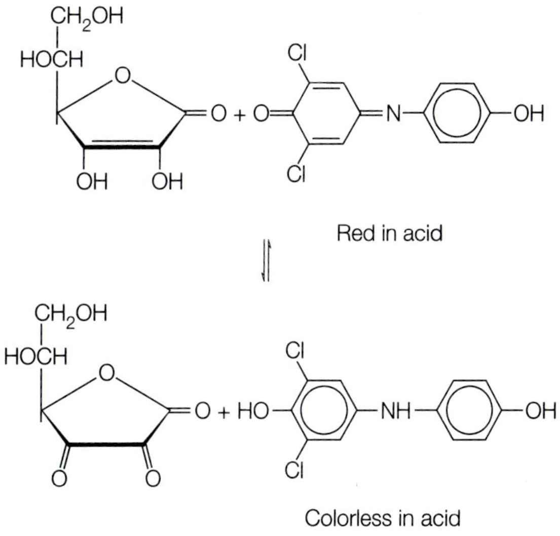

second titration allows determination of reducing substances other than ascorbic acid. A second way to reduce the effect of other reducing substances is to perform the titration in the pH range 1 to 3. The interfering agents react very slowly with DCIP under these conditions.

Determination of vitamin C in biological fluids such as blood and urine is more difficult because only small amounts of the vitamin are present and many interfering reducing agents are present. Substances containing sulfhydryl groups, sulfite, and thiosulfate are common in biological fluids and react with DCIP, but much more slowly than ascorbic acid. The interference by sulfhydryl groups is often minimized by the addition of *p*-chloromercuribenzoic acid.

C. Overview of the Experiment

Cholesterol

Clinical methods based on enzyme reactions are now used routinely for the determination of glucose, cholesterol, urea, uric acid, and many other metabolites in blood, urine, and other biological fluids as well as tissue specimens. Diagnostic kits containing all the required reaction components are commercially available at reasonable cost. These kits make it possible for measurements to be made in a simple, accurate, rapid, and reproducible manner. In modern hospital laboratories, much of this testing is completely automated, which greatly improves the speed and reproducibility, but not necessarily the accuracy. The procedures used to measure total cholesterol and HDL-cholesterol are outlined in Figures E11.4 and E11.5. Depending

Figure E11.4

The flowchart for measurement of total serum cholesterol.

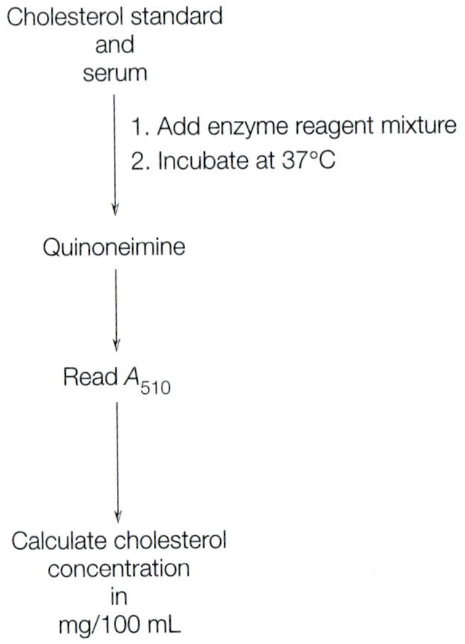

Cholesterol standard
and
serum

    1. Add enzyme reagent mixture
    2. Incubate at 37°C

Quinoneimine

Read A_{510}

Calculate cholesterol
concentration
in
mg/100 mL

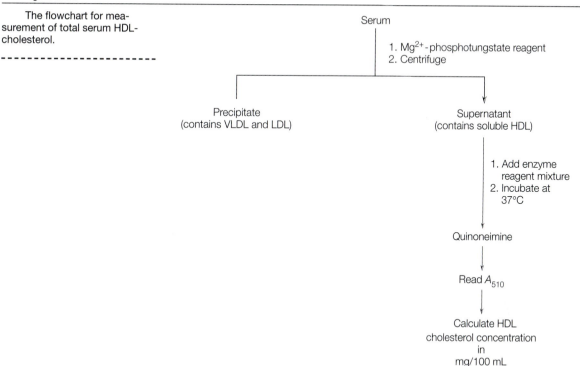

Serum

1. Mg^{2+}-phosphotungstate reagent
2. Centrifuge

Precipitate
(contains VLDL and LDL)

Supernatant
(contains soluble HDL)

1. Add enzyme
   reagent mixture
2. Incubate at
   37°C

Quinoneimine

Read A_{510}

Calculate HDL
cholesterol concentration
in
mg/100 mL

on the number of serum samples analyzed, this part of the experiment will require 1–2 hours to complete.

Because of the possible transmission of infectious diseases during the use of human blood samples, it is recommended that students use bovine, porcine, or other animal serum samples. Sera from a wide variety of animals, including humans, are available in a lyophilized form from Sigma Chemical Co. When reconstituted with water, these provide convenient unknowns.

Vitamin C

In this experiment several samples will be prepared, and the amount of vitamin C in each will be determined. Each sample will be titrated with standardized redox indicator, DCIP. The vitamin C content will be calculated in terms of milligrams per milliliter (mg/mL) of sample.

This experiment is flexible in that students may choose the samples they wish to analyze. It is suggested that several fruits and vegetables as well as an unknown ascorbic acid sample be available for analysis. The vitamin C content of raw and cooked vegetables can also be measured and compared. Commercially available vitamin C tablets or multivitamin pills provide interesting samples for analysis. The time required for this experiment can be adjusted by controlling the number and type of samples.

II.   MATERIALS AND SUPPLIES

A. Measurement of Cholesterol

- Serum samples. Bovine or porcine blood samples can be obtained from a local slaughterhouse or serum samples may be purchased.
- Cholesterol aqueous standard I. This contains cholesterol (200 mg/100 mL) in water containing stabilizers and sodium azide as preservative.
- Cholesterol aqueous standard II. This contains cholesterol (50 mg/100 mL) in water solution containing stabilizers and sodium azide as preservative.
- LDL precipitating reagent. This solution contains phosphotungstate, magnesium ions, and sodium azide.
- Cholesterol assay solution. The stock reagent contains pancreatic cholesterol esterase, microbial cholesterol oxidase, horseradish peroxidase, 4-aminoantipyrine, and phenol. Your instructor will reconstitute the stock reagent by addition of water.
- Spectrophotometer and cuvettes. Any spectrometer that measures in the range 510 ± 10 nm is suitable.
- Constant-temperature bath at 37°C
- Centrifuge, capable of speeds up to 2000 rpm
- Centrifuge tubes, 10-mL, conical
- Saline solution, 0.15 M NaCl in water
- Hydrocarbon foil

B. Measurement of Vitamin C

- Fruit juices—orange, grapefruit, lemon, or lime
- Whole fruits and vegetables—oranges, grapefruit, lemons, limes, green peppers, tomatoes, potatoes, spinach, lettuce, cabbage, and others
- Vitamin C tablets or multivitamin pills
- Metaphosphoric acid/acetic acid solution, 4%
- Unknown ascorbic acid in metaphosphoric acid/acetic acid solution, 0.5 to 3.0 mg/mL
- Standard ascorbic acid in metaphosphoric acid/acetic acid solution, 0.50 mg/mL
- 2,6-Dichlorophenolindophenol solution in H_2O, 25 mg/100 mL
- Ascorbic acid oxidase, lyophilized powder
- Buret; use a 10-mL microburet
- Mortar and pestle
- Knife

- Filter paper, fast flow
- Glass funnel

III.   EXPERIMENTAL PROCEDURE

> **CAUTION**
>
> Several reagents used in this experiment contain sodium azide, a deadly poison. **Never pipet any solutions by mouth.**
>
> Sodium azide may react with lead and copper plumbing to form highly explosive metal azides. If allowed in your region, flush with a large volume of water to prevent azide accumulation. **Because of possible pathogenic contamination, use of human blood serum is not recommended.**

A. Measurement of Cholesterol

Measurement of Total Cholesterol

The procedure will be described for the analysis of two different serum samples. Turn on the spectrometer and allow warm-up for 15 to 20 minutes. Set the wavelength to 510 nm. Obtain two 3-mL cuvettes or colorimeter tubes for the blank and standard. In addition, one cuvette will be needed for each serum sample to be tested. Label the cuvettes Blank, Standard, Serum$_1$, and Serum$_2$. Make the following additions:

Blank: 0.02 mL of water

Standard: 0.02 mL of cholesterol standard I

Serum$_1$: 0.02 mL of Serum$_1$

Serum$_2$: 0.02 mL of Serum$_2$

Pipet 1.0 mL of cholesterol enzyme reagent into each tube. Mix well, but do not shake. Shaking will cause foaming and protein denaturation. Mixing is best done by tightly covering the cuvette with hydrocarbon foil and gently inverting 3 to 5 times. Incubate the cuvettes at 37°C in a constant-temperature water bath for 10–15 minutes. Remove all the cuvettes from the bath and wipe dry with a tissue. Add 2.0 mL of saline water to each cuvette and mix well. Place the blank in the spectrometer and adjust A_{510} to 0.0. Read and record in your notebook A_{510} for the Standard, Serum$_1$, and Serum$_2$. Take the absorbance readings within 30 minutes after removal from the bath to avoid color fading.

This procedure does not take into account any absorbance due to the serum. If the serum is turbid, a correction should be made by measuring the absorbance at 510 nm of a 0.02-mL sample of blood serum in 3.0 mL of saline water. Read the A_{510} of this solution using saline water as blank. Record this reading in your notebook as A_c for correction. The calculation for cholesterol concentration will be described in the Analysis of Results.

Measurement of HDL-Cholesterol

Obtain a conical centrifuge tube for each serum sample you wish to analyze. To each tube add 0.4 mL of serum and 0.05 mL of phosphotungstate precipitating reagent. Mix well. Centrifuge the tubes for 10 minutes at 2000 rpm. Obtain four cuvettes, one each for Blank, Standard, $Serum_1$, and $Serum_2$. Label the cuvettes and add the following reagents:

Blank: 0.05 mL of water

Standard: 0.05 mL of cholesterol aqueous standard II (50 mg/100 mL)

$Serum_1$: 0.05 mL of $supernatant_1$ from centrifugation

$Serum_2$: 0.05 mL of $supernatant_2$ from centrifugation

Add 1.0 mL of cholesterol enzymatic reagent to each cuvette, cover with hydrocarbon foil, and mix well. Incubate for 10–15 minutes at 37°C. Add 2.0 mL of saline water to each cuvette. Adjust spectrometer to zero with Blank and read A_{510} for each sample. Record in your notebook for further calculations.

B. Measurement of Vitamin C

Standard Ascorbic Acid Solution

Fill a microburet with DCIP solution. The top of the solution should be at or slightly below the zero mark of the buret, and the glass tip must be full of solution. Using a pipet, transfer 1.0 mL of the ascorbic acid standard solution to a 50-mL Erlenmeyer flask containing 5 mL of 4% metaphosphoric acid/acetic acid solution. Read and record the initial reading on the buret. Titrate by rapid, dropwise addition of DCIP from the buret while mixing the contents of the flask. Add DCIP solution until a distinct rose-pink color persists for 15 to 20 seconds. Record the final reading on the buret. Repeat this procedure twice more, each time with a fresh 1.0-mL sample of ascorbic acid standard. In a similar fashion, titrate three blanks, each containing 5.0 mL of 4% metaphosphoric acid/acetic acid solution and 1.0 mL of water. Average the results for each series of measurements.

Unknown Ascorbic Acid Solution

Obtain a sample containing an unknown amount of ascorbic acid from your instructor. Place 1.0 mL of the unknown in a 50-mL Erlenmeyer flask. Add 5.0 mL of metaphosphoric acid solution to the flask and titrate as before with DCIP. Repeat with two more samples of the unknown.

Fruit Juice

Mix the stock solution of juice well and pour about 15 mL into a small beaker. Dilute 10.0 mL of this juice to 50 mL with the metaphosphoric acid solution. Filter this solution through a rapid flow, fluted filter paper. Pipet 10.0 mL of the filtrate into a 50-mL Erlenmeyer and titrate rapidly to a persistent rose-pink color with the standard DCIP solution as previously described. Repeat the titration with two 10.0-mL samples of the diluted juice. For a blank, add 10.0 mL of water to 40 mL of metaphosphoric acid solution and titrate three 10.0-mL samples with DCIP. Record the volume of DCIP necessary to titrate each juice sample and each blank.

To test for the presence of interfering substances in the juice, pipet a 10.0-mL sample of the fresh, undiluted juice into a 50-mL Erlenmeyer. Add a few crystals of ascorbic acid oxidase to destroy the ascorbic acid. Let stand, after gentle mixing, for 10 minutes. Add 40 mL of metaphosphoric acid solution. Titrate three 10.0-mL portions of this diluted juice with DCIP.

Whole Fruit

Peel the fruit, weigh to the nearest 0.1 g, and cut a sample of 25 to 50 g from the edible portion. Weigh the sample to the nearest 0.1 g. Slice the sample into many small pieces, transfer to a mortar, and grind with 25 mL of the metaphosphoric acid solution. Pour the liquid from the extract into a 100-mL volumetric flask. Grind the solid residue in the mortar twice using 25 mL of the metaphosphoric acid solution. Add the liquid extract each time to the 100-mL volumetric flask. Add metaphosphoric acid solution to the volumetric flask to a total volume of 100 mL. Filter the solution through a rapid flow, fluted filter paper. Transfer 10.0 mL of the filtrate into a 50-mL Erlenmeyer flask. Titrate rapidly with the DCIP solution. Repeat the titration on two more 10.0-mL samples of the filtrate. Titrate 10.0 mL of metaphosphoric acid solution as a blank.

Vitamin C in Raw and Boiled Vegetable

Cut two samples of a vegetable (green pepper, lettuce, cabbage, etc.) so that each weighs exactly the same amount (10–15g). Cut one sample into smaller pieces and grind in a mortar with 25 mL of the metaphosphoric acid solution. Pour the liquid from the extract into a 100-mL volumetric flask. Grind the solid residue in the mortar two more times using 25 mL of the metaphosphoric acid solution. Each time, combine all the extracts in the 100-mL flask. Add metaphosphoric acid solution to the volumetric flask to a total volume of 100 mL. Filter the solution through a rapid flow, fluted filter paper. Transfer 10.0 mL of the filtrate into a 50-mL Erlenmeyer flask and titrate, as before, with the DCIP solution. Repeat the titration on two more 10.0-mL samples of the filtrate. Titrate 10.0 mL of metaphosphoric acid solution as a blank.

Place the other vegetable sample in a beaker containing 25 mL of water. Heat the water to boiling and continue to boil for 10 minutes. Pour off and save the water, cut vegetable into small pieces, and transfer to a mortar. Grind with three 25-mL portions of metaphosphoric acid solution as before. Combine all the metaphosphoric acid extracts in a 100-mL volumetric flask. Filter the extract and titrate with DCIP as before. Mix 5.0 mL of the cooking water with 5.0 mL of metaphosphoric acid solution. Titrate with DCIP.

Vitamin C Tablets or Multivitamin Pills

Design your own procedure for analysis.

IV. ANALYSIS OF RESULTS

A. Measurement of Cholesterol

Calculation of Total Serum Cholesterol

The cholesterol assay as used in this experiment is linear up to 500 mg/100 mL serum. A calibration curve is not essential and a single standard can be used. Use Equation E11.8 to calculate the cholesterol concentration. If the serum sample was turbid, you should have determined A_{510} of serum in saline water, A_c. If so, subtract this from $A_{510(x)}$

$$\text{Total cholesterol concentration (mg/100 mL)} = \frac{A_{510(x)} - A_c}{A_{510(s)}} \times C_s$$

where A_c is the absorbance of diluted serum at 510 nm. The concentration of standard cholesterol (C_s) is 200 mg/100 mL. Calculate total cholesterol in the serum samples you tested. Compare your results with normal values of serum cholesterol.

Calculation of HDL-Cholesterol Concentration

The calculation is identical to that of total serum cholesterol except that the standard cholesterol solution has a concentration of 50 mg/100 mL. Calculate the concentration of serum HDL-cholesterol in the samples you tested. Also, calculate the concentration of cholesterol associated with LDL for each serum sample.

B. Measurement of Vitamin C

Standard Ascorbic Acid Solution

Subtract the volume of titrant required for the blank from the titrant required for the sample. The difference represents the volume of DCIP that is equivalent to 0.50 mg of vitamin C. Calculate the standard deviation of your answer. What are the limits for 95% confidence? Review Chapter 1, Section F for statistical analysis.

Unknown Ascorbic Acid Solution

Calculate the concentration of ascorbic acid in the unknown sample in units of mg/mL. Express your answer with confidence limits at the 95% confidence level.

Fruit Juice

Calculate the amount of ascorbic acid in the juices and report in terms of mg/100 mL of pure juice. Again, calculate confidence limits at the 95% confidence level. Remember that you started with 10 mL of pure juice, diluted it with 40 mL of metaphosphoric acid solution, and titrated 10 mL of the diluted juice. Are there reducing substances in addition to ascorbic acid in the juice?

Whole Fruit

Calculate the total amount of ascorbic acid in the whole fruit. Use proper statistical analysis.

Vitamin C in Raw and Boiled Vegetable

Calculate the amount of ascorbic acid in each sample. Was any lost during the boiling process? Explain. Could vitamin C be detected in the boiled water? Typical values for vitamin C content are shown below.

Orange juice	20–80 mg/100 mL Average = 45 mg/100 mL
Grapefruit juice	35–65 mg/100 mL Average = 40 mg/100 mL
Lemon juice	30–70 mg/100 mL Average = 45 mg/100 mL
Lime juice	5–40 mg/100 mL Average = 15 mg/100 mL
Plasma	0.2–2 mg/100 mL

Vitamin C Tablets or Multivitamin Pills

Calculate the amount of ascorbic acid in each sample. Does it agree with the label on the bottle?

Study Problems

▶ *1.* Assume you are to determine the blood serum level of cholesterol by the single standard method. The absorbance data obtained are: $A_{standard}$ = 0.350 (250 mg/100 mL); A_{sample} = 0.390; and $A_{correction}$ = 0.02. Calculate the concentration of cholesterol in mg/100 mL. What would be the % error if $A_{correction}$ were not used?

▶ *2.* What is the primary assumption made in the use of a single standard as in Problem 1?

▶ *3.* The cholesterol assay is linear to a level of 500 mg/100 mL. What would you do if a determination for cholesterol showed the level to be about 650 mg/100 mL?

    4. Write the reaction for the oxidation of ascorbic acid catalyzed by ascorbic acid oxidase.

▶ *5.* A 10-mL sample of pure orange juice was diluted with 40 mL of metaphosphoric acid solution. Then 10 mL of the diluted juice was titrated to an end point with 9.27 mL of DCIP. A blank required 0.25 mL of DCIP. A 1-mg sample of pure ascorbic acid required 6.52 mL (after blank correction) of DCIP. What is the concentration of ascorbic acid in the orange juice in mg/100 mL?

▶ *6.* How does *p*-chloromercuribenzoic acid function to remove sulfhydryl group interference during the DCIP titration?

▶ *7.* A whole, peeled orange weighed 100 g. Three sections (30 g) were extracted with metaphosphoric acid and the total extract filtered and diluted to 100 mL. A sample of 10 mL of the filtered extract required 4.10 mL of DCIP after blank correction. Also, 1 mg of standard ascorbic acid required 7.2 mL of DCIP (blank corrected). How many milligrams of ascorbic acid are present in the whole, peeled orange?

▶ *8.* To what class of biomolecules does vitamin C belong?

▶ *9.* Explain the principles behind the use of the phosphotungstate reagent to measure HDL-cholesterol.

▶ *10.* What is the reaction class for each of the three enzymes used in the cholesterol assay? Choose from oxidoreductase, hydrolase, transferase, and ligase.

Further Reading

J. T. Baker Chemical Co., ***Product Information Bulletin***, Commodity No. H 114, (1971), Phillipsburg, NJ. Analytical uses of 2,6-DCIP.

R. Boyer, ***Concepts in Biochemistry*** (1999), Brooks/Cole (Pacific Grove, CA), pp. 123, 181 vitamin C; pp. 251–252, 577–589 cholesterol.

T. Devlin, Editor, ***Textbook of Biochemistry with Clinical Correlations***, 4th ed. (1997), John Wiley & Sons (New York), pp. 1127, 1098. Vitamin C and cholesterol.

R. Eitenmiller and W. Landen, ***Vitamin Analysis for the Health and Food Sciences*** (1998), CRC Press (Boca Raton, FL). Review of vitamin measurement.

R. Garrett and C. Grisham, ***Biochemistry***, 2nd ed. (1999), W. B. Saunders (Orlando, FL), pp. 599–600; 840–846. An introduction to vitamin C and cholesterol.

B. Halliwell, ***Trends Biochem. Sci.*** **24,** 255–259 (1999). "Vitamin C: Poison, Prophylactic or Panacea?"

L. Kaplan and A. Pesce, ***Clinical Chemistry—Theory, Analysis and Correlation*** (1989), C. V. Mosby (St. Louis). A teaching text for medical technologists and laboratory technicians.

L. Machlin, ***Handbook of Vitamins***, 2nd ed. (1991), Marcel Dekker (New York). A complete guide to the vitamins.

S. Margolis and R. Schapira, *J. Chromatogr.* **690,** 25–33 (1996). "The Measurement of L-Ascorbic Acid and D-Ascorbic Acid in Biological Samples."

C. Mathews, K. van Holde, and K. Ahern, *Biochemistry,* 3rd ed. (2000), Benjamin/ Cummings (San Francisco), pp. 554–555; 323–324. An introduction to vitamin C and cholesterol.

Sigma Chemical Co., Cholesterol, Total and HDL, Technical Bulletins, Procedure No. 351 and Procedure No. 352–3 (1985), Sigma Diagnostics (P.O. Box 14508, St. Louis, MO 63178).

L. Stryer, *Biochemistry,* 4th ed. (1995), W. H. Freeman (New York), pp. 267; 455, 691–702. Introduction to vitamin C and cholesterol.

N. Tietz, Editor, *Fundamentals of Clinical Chemistry,* 2nd ed. (1976), W. B. Saunders (Philadelphia), pp. 506–514; 547–551; 999–1002. A classic reference in the area of clinical chemistry.

D. Voet, J. Voet, and C. Pratt, *Fundamentals of Biochemistry* (1999), John Wiley & Sons (New York), pp. 135; 228–229; 260–264. Introduction to vitamin C and cholesterol.

Cholesterol on the Web

http://www.ktl.fi/monica/public/publications/manual/part3/iii-2.htm
> Click on Recommended analytical procedures for total cholesterol and HDL-cholesterol measurement. Review procedure details.

12

ACTIVITY AND THERMAL STABILITY OF GEL–IMMOBILIZED PEROXIDASE

• **Recommended Reading**

Chapter 4, Section B; *Experiment 5.*

• **Synopsis**

Immobilized enzymes are becoming increasingly important in commercial processes. In this experiment, students will trap molecules of the enzyme horseradish peroxidase within a polyacrylamide gel matrix. The reaction kinetics and thermal stability of the immobilized enzyme will be measured. This experiment introduces students to the use of enzymes in biotechnology.

I. INTRODUCTION AND THEORY

The principles of biotechnology, the application of biological cells or cell components to technically useful operations, are seldom introduced to students in today's biology, chemistry, and biochemistry classes. With the increasing impact of modern biology on all areas of science and industry, it is imperative that students have some understanding of biotechnology operations.

The use of immobilized enzymes in biotechnology is becoming widespread and encompasses several areas, including the chemical industry, pharmaceuticals, food production, medicine, environmental waste, clinical chemistry, and agriculture. Specific applications include wine making, cheese making, production of alcohol for fuel, treatment of wastewater, metal mining, measurement of glucose and other constituents in body fluids, and enzyme replacement in individuals with genetic disorders.

Immobilized Enzymes

Enzymes are valuable and versatile commercial reagents for the following reasons: (1) they have high catalytic activities, (2) they catalyze a great variety of reactions, (3) they provide the potential for stereospecific reactions, (4) they function under generally mild reaction conditions, and (5) side reactions and secondary products are rare. However, along with these advantages as catalysts, enzymes also have some disadvantages: (1) they are generally available only in small quantities, (2) they are fragile, unstable molecules, and (3) they are expensive from a commercial point of view. Therefore, their commercial use is practical only if methods can be found to increase their stability and if enzymes can be recovered and reused after the desired process is complete. In other words, enzymes used in commercial processes must be recycled. One of the best methods for ensuring recovery of active enzyme is immobilization of the catalyst molecules. Immobilization may be defined as any process that limits the movement or free diffusion of the enzyme molecule and usually involves attachment of the enzyme to an inert, water-insoluble support or entrapment within a water-insoluble matrix or microsphere.

In addition to recovery and reuse of immobilized enzymes, there are several other advantages to their use: (1) since they are in an insoluble form, they can be separated from the reaction mixture and do not contaminate the product; (2) an immobilized enzyme is often more stable to heat, pH change, organic solvents, and other adverse reaction conditions than a free enzyme; (3) they can be used in a continuous manner, for example, in a column flow reactor; and (4) the end point of the reaction can be controlled simply by physically separating the enzyme from the solution. Immobilized enzymes are also becoming important in basic biochemical research. Many enzymes in their natural environment in organisms are not free but are immobilized in cells, tissues, and membranes. Even cytoplasmic enzymes are held somewhat rigidly in a gelatinous matrix or cytoskeleton. Therefore, studying the characteristics of immobilized enzymes will provide additional insight into how enzymes function in the living organism.

Methods of Immobilization

Four methods have been developed for enzyme immobilization: (1) physical adsorption onto an inert, insoluble, solid support such as a polymer; (2) chemical covalent attachment to an insoluble polymeric support; (3) encapsulation within a membranous microsphere such as a liposome; and (4) entrapment within a gel matrix. The choice of immobilization method is dependent on several factors, including the enzyme used, the process to be carried out, and the reaction conditions. In this experiment, an enzyme, horseradish peroxidase (donor: H_2O_2 oxidoreductase; EC 1.11.1.7), will be imprisoned within a polyacrylamide gel matrix. This method of entrapment has been chosen because it is rapid, inexpensive, and allows kinetic characterization of the immobilized enzyme. Immobilized peroxidase catalyzes a reaction that has commercial potential and interest, the reductive cleavage of hydrogen peroxide, H_2O_2, by an electron donor, AH_2:

$$>> \quad H_2O_2 + AH_2 \underset{}{\overset{peroxidase}{\rightleftharpoons}} 2H_2O + A \qquad \textit{Equation E12.1}$$

The most commonly used matrix for entrapment of enzymes is cross-linked polyacrylamide. When acrylamide and methylene bisacrylamide (a cross-linking agent) are allowed to polymerize, a cross-linked polymer is produced (Figure E12.1). If the proper ratio of acrylamide and methylene bisacrylamide is used, the extent of cross-linking is such that relatively large enzyme molecules can be trapped within the polymer cage, while still allowing relatively small substrate and product molecules to diffuse readily in and out of the matrix. Acrylamide gel has another advantage in that it is non-ionic; therefore, charged reactant and product molecules are not retained inside the gel. To produce an entrapped enzyme, a solution of the desired enzyme is mixed with the monomer acrylamide, cross-linking agent methylene bisacrylamide, and a catalyst added for polymer initiation. Within a few minutes, a solid gel mass is produced, which is washed to remove free enzyme that has not been trapped. The gel containing the trapped enzyme is now ready for further investigation and analysis.

Characterization of Immobilized Enzymes

Before an immobilized enzyme can be used for an industrial process, it is essential to characterize it in terms of its catalytic and kinetic properties. A quantitative assay must be developed to measure the activity, kinetic parameters, and stability of the enzyme. In a coupling reaction, H_2O_2 rapidly reacts with phenol and 4-aminoantipyrine (electron donor) in the presence of peroxidase to produce a quinoneimine chromogen (Equation E12.2, Figure E11.2), which is intensely colored with a maximum absorbance at 510 nm. (This is the same as the product formed in the analysis of cholesterol in Experiment 11.)

$$>> \quad H_2O_2 + phenol + 4\text{-aminoantipyrine} \underset{}{\overset{peroxidase}{\rightleftharpoons}} quinoneimine + 2H_2O \quad \textit{Equation E12.2}$$

The amount of peroxidase present will influence the amount of quinoneimine product formed. In fact, there is a direct and linear relationship between the

Figure E12.1

Entrapment of an enzyme molecule (◯) by cross-linked polyacrylamide gel prepared by mixing enzyme with acrylamide, methylene bisacrylamide, and catalyst.

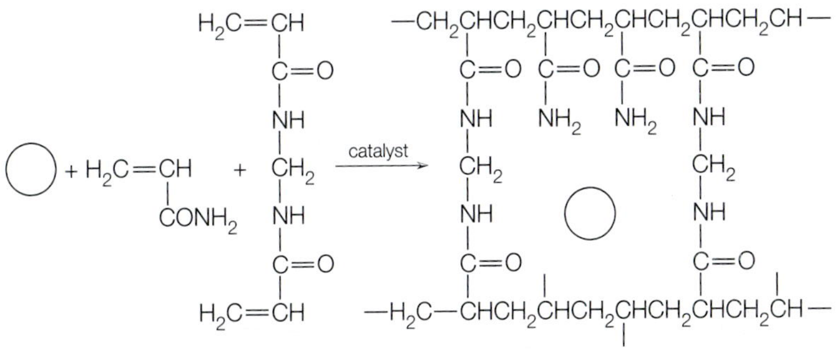

quantity of peroxidase present in the gel or solution and the intensity of the color produced. The intensity of color or absorbance can be measured in a spectrophotometer. Therefore, this assay can be used to measure the rate of the enzyme-catalyzed reaction, which allows calculation of important kinetic constants including the number of enzyme activity units (U), the Michaelis constant (K_M), and the maximum velocity (V_{max}). An enzyme supported within a matrix may not behave kinetically in the same way as one free in solution. There are several reasons for this: (1) immobilization may force the enzyme molecule to take on a different conformation, (2) the chemical microenvironment around the immobilized enzyme may be different, and (3) the rate of the reaction may be more greatly influenced by the diffusion of substrate molecules into the gel. The stability of an enzyme is also often changed upon immobilization. This characteristic may be increased in some instances due to stabilizing effects of the surrounding matrix environment, or it may be decreased because of a denaturing microenvironment inside the gel.

Overview of the Experiment

Several reaction characteristics of gel-immobilized peroxidase, including activity and thermal stability, will be examined in this experiment. A spectrophotometric assay for peroxidase is introduced. The literature reports many assays for measurement of peroxidase activity. The aminoantipyrine-phenol assay is selected because it is rapid, convenient, and accurate and requires less toxic reagents than other assays.

Time requirements for this experiment are:

Part A: Preparation of immobilized peroxidase–45 minutes.
Part B: Assay of immobilized peroxidase–1 hour.
Part C: Thermal stability of immobilized peroxidase–1 hour.

II. MATERIALS AND SUPPLIES

CAUTION

Phenol is a chronic poison. It may be fatal if inhaled, swallowed, or absorbed through skin.

First Aid: In case of contact, immediately flush eyes or skin with copious amounts of water for at least 15 minutes. If inhaled, remove to fresh air.

Acrylamide in the unpolymerized form is a skin irritant and an accumulative neurotoxin. Wear gloves and a mask while weighing the dry powder. Do not breathe the dust. Prepare all acrylamide solutions in the hood. Do not mouth pipet any solutions used for gel formation.

First Aid: In case of contact, immediately flush with copious amounts of water for at least 15 minutes. If inhaled, remove to fresh air.

Disposal of Gels: Allow to dry in air and throw in a special container provided by the instructor.

- Potassium phosphate buffer, 0.2 M, pH 7.0
- Screw cap scintillation or similar type of vials, 20 mL
- Reagents for gel preparation:

  A. 30% acrylamide plus 0.8% methylene bisacrylamide in potassium phosphate buffer

  B. 10% ammonium persulfate in potassium phosphate buffer

  C. N,N,N',N'-tetramethylethylenediamine (TEMED)
- Ammonium persulfate solution, 1.0 g/ 10 mL of phosphate buffer (prepare fresh just before use)
- Horseradish peroxidase, 0.1 mg/mL in glass-distilled H_2O (150–200 units/mg)
- Spectrophotometer and cuvettes
- Clinical centrifuge
- 4-Aminoantipyrine–phenol solution. Prepare by dissolving 810 mg phenol (**Caution**) in 40 mL of glass-distilled water and adding 25 mg of 4-aminoantipyrine. Dilute to a final volume of 50 mL with glass-distilled water.
- Hydrogen peroxide, 0.0017 M in water. Prepare by mixing 1 mL of 30% H_2O_2 with 99 mL of glass-distilled water. Further dilute 1 mL of this solution to 50 mL with 0.2 M phosphate buffer, pH 7.0 (prepare fresh just before use).
- Syringe (5 mL) with 0.8 μ filter system
- Constant-temperature water bath, 60°C

III. EXPERIMENTAL PROCEDURE

A. Preparation of Immobilized Peroxidase

To prepare the acrylamide gel, add the following to a 20-mL vial: 3.25 mL potassium phosphate buffer, 2.7 mL of solution A (acrylamide-bisacrylamide), and 80 μL of solution B (ammonium persulfate). Mix and add 1.0 mL of 0.1 mg/mL peroxidase solution. Add 10 μL of reagent TEMED and mix with a vortex mixer. The solution should become opaque within a few minutes and completely polymerized within 20–30 minutes. With a spatula, transfer the gel to a vacuum filtration system to remove most of the solution. To wash the gel, transfer it to a test tube containing 5 mL of water. Break up the gel by aspirating with a Pasteur pipet. Centrifuge the gel mixture for 5 minutes at 1000–1200 rpm. Decant supernatant and add 10 mL of water to the gel. Again, break up the gel by aspirating with a Pasteur pipet and centrifuge as before. Repeat this washing process two more times. The final centrifugation should be done for 5 minutes at 1200–1500 rpm. Dry the gel by vacuum filtration for a few minutes and weigh on a balance. Approximately 2–3 g of semiwet gel will be obtained. Proceed directly to part B.

B. Assay of Immobilized Peroxidase

A 3-minute fixed-time assay will be used to measure peroxidase activity. Two tubes will be set up for each assay, one tube for the zero point and the other tube for the 3-minute assay. The amount of gel-immobilized peroxidase will be different for each assay. Obtain six test tubes for reaction vessels and label them 1–6. Weigh two samples of 0.05 g of immobilized peroxidase and transfer one sample to tube 1 and the other sample to tube 2. To tubes 1 and 2, add 2.5 mL of the aminoantipyrine-phenol solution. Mix well. To tube 1 (zero point), add 2.5 mL of the H_2O_2 solution, rapidly mix well, and immediately quench the reaction by pouring into the barrel of a syringe with filter system. Use the plunger to force the solution through the filter into a glass cuvette. This whole procedure from mixing to transfer should take no more than 10 seconds. Read and record in your notebook the absorbance of the reaction mixture at 510 nm.

To tube 2 (3-minute point), add 2.5 mL of the H_2O_2 solution, immediately mix, and note the time. Gently and continuously mix the reaction mixture. At the end of exactly 3 minutes, transfer the reaction mixture to the barrel of a syringe with filter system. Force the solution through the filter into a glass cuvette. Read the absorbance of the reaction mixture at 510 nm and record in your notebook.

In tubes 3 and 4, 0.1 g of immobilized peroxidase will be used. Tube 3 will represent the zero point and tube 4 the 3-minute point. Weigh two samples of 0.1 g of immobilized peroxidase and transfer to tubes 3 and 4. Add 2.5 mL of the aminoantipyrine-phenol solution to each tube and mix well. Measure the peroxidase activity in each tube by adding 2.5 mL of the H_2O_2 solution as before; tube 3 will be treated like tube 1 and tube 4 like tube 2. Read and record the A_{510} for tubes 3 and 4. Repeat the procedure with tubes 5 (zero point) and 6 (3-minute point), each containing 0.2 g of immobilized peroxidase.

C. Thermal Stability of Immobilized Peroxidase

In this section, the thermal stability of acrylamide gel–immobilized peroxidase will be compared to that of the free enzyme. The free enzyme is assayed in the following manner: Dilute 1 mL of the stock horseradish peroxidase (0.1 mg/mL, 15 units/mL) with 299 mL of glass-distilled water. Add 1.0 mL of this diluted enzyme to each of two test tubes. Place one of the tubes in a 60°C water bath for exactly 4 minutes. Allow the other tube to sit at room temperature for the same time interval. Cool the higher-temperature tube to room temperature by placing in a water bath. To each tube add 2.0 mL of the aminoantipyrine-phenol stock solution and 2.0 mL of the H_2O_2 solution and mix well. Allow the tubes to sit at room temperature for exactly 3 minutes; then immediately read the A_{510} for each. Record the results in your notebook.

The immobilized enzyme is assayed in the following manner: Weigh out two samples of 0.1 g of the gel and transfer to two separate test tubes each containing 0.5 mL of phosphate buffer. Place one of the tubes in the 60°C water bath and allow the other to remain at room temperature. At the end of 4 minutes, remove the tube from the 60°C bath, cool to room temperature in a water bath, and add 2.25 mL of aminoantipyrine-phenol solution and 2.25 mL of H_2O_2 solution. Gently mix the reaction mixture for exactly 3 minutes and remove the gel by passing the mixture through the syringe-filter system. Read the A_{510}. To the room-temperature tube, add 2.25 mL of aminoantipyrine-phenol solution and 2.25 mL of H_2O_2. Gently mix for exactly 3 minutes and separate the gel with a syringe-filter system. Read the A_{510} of the reaction mixture.

IV.   ANALYSIS OF RESULTS

A. Preparation of Immobilized Peroxidase

Describe your observations of the polymerization process. How many grams of acrylamide gel–immobilized peroxidase did you obtain?

B. Assay of Immobilized Peroxidase

The activity of the immobilized peroxidase can be calculated from the absorbance change for each reaction mixture. The absorbance change is calculated as follows:

$$\Delta A = A_{3\ min} - A_{0\ min}$$

where

ΔA = overall absorbance change

$A_{3\ min}$ = absorbance at 510 nm of tubes 2, 4, and 6

$A_{0\ min}$ = absorbance at 510 nm of tubes 1, 3, and 5

Calculate the ΔA/min for each of the three sets of conditions (0.05 g, 0.10 g, and 0.2 g of gel). Prepare a plot of ΔA/min (y axis) vs. mg of gel (x axis). Describe and explain the shape of the graph.

Use the equation below to calculate the units of activity per mg of gel. The number 6.58 in the denominator is the absorption coefficient for the quinoneimine chromogen assay product.

$$\text{Units/mg} = \frac{\Delta A/\text{min}}{6.58 \ \times \ \text{mg of gel}}$$

Repeat this calculation for all three amounts of gel (0.05, 0.10, 0.20 g). Compare the three results.

C. Thermal Stability of Immobilized Peroxidase

Determine the reaction rate (ΔA/min) for each of the four reaction conditions. Assume that $A_{0\ min} = 0$ for each assay.

ΔA_1 = absorbance change for free peroxidase at room temperature

ΔA_2 = absorbance change for free peroxidase at 60°C

ΔA_3 = absorbance change for immobilized peroxidase at room temperature

ΔA_4 = absorbance change for immobilized peroxidase at 60°C

Calculate the % activity remaining after heating the free and immobilized enzyme:

$$\% \text{ Activity remaining}_{free} = \frac{\Delta A_2}{\Delta A_1} \times 100$$

$$\% \text{ Activity remaining}_{immobilized} = \frac{\Delta A_4}{\Delta A_3} \times 100$$

Compare the two final results. Which is more stable to heat, free or immobilized enzyme?

Calculate the specific activity of the free enzyme in units/mg using the following equation:

$$\text{Units/mg} = \frac{\Delta A_{510}/\text{min}}{6.58 \ \times \text{mg enzyme/mL reaction mixture}}$$

Compare this number with known specific activity of the enzyme listed on the reagent bottle.

Study Problems

> *1.* What is the purpose of washing the acrylamide gel several times with water as described in the experimental section, part A? How could you experimentally determine when washing is complete?

> *2.* Design an experiment using the immobilized peroxidase to calculate the Michaelis constant, K_M, and the maximum velocity, V_{max}.

▶ 3. How could you determine if the gel is "leaking" peroxidase? What experimental modifications could be made to prevent this?

▶ 4. What should be the shape of a plot of ΔA/min (immobilized enzyme activity) vs. milligrams of gel? Explain.

▶ 5. What is the purpose of riboflavin, ammonium persulfate, and sunlight in this experiment?

▶ 6. The data below were collected by students completing this lab. From their data, calculate the units of enzyme activity per mg of gel.

Amount of gel = 100 mg

ΔA/min = 0.03 per minute

7. What are some of the benefits of using immobilized enzymes in a commercial process?

▶ 8. The breakdown of hydrogen peroxide catalyzed by peroxidase is an oxidation-reduction reaction. Study Equation E12.1 and answer the following questions.

(a) What is the reducing agent?

(b) Is H_2O_2 oxidized or reduced?

(c) What is the cofactor for peroxidase?

9. How would you change the experimental conditions in order to trap an enzyme that has a molecular weight over 1 million? Horseradish peroxidase has a molecular weight of 40,000.

10. What is the purpose of filtering the assay tube in part B before reading A_{510}?

Further Reading

M. Allison and C. Bering, *J. Chem. Educ.* **75,** 1278–1280 (1998). Immobilization of lactase.

G. Bickerstaff, Editor, *Immobilization of Enzymes and Cells* (1997), Humana (Totowa, NJ). A complete book on enzyme immobilization.

R. Boyer and J. McArthur, *Biotech. Educ.* **2,** 17–20 (1991). "Activity and Thermal Stability of an Immobilized Enzyme."

M. Busto, *Biochem. Educ.* **26,** 304–308 (1998). Immobilization of β-glucosidase.

K. Mosbach, Editor, *Methods in Enzymology,* Vol. 137 (1988), Academic Press (San Diego). An entire volume on immobilized enzymes and cells.

C. Worthington, Editor, *Worthington Enzyme Manual* (1988), Worthington Biochemical Corporation (Freehold, NJ 07728), pp. 254–260. A book that provides data on many enzymes, including peroxidase.

13

EXTRACTION AND CHARACTERIZATION OF BACTERIAL DNA

- **Recommended Reading**

Chapter 2, Section C; *Chapter 5,* Sections A and B.

- **Synopsis**

DNA, the genetic substance in biological cells, is a large polymer composed of deoxyribonucleotide monomers. This experiment introduces the student to a general method for isolation and partial purification of DNA from microorganisms. The procedure consists of disrupting the cell wall or membrane, dissociating bound proteins, and separating the DNA from other soluble compounds. The isolated DNA is characterized and quantified by ultraviolet spectroscopy under native and denaturing conditions. In addition, ethidium bromide interaction with duplex DNA results in enhanced fluorescence of the dye, which may be used to (1) measure the concentration of DNA solutions and (2) study the binding of a polyamine, spermine.

I. INTRODUCTION AND THEORY

DNA Structure and Function

DNA in all forms of life is a polymer made up of nucleotides containing four major types of heterocyclic nitrogenous bases, adenine, thymine, guanine, and cytosine. The nucleotides are held together by 3′, 5′-phosphodiester bonds (Figure E13.1). The quantitative ratio and sequence of bases vary with the source of the DNA. Native DNA exists as two complementary strands held together by hydrogen bonds and arranged in a double helix. DNA in prokaryotic cells (simple cells with no major organelles and a single chromosome) exists as a single molecule in a circular, double-stranded form with a molecular weight of at least 2×10^9. Eukaryotic cells (cells with major organelles) contain several chromosomes and, thus, several very large DNA molecules.

DNA was first isolated from biological material in 1869, but its participation in the transfer of genetic information was not recognized until the mid-1940s. Since that time, DNA has been the subject of thousands of physical, chemical, and biological investigations. A landmark discovery was the elucidation of the three-dimensional structure of DNA by Watson and Crick in 1953. The **double helix** as envisioned by Watson and Crick is now recognized as a significant structural form of native DNA (Figure E13.2). Other discoveries of importance include methods for sequential analysis of the nucleotide bases, regulation of gene expression, cleavage of DNA by specific restriction endonucleases, and the production of recombinant DNA.

The nucleic acids are among the most complex molecules that you will encounter in your biochemical studies. When the dynamic role that is played by DNA in the life of a cell is realized, the complexity is understandable. It is difficult to comprehend all the structural characteristics that are inherent in the DNA molecules, but most biochemistry students are familiar with the double-helix model of Watson and Crick. The discovery of the double-helical structure of DNA is one of the most significant breakthroughs in our understanding of the chemistry of life. This experiment will introduce you to the basic structural characteristics of the DNA molecule and to the forces that help establish the complementary interactions between the two polynucleotide strands.

Isolation of DNA

Because of the large size and the fragile nature of chromosomal DNA, it is very difficult to isolate in an intact, undamaged form. Several isolation procedures have been developed that provide DNA in a biologically active form, but this does not mean it is completely undamaged. These preparations yield DNA that is stable, of high molecular weight, and relatively free of RNA and protein. Here, a general method will be described for the isolation of DNA in a relatively pure form from microorganisms. The procedure outlined is applicable to many microorganisms and can be modified as necessary.

Designing an isolation procedure for DNA requires extensive knowledge of the chemical stability of DNA as well as its condition in the cellular environment. Figures E13.1 and E13.2 illustrate several chemical bonds in DNA that may be susceptible to cleavage during the extraction process. The experimental factors that must be considered and their effects on various structural aspects of intact DNA are outlined below.

1. pH

    (a) Hydrogen bonding between the complementary strands is stable between pH 4 and 10.

    (b) The phosphodiester linkages in the DNA backbone are stable between pH 3 and 12.

Figure E13.1

The covalent structure of DNA.

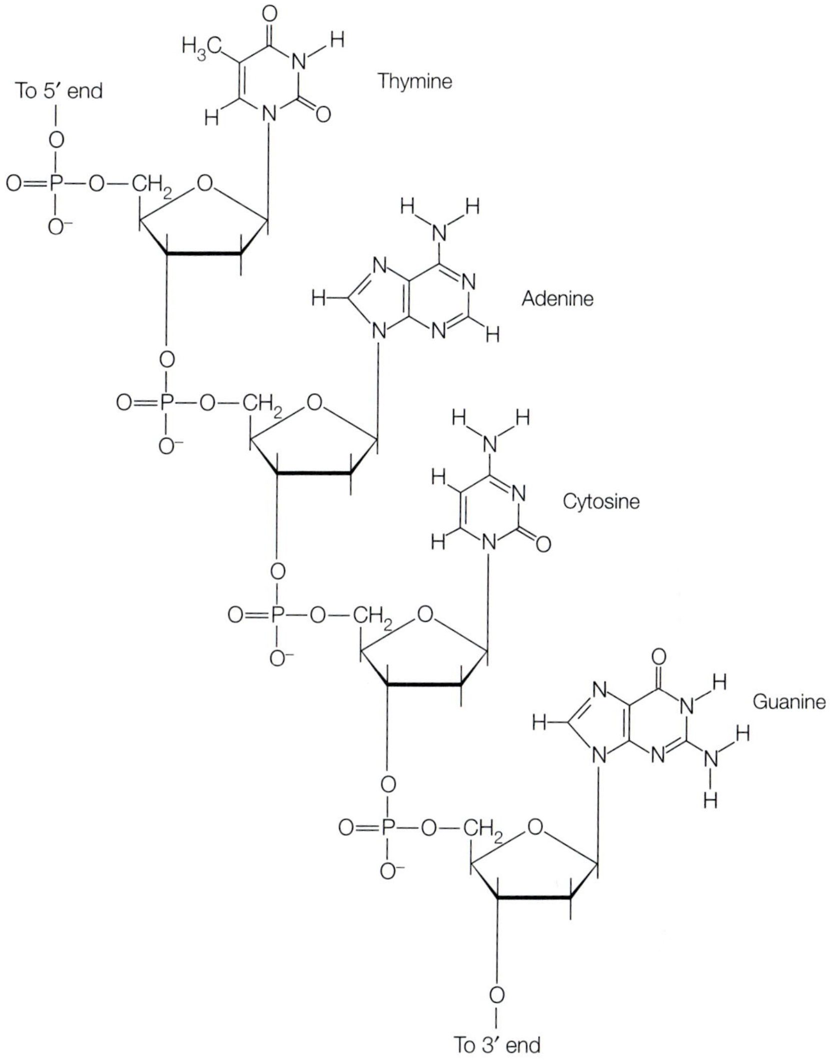

(c) N-Glycosidic bonds to purine bases (adenine and guanine) are hydrolyzed at pH value of 3 and less.

2. Temperature

(a) There is considerable variation in the temperature stability of the hydrogen bonds in the double helix, but most DNA begins to unwind in the range of 80–90°C.

(b) Phosphodiester linkages and N-glycosidic bonds are stable up to 100°C.

Figure E13.2

The helix conformation of
DNA showing base pairs.

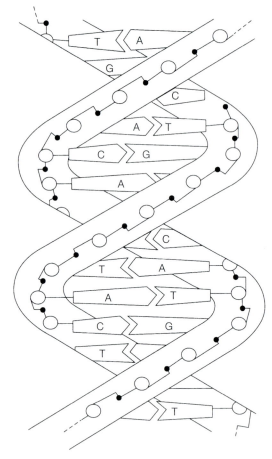

3. Ionic strength

    (a) DNA is most stable and soluble in salt solutions. Salt concentrations of less than 0.05 *M* weaken the hydrogen bonding between complementary strands.

4. Cellular conditions

    (a) Before the DNA can be released, the cell wall of the organism must by lysed. The ease with which the cell wall is disrupted varies from organism to organism. In some cases (yeast), extensive grinding or sonic treatment is required, whereas in others (*Bacillus subtilis*), enzymatic hydrolysis of the cell wall is possible.

    (b) Several enzymes are present in the cell that may act to degrade DNA, but the most serious damage is caused by the deoxyribonucleases. These enzymes catalyze the hydrolysis of phosphodiester linkages.

    (c) Native DNA is present in the cell as DNA-protein complexes. Basic proteins called **histones** must be dissociated from the DNA during the extraction process.

5. Mechanical stress on the DNA

   (a) Gentle manipulations may not always be possible during the isolation process. Grinding, shaking, stirring, and other disruptive procedures may cause cleavage (shearing or scission) of the DNA chains. This usually does not cause damage to the secondary structure of the DNA, but it does reduce the length of the molecules.

Now that these factors are understood, a general procedure of DNA extraction will be outlined.

Step 1. Disruption of the cell wall and release of the DNA into a medium in which it is soluble and protected from degradation. The isolation procedure described here calls for the use of an enzyme, lysozyme, to disrupt the cell wall. Lysozyme catalyzes the hydrolysis of glycosidic bonds in cell wall peptidoglycans, thus causing destruction of the cell wall and allowing the release of DNA and other cellular components. The medium for solution of DNA is a buffered saline solution containing EDTA. DNA, which is ionic, is more soluble and stable in salt solution than in distilled water. The EDTA serves at least two purposes. First, it binds divalent metal ions (Ca^{2+}, Mg^{2+}, Mn^{2+}) that could form salts with the anionic phosphate groups of the DNA. Second, it inhibits deoxyribonucleases that have a requirement for Mg^{2+} or Mn^{2+}. The mildly alkaline medium (pH 8) acts to reduce electrostatic interaction between DNA and the basic histones and the polycationic amines, spermine and spermidine. The relatively high pH also tends to diminish nuclease activity and denature other proteins.

Step 2. Dissociation of the protein-DNA complexes. Detergents are used at this stage to solubilize the inner membrane and disrupt the ionic interactions between positively charged histones and the negatively charged backbone of DNA. Sodium dodecyl sulfate (SDS), an anionic detergent, binds to proteins and gives them extensive anionic character. A secondary action of SDS is to act as a denaturant of deoxyribonucleases and other proteins. Also favoring dissociation of protein-DNA complexes is the alkaline pH, which reduces the positive character of the histones. To ensure complete dissociation of the DNA-protein complex and to remove bound cationic amines, a high concentration of a salt (NaCl or sodium perchlorate) is added. The salt acts by diminishing the ionic interactions between DNA and cations.

Step 3. Separation of the DNA from other soluble cellular components. Before DNA is precipitated, the solution must be deproteinized. This is brought about by treatment with chloroform–isoamyl alcohol followed by centrifugation. Upon centrifugation, three layers are produced: an upper aqueous phase, a lower organic layer, and a compact band of denatured protein at the interface between the aqueous and organic phases. Chloroform causes surface denaturation of proteins. Isoamyl alcohol reduces foaming and stabilizes the interface between the aqueous phase and the organic phases where the protein collects.

The upper aqueous phase containing nucleic acids is then separated and the DNA precipitated by addition of ethanol. Because of the ionic nature of DNA, it becomes insoluble if the aqueous medium is made less polar by addition of an organic solvent. The DNA forms a threadlike precipitate that can be collected by "spooling" onto a glass rod. The isolated DNA may still be contaminated with protein and RNA. Protein can be removed by dissolving the spooled DNA in saline medium and repeating the chloroform–isoamyl alcohol treatment until no more denatured protein collects at the interface.

RNA does not normally precipitate like DNA, but it could still be a minor contaminant. RNA may be degraded during the procedure by treatment with ribonuclease after the first or second deproteinization step. Removal of RNA sometimes makes it possible to denature more protein using chloroform–isoamyl alcohol. If DNA in a highly purified state is required, several deproteinization and alcohol precipitation steps may be carried out. It is estimated that up to 50% of the cellular DNA is isolated by this procedure. The average yield is 1 to 2 mg per gram of wet packed cells.

Characterization of DNA

Ultraviolet Absorption

A complete understanding of the biochemical functions of DNA requires a clear picture of its structural and physical characteristics. DNA has significant absorption in the UV range because of the presence of the aromatic bases adenine, guanine, cytosine, and thymine. This provides a useful probe into DNA structure because structural changes such as helix unwinding affect the extent of absorption. In addition, absorption measurements are used as an indication of DNA purity. The major absorption band for purified DNA peaks at about 260 nm. Protein material, the primary contaminant in DNA, has a peak absorption at 280 nm. The ratio A_{260}/A_{280} is often used as a relative measure of the nucleic acid/protein content of a DNA sample. The typical A_{260}/A_{280} for isolated DNA is about 1.8. A smaller ratio indicates increased contamination by protein.

Thermal Denaturation

If DNA solutions are treated with denaturing agents (heat, alkali, organic solvents) their ultraviolet-absorbing properties are strikingly increased. Figure E13.3 shows the effect of temperature on the UV absorption of DNA. The curve is obtained by plotting $A_{260(T)}/A_{280(25°)}$ vs. temperature (T). Heating through the temperature range of 25°C to about 80°C results in only minor increases in absorption. However, as the temperature is further increased, there is a sudden increase in UV absorption followed by a constant A_{260}. The total increase in absorption is usually on the order of 40% and occurs over a small temperature range. Figure E13.3 is called a **thermal denaturation curve**, a **temperature profile**, or a **melting curve**. The temperature corresponding to the midpoint of each absorption increase is defined as T_m, the **transition temperature** or **melting temperature**. (This should not

Figure E13.3

A typical thermal denaturation curve for DNA. T_m is measured as the temperature at the midpoint of the absorbance increase.

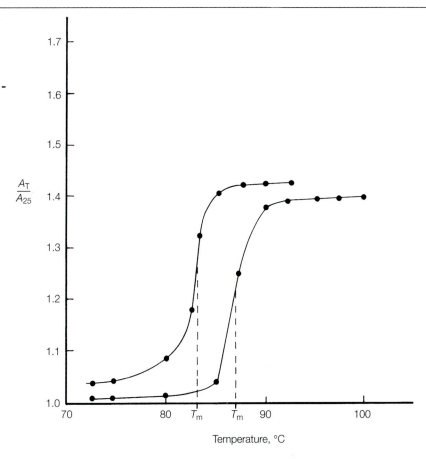

be confused with melting point, transformation of a substance from solid to liquid, as was routinely studied in organic laboratory.) Each species of DNA has a characteristic T_m value that can be used for identification and characterization purposes.

The origin of the absorption increase, called a **hyperchromic effect,** is well understood. The absorption changes are those that result from the transition of an ordered double-helix DNA structure to a denatured state or random, unpaired DNA strands. Native DNA in solution exists in the double helix held together primarily by hydrogen bonding between complementary base pairs on each strand (see Figure E13.2). Hydrophobic and π-π interactions between stacked base pairs also strengthen the double helix. Agents that disrupt these forces (hydrogen bonding, hydrophobic and π-π interactions) cause dissociation or unwinding of the double helix. In a random coil arrangement, base-base interactions are at a minimum; this alters the resonance behavior of the aromatic rings, causing an increase in absorption. The process of DNA dissociation can, therefore, be characterized by monitoring the UV-absorbing properties of DNA under various conditions.

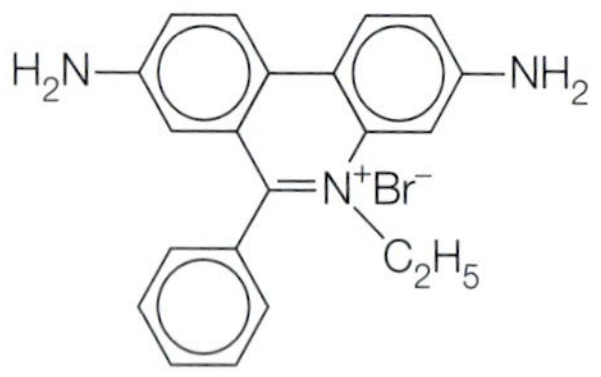

Figure E13.4

Structure of the fluorescent intercalation dye ethidium bromide.

- -

Binding and Fluorescence of Ethidium Bromide

Fluorescence assays are considered among the most convenient, sensitive, and versatile of all laboratory techniques. However, the purine and pyrimidine bases yield only weak fluorescence spectra. Le Pecq and Paoletti (1967) showed that the fluorescence of a dye, ethidium bromide, is enhanced about 25-fold when it interacts with DNA. Ethidium bromide, which is a relatively small planar molecule (Figure E13.4), binds to DNA by insertion between stacked base pairs (intercalation). The process of intercalation is especially significant for aromatic dyes, antibiotics, and other drugs. Some dyes, when intercalated into DNA, show an enhanced fluorescence that can be used to detect DNA molecules after gel electrophoresis measurements (see Chapter 4 and Experiments 14 and 15) and to characterize the physical structure of DNA. Two analyses of DNA will be completed in this experiment:

1. Measurement of the Concentration of DNA and RNA in Solution

Most assays measure both double-stranded and single-stranded DNA. Ethidium interaction with single-stranded DNA does not lead to increased fluorescence, so duplex DNA can be quantified in the presence of dissociated DNA. Solutions of purified DNA are commonly contaminated with RNA. Since single-stranded RNA can form hairpin loops with base pairing and duplex formation (as in tRNA), ethidium also binds with enhanced fluorescence to these duplex regions of RNA. Addition of ribonuclease A results in digestion of RNA and loss of fluorescence due to ethidium binding of RNA. The amount of fluorescence lost is proportional to the concentration of RNA. The ethidium fluorescence remaining after ribonuclease treatment is directly proportional to the concentration of duplex DNA (Equation E13.1).

$$F_{total} = F_{DNA} + F_{RNA}$$

Equation E13.1

where

F = fluorescence yield due to each type of nucleic acid

When the F_{RNA} term is reduced to zero, the total fluorescence is a direct measure of DNA concentration. The actual concentration of DNA in solution can be calculated by use of a standard solution of DNA or from a standard curve.

2. Polyamine Binding to DNA

The site of action of many antibiotics is the DNA molecule of an invading organism. In fact, the physiological action of many drugs, synthetic and natural, depends on interaction with DNA, which often leads to an intercalation complex. It is possible to demonstrate binding and to estimate the association constants for ligand-DNA complexes using the ethidium assay. A solution of DNA treated with excess ethidium bromide gives a maximum fluorescence yield. If a ligand is added that can compete with ethidium for the intercalation sites on DNA, the ethidium dissociates from the DNA. The fluorescence yield of ethidium decreases at an amount that is directly proportional to the concentration of added ligand. That is, the greater the concentration of bound ligand, the greater the extent of ethidium dissociation and the greater the decrease in ethidium fluorescence. Of course, it must be shown that the ligand does not fluoresce under the conditions of the experiment and that it does not interfere with ethidium fluorescence. An estimate of the binding constant can be obtained from the binding curve of % fluorescence vs. ligand concentration. The ligand concentration that causes a 50% decrease in ethidium concentration is approximately inversely proportional to the binding constant, k_a.

There is particularly strong interest in the interactions between polyamines and DNA. Four polyamines, 1,4-diaminobutane (putrescine), 1,5-diaminopentane (cadaverine), spermidine, and spermine (Figure E13.5), are metabolic products in many cells and are found in relatively high concentrations. Putrescine and cadaverine are derived from the decarboxylation of ornithine and lysine, respectively. The other two polyamines, spermidine and spermine, are synthesized in rather complex processes requiring S-adenosylmethionine. Polyamines are cationic, so they bind strongly to nucleic acids. They interact by forming salt bridges between their positively charged amino groups and negatively charged phosphate anions of the nucleotides. The polyamines may help control the progression of cells through the cell cycle by regulating the rates of nucleic acid and protein synthesis. The ethidium assay is particularly convenient for demonstrating polyamine binding to DNA because the analysis is rapid; also, since the ligands are not aromatic, they display no measurable interference of ethidium fluorescence.

Overview of the Experiment

This experiment outlines an isolation procedure for DNA that may be applied to most microorganisms. Although this isolation method was first

---

Figure E13.5

Structures of several significant polyamines.

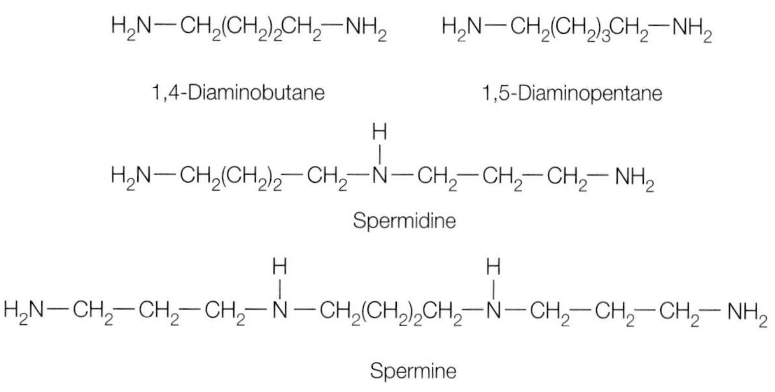

- -

$$H_2N-CH_2(CH_2)_2CH_2-NH_2 \qquad H_2N-CH_2(CH_2)_3CH_2-NH_2$$

1,4-Diaminobutane   1,5-Diaminopentane

$$H_2N-CH_2(CH_2)_2-CH_2-\overset{\overset{\displaystyle H}{|}}{N}-CH_2-CH_2-CH_2-NH_2$$

Spermidine

$$H_2N-CH_2-CH_2-CH_2-\overset{\overset{\displaystyle H}{|}}{N}-CH_2(CH_2)_2CH_2-\overset{\overset{\displaystyle H}{|}}{N}-CH_2-CH_2-CH_2-NH_2$$

Spermine

described nearly 40 years ago, it is still one of the best for extracting larger quantities (mg) of DNA (Marmur, 1961, 1963). DNA may be isolated from *B. subtilis, E. coli,* and *C. welchii.* Other microorganisms may be used, but modifications in the procedure may have to be made as described by Marmur. The isolated DNA is usually recovered in relatively good yield (about 50%) and there is only limited shearing of the product. After extraction, the DNA will be analyzed by UV measurement to obtain an estimate of DNA purity and to study thermal denaturation. This experiment will also illustrate two of the many applications of the ethidium fluorescence assay to DNA study. The isolation and analysis of DNA may be completed during a 3 to 4-hour laboratory period. The actual isolation requires about 1½ hours and UV analysis about 30 minutes. The measurement of DNA solutions using the ethidium assay requires about 1 hour and the spermine binding 1–2 hours.

II.   MATERIALS AND SUPPLIES

– Bacterial cells, wet packed, 2 to 3 g. Use *B. subtilis, Escherichia coli,* or *Clostridium welchii.* If lyophilized cells are available, use 0.5 g.

– Saline-EDTA, 0.15 M NaCl plus 0.1 M ethylenediaminetetraacetate, pH 8

– Sodium dodecyl sulfate (SDS), 25% in water

– Lysozyme solution, 10 mg/mL in water

– Sodium perchlorate, 5 M

– Chloroform–isoamyl alcohol, 24:1

– 95% ethanol

– Tris buffer I; 0.01 M Tris-HCl plus 0.05 M NaCl, pH 7.5

– Quartz cuvettes

– Spectrophotometer for UV measurements

– Magnetic stirrer

– Water bath at 60°C

– Water bath at 90°C

– DNA standard solution, 50 μg/mL in Tris buffer I

– Ethidium bromide solution, pH 7.5, 0.005 M Tris-HCl, 0.5 μg/mL ethidium bromide, and 0.5 mM EDTA

– Ribonuclease A, 20 mg/mL in 0.10 M Tris-HCl, pH 7.5

– Water bath at 37°C

– Spermine solution, 1 × 10^{-4} M in Tris buffer I

– Spectrofluorimeter and fluorescence cuvettes:

Excitation wavelength	525 nm
Emission wavelength	590 nm
Slit width	20 nm
Scale	× 100

III. EXPERIMENTAL PROCEDURE

A. Isolation of DNA

> **CAUTION**
> • Experimental work with bacterial cells presents a potential biohazard. Some strains of the bacteria recommended for use may cause gastroenteritis (abdominal pain and diarrhea). Do not pipet any solutions by mouth. Always wear gloves and wash your hands well with hot water and soap before eating or drinking.
> • Ethanol is volatile and flammable. There should be no flames in the laboratory during this experiment.

Suspend 2 to 3 g of wet packed bacterial cells or 0.5 g of lyophilized cells in 25 mL of saline–EDTA solution in a 125-mL Erlenmeyer flask. Add 1.0 mL of the lysozyme solution (10 mg of lysozyme) and incubate the mixture at 37°C for 30 to 45 minutes. To bring about complete cell lysis, add 2.0 mL of 25% SDS solution and heat the mixture in a 60°C water bath for 10 minutes. To avoid excessive foaming, stir the mixture very gently. As the nucleic acid is released from the lysed cells, the solution will show an increase in viscosity and a decrease in cloudiness.

After the heating period, allow the mixture to cool to room temperature or cool the flask in cold running water. Add 9.0 mL of sodium perchlorate solution (5 M), which brings the final salt concentration to about 1 M. Mix well, but gently, and then add a volume of chloroform–isoamyl alcohol (24:1) equal to the total volume of the extraction mixture. Shake the solution in a separatory funnel, a tightly stoppered flask, or a screwcap centrifuge tube for 10 to 15 minutes. Then, centrifuge the emulsion for 5 minutes at 10,000 × g. Three layers should be visible: a bottom organic phase, a middle band of denatured protein at the interface, and an upper aqueous layer containing the nucleic acids. Carefully transfer the aqueous layer to a 125-mL beaker or large test tube. Precipitate the nucleic acids by carefully layering about 2 volumes of 95% ethanol over the aqueous phase. The ethanol should be poured down the side of the beaker or tube using a glass stirring rod. Mix the two layers very gently with a circular motion of the stirring rod and "spool" all the fibers of nucleic acid onto the rod. Press the spooled DNA against the inside of the glass vessel to remove solvent. Dissolve the spooled nucleic acid in 10 to 15 mL of Tris buffer I. Save the DNA solution for parts B, C, and D.

B. Ultraviolet Measurement and Denaturation of Isolated DNA

Dilute a 0.5-mL aliquot of the DNA solution with 4.5 mL of Tris buffer I. Transfer some of the solution to a quartz cuvette and determine the A_{260}, using a cuvette containing Tris buffer I as reference. If the A_{260} is greater than 1, quantitatively dilute the sample until the absorbance reading is between 0.5 and 1.0. Determine and record the A_{280} on the same DNA sample.

Prepare 10 mL of a DNA solution in Tris buffer I at a DNA concentration of about 20 μg/mL. Measure and record the A_{260}. It should be around 0.4 absorbance units. Transfer 3.0 mL of the DNA solution into each of three test tubes. Place a marble over the top of each tube. Maintain one tube at room temperature and place the other two in a 90°C water bath for 15 minutes. After the incubation period, remove the tubes. Quick-cool one heated tube in an ice bath and allow the other heated tube to cool slowly to room temperature over a period of about 1 hour. Measure and record final A_{260} readings on each of the three tubes.

C. Ethidium Bromide Assay

A relative fluorescence intensity scale will be defined in part 1 of the experiment. The cuvette holder should be provided with a water jacket to maintain the temperature within ± 0.5°. It is recommended that you use the same cuvette for all measurements unless a pair is available that is very well matched. Fluorescence measurements are temperature dependent. It is necessary to maintain all reagents, except the ribonuclease solution, at 37°C, the same as the setting for the cuvette holder in the fluorimeter.

> **CAUTION**
>
> • Ethidium bromide is a mutagen and a potential carcinogen. Always wear gloves and a face mask while weighing or handling the pure chemical. Gloves should always be worn while using solutions of the dye. Do not allow the solutions of ethidium bromide to come into contact with your skin. **Do not mouth pipet any ethidium bromide solutions.** Save all assay mixtures and excess ethidium bromide solutions after measurements are taken, and pour them into a container marked "Waste Ethidium."

1. Concentration of DNA Solutions

Turn on the fluorimeter lamp and allow it to warm up for 15 minutes. Set the fluorescence scale for the standard DNA in the following manner. Pipet 3.0 mL of pH 7.5, ethidium bromide–Tris buffer solution into a fluorescence cuvette. With a micropipet, transfer 15 μL of the standard DNA into the cuvette and mix well. Place the cuvette in the spectrofluorimeter and adjust the fluorescence intensity to "100", F_{std}. This is, of course, an arbitrary setting. The actual setting that is used is not especially important, but you must know what the setting is. The number 100 is convenient for comparison purposes and is customary in fluorescence measurements.

Pour the contents of the cuvette into the waste ethidium container provided by your instructor. Clean the cuvette carefully with warm water, rinse several times with distilled water, and dry.

Again, transfer 3.0 mL of the pH 7.5, ethidium bromide–Tris buffer solution into the cuvette. Add 15 μL of the unknown DNA solution. Mix well and place in the spectrofluorimeter. Record the fluorescence intensity, F_{total}. The measured fluorescence is due to both DNA and RNA (if present). The amount of fluorescence due to ethidium-RNA can be eliminated by digesting the RNA with ribonuclease. To the cuvette containing pH 7.5, ethidium bromide–Tris buffer and unknown DNA, add 2 μL of ribonuclease solution. Mix well and incubate the cuvette at 37°C for 20 minutes in a water bath. After incubation, measure the fluorescence intensity, (F_{total}). Compare this with the original measurement taken on the unknown DNA (F_{total}). Is RNA present in your DNA sample? Calculate the concentration of DNA in your unknown solution as described in the Analysis of Results. The possibility exists that the solution of ribonuclease may contain a fluorescent impurity. What control experiment should be done to correct this situation?

2. Binding of Spermine to DNA

Standardize the spectrofluorimeter in the following way. Pipet 2.0 mL of the pH 7.5, ethidium bromide–Tris buffer into a cuvette. Add 1.0 mL of Tris buffer I and 20 μL of standard DNA solution. Mix and place in the fluorimeter. Adjust the fluorescence intensity to 100. Clean the cuvette as described in part A and repeat the assay using various concentrations of spermine. Prepare a table displaying the amount of each component to be added. Four reagents must be in the table: pH 7.5, ethidium bromide–Tris buffer, DNA solution, spermine, and Tris buffer I. Maintain the volume of DNA at 20 μL and ethidium bromide solution at 2.0 mL for all assays. Use 0.05, 0.1, 0.2, 0.4, 0.6, 0.8, and 1.0 mL of spermine in the assays. Remember that the total volume of all constituents in the cuvette must remain constant at 3.02 mL for all the assays. Therefore, the amount of Tris buffer I must change with the amount of spermine added. Prepare each assay separately by adding the proper amount of each component to the cuvette. Mix well and record the fluorescence intensity of each cuvette.

IV.   ANALYSIS OF RESULTS

A. Isolation of DNA

Write all observations of your nucleic acid preparation in your notebook. Prepare a flowchart for the isolation procedure. Explain the purpose of each step and the role played by each reagent. Speculate on any damage that may have altered the molecular structure of the DNA.

B. Ultraviolet Measurement and Denaturation of Isolated DNA

Calculate the A_{260}/A_{280} ratio. Extensively purified DNA has a ratio of about 1.8. Comment on the protein content of your isolated nucleic acid. Calculate the DNA concentration ($\mu g/mL$) in the Tris buffer I solution and determine the total yield of isolated DNA. Assume the $E_{260}^{1\%}$ of native DNA is about 200. Information available in Chapter 2, Section C may also be used to calculate the concentration of DNA.

Calculate the ratio $A_{260(T)}/A_{260(25°)}$ for each of the three tubes. Why are the three ratios not the same? Explain the experimental observation in terms of the molecular structure of DNA. What was the percentage increase in the overall hyperchromic effect? Give a qualitative answer regarding the purity of the DNA sample you used.

C. Concentration of DNA Solutions

The concentration of the unknown DNA solution can be calculated by a simple comparison of the fluorescence intensity obtained from the standard DNA reaction mixture (F_{std}) and the fluorescence of the unknown DNA mixture (Equation E13.2).

>>

$$[DNA]_x = \frac{[DNA]_{std} F_{DNA-x}}{F_{DNA-std}}$$

Equation E13.2

where

$[DNA]_x$ = concentration of the unknown solution of DNA in $\mu g/mL$

$[DNA]_{std}$ = concentration of the standard solution of DNA in $\mu g/mL$

$F_{DNA-std}$ = fluorescence yield of the standard DNA solution

F_{DNA-x} = fluorescence yield of the standard DNA solution after incubation with ribonuclease

To correct for possible fluorescent contamination in the RNAase solution, place 3.0 mL of pH 7.5, ethidium bromide–Tris buffer in a cuvette. Add 15 μL of Tris buffer I and 2 μL of ribonuclease. Mix well and incubate at 37°C for 20 minutes. Record the fluorescence intensity, if any. If the fluorescence is significant (> 1 intensity unit), subtract from F_{DNA-x} before calculation of unknown DNA. Calculate the concentration of DNA in your sample in units of $\mu g/mL$. Was RNA present in your unknown DNA solution?

D. Binding of Spermine

Prepare a table of fluorescence reading vs. spermine concentration in the cuvette in μM. If the fluorescence of untreated DNA was set to 100 on the fluorimeter, each fluorescence reading taken can be considered a percentage of the maximum possible. Plot the fluorescence reading (y axis) against the spermine concentration (x axis) that caused that particular fluorescence reading. Connect the points with a straight line. The binding constant, k_a, is estimated as the reciprocal of the spermine concentration that produces 50% of the fluorescence of untreated DNA.

Study Problems

> 1. Why does the solution become more viscous during lysozyme and SDS treatment?

> 2. How could RNA be removed from the nucleic acid preparation?

> 3. How does SDS help disrupt the cell membrane?

> 4. If extremely high purity DNA is desired, what further steps could be followed in addition to those described here?

> 5. A solution of purified DNA gave an A_{260} of 0.55 when measured in a quartz cuvette of 1-cm path length. What is the concentration of DNA in μg/mL?

> 6. The concentration of a purified DNA solution is 35 μg/mL. Predict the value of A_{260}.

> 7. Predict and explain the action of each of the following conditions on the T_m of native DNA.
> (a) Measure T_m in pH 12 buffer.
> (b) Measure T_m in distilled water.
> (c) Measure T_m in 50% methanol water.
> (d) Measure T_m in standard Tris-HCl buffer solution containing sodium dodecyl sulfate.

> 8. A sample of highly purified DNA gave an A_{260} at 25°C of 0.45. The sample was divided into two samples and each was treated as follows:

Sample I		Sample II	
1. Heat to 100°C		1. Heat to 100°C	
$A_{260(100°C)}$	= 0.65	$A_{260(100°C)}$	= 0.65
2. Quick-cool in ice water		2. Very slow cool to 25°C	
$A_{260(25°C)}$	= 0.55	$A_{260(25°C)}$	= 0.46

Explain why different final A_{260} readings at 25°C are obtained.

> 9. Compare the structure of ethidium bromide (Figure E13.4) with those of the polyamines (Figure E13.5). Would you expect the binding of spermine or spermidine to DNA to be competitive with the binding of ethidium? Explain.

> 10. Explain how this experiment could be modified to measure the concentration of an RNA solution.

Further Reading

W. Bauer and J. Vinograd, *J. Mol. Biol.* **33,** 141 (1968). "The Interaction of Closed Circular DNA with Intercalative Dyes."

G. Blackburn and M. Gait, *Nucleic Acids in Chemistry and Biology* (1991), Oxford University Press (New York). An excellent book on general aspects of nucleic acids.

R. Boyer, *Concepts in Biochemistry* (1999), Brooks/Cole (Pacific Grove, CA), pp. 30–44; 280–310. Structure and function of DNA.

J. Day, *Biochem. Educ.* **25,** 41–43 (1997). "Isolation of Nuclear, Chloroplast, and Mitochondrial DNA from Plants."

R. Garrett and C. Grisham, *Biochemistry,* 2nd ed (1999) W. B. Saunders (Orlando, FL), pp. 327–394. Structure and function of DNA.

A. Lehninger, D. Nelson, and M. Cox, *Principles of Biochemistry,* 3rd ed. (1999), Worth Publishers (New York), pp. 907–930. DNA structure and function.

J. Le Pecq and C. Paoletti, *J. Mol. Biol.* **27,** 87 (1967). "A Fluorescent Complex Between Ethidium Bromide and Nucleic Acids."

J. Marmur, in *Methods in Enzymology,* Vol. VI, S. Colowick and N. Kaplan, Editors (1963), Academic Press (New York), pp. 726–738. "A Procedure for the Isolation of Deoxyribonucleic Acid from Microorganisms."

J. Marmur, *J. Mol Biol.* **3,** 208 (1961). "A Procedure for the Isolation of Deoxyribonucleic Acid from Microorganisms."

C. Mathews, K. van Holde, and K. Ahern, *Biochemistry,* 3rd ed. (2000), Benjamin/Cummings (San Francisco), pp. 84–125. DNA structure and function.

K. Strothkamp and R. Strothkamp, *J. Chem. Educ.* **71,** 77–79 (1994). "Fluorescence Measurements of Ethidium Binding to DNA."

L. Stryer, *Biochemistry,* 4th ed. (1995), W. H. Freeman (New York), pp. 75–144; 785–818. Structure and function of DNA.

D. Voet and J. Voet, *Biochemistry,* 2nd ed. (1995) John Wiley & Sons (New York), pp. 830–914. DNA structure and function.

D. Voet, J. Voet, and C. Pratt, *Fundamentals of Biochemistry,* (1999) John Wiley & Sons (New York), pp. 41–76; 725–771. DNA structure and function.

DNA Purification on the Web

http://research.nwfsc.noaa.gov/protocols.html
    Various procedures for DNA isolation.

http://www.tgbiotech.com/methods.html
    Methods for DNA preparation.

http://res.agr.ca/winn/ethidium.htm
    Decontamination of dilute solutions of ethidium bromide.

14

PLASMID DNA ISOLATION AND CHARACTERIZATION BY ELECTROPHORESIS

- **Recommended Reading**

Chapter 4, Sections A, B, C; *Experiments 13* and *15.*

- **Synopsis**

Extrachromosomal DNA molecules called plasmids are harbored in some strains of *E. coli.* The normal copy number of the plasmids is small, between 2 and 10; however, if these strains of *E. coli* are grown in the presence of chloramphenicol, up to 3000 copies may be replicated per cell. Plasmid DNA has been demonstrated to be a useful vehicle in molecular cloning. This experiment describes a method for the growth of *E. coli* and amplification of the ColE1 plasmids. The plasmids will be isolated from *E. coli* cells by one of two methods, a large-scale boiling method or a microscale alkaline lysis method. The DNA plasmids will be measured for molecular size by agarose electrophoresis.

I. INTRODUCTION AND THEORY

Our knowledge of DNA structure and function has increased at a gradual pace over the past 50 years, but some discoveries have had a special impact on the progress and direction of DNA research. Although DNA was first discovered in cell nuclei in 1869, it was not confirmed as the carrier of genetic information until 1944. This major discovery was closely followed by the announcement of the double-helix model of DNA in 1953. The more recent development of technology that allows manipulation by insertion of "foreign" DNA fragments into the natural, replicating DNA of an organism may well have a greater impact on the direction of DNA research than the previous two discoveries. In the short time since the first construction and replication of plasmid "recombinant DNA," several scientific and medical applications of the new technology have been announced. This new era of "genetic engineering" has captivated the general public and students alike. Some of the predicted achievements in recombinant DNA research have been slow in coming; however, future workers in biochemistry and related fields must be familiar with the principles and techniques of DNA recombination.

Recombinant DNA

DNA recombination or **molecular cloning** consists of the covalent insertion of DNA fragments from one type of cell or organism into the replicating DNA of another type of cell. Many copies of the hybrid DNA may be produced by the progeny of the recipient cells; hence, the DNA molecule is **cloned.** If the inserted fragment is a functional gene carrying the code for a specific protein, many copies of that gene and translated protein could be produced in the host cell. This process has become important for the large-scale production of proteins (insulin, somatostatin, and other hormones) that are of value in medicine and basic science but are difficult and expensive to obtain by other methods.

The basic steps involved in performing a recombinant DNA experiment are listed below and are outlined in Figure E14.1.

1. Select and isolate a natural DNA to serve as the carrier (sometimes called **vehicle** or **vector**) for the foreign DNA.

2. Cleave both strands of the DNA carrier with a restriction endonuclease.

3. Prepare and insert the foreign DNA fragment into the vehicle. This produces a **recombinant** or **hybrid** DNA molecule.

4. Introduce the hybrid DNA into a host organism (usually a bacterial cell), where it can be replicated. This process is called **transformation.**

5. Develop a method for identifying and screening the host cells that have accepted and are replicating the hybrid DNA.

Our current state of knowledge of recombinant DNA is the result of several recent technological advances. The first major breakthrough was the isolation of mutant strains of *E. coli* that are not able to degrade or restrict foreign DNA. These strains are now used as host organisms for the replication of recombinant DNA. The second advance was the development of bacterial extrachromosomal DNA (plasmids) and bacteriophage DNA as cloning vehicles. The final, necessary advance was the development of methods for insertion of the foreign DNA into the natural vehicle. The discovery of **restriction endonucleases,** enzymes that catalyze the hydrolytic cleavage of DNA at selective sites, provided a gentle and specific method for opening (linearizing) circular vehicles. (See Experiment 15.) Methods for covalent joining of the foreign DNA to the vehicle ends and final closure of the circular hybrid plasmid were then developed. The enzyme **DNA ligase,** which can catalyze the ATP-dependent formation of

Figure E14.1

Outline of a typical procedure for a DNA recombination
experiment.

- -

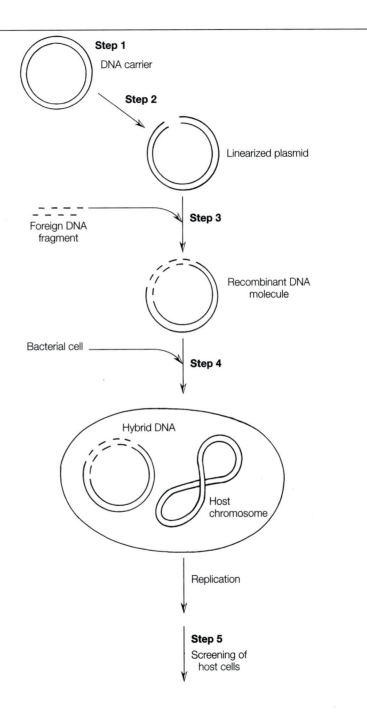

Step 1

DNA carrier

Step 2

Linearized plasmid

Foreign DNA
fragment

Step 3

Recombinant DNA
molecule

Bacterial cell

Step 4

Hybrid DNA

Host
chromosome

Replication

Step 5

Screening of
host cells

phosphodiester linkages at the insertion sites, is used for final closure (Figure E14.2).

Plasmids as Cloning Vehicles

Many bacterial cells contain self-replicating, extrachromosomal DNA molecules called **plasmids.** This form of DNA is closed circular, double-stranded, and much smaller than chromosomal DNA; its molecular weight ranges from 2×10^6 to 20×10^6, which corresponds to between 3000 and 30,000 base pairs. Bacterial plasmids normally contain genetic information for the translation of proteins that confer a specialized and sometimes protective characteristic (phenotype) on the organism. Examples of these characteristics are enzyme systems necessary for the production of antibiotics, enzymes that degrade antibiotics, and enzymes for the production of toxins. Plasmids are replicated in the cell by one of two possible modes. **Stringent replicated** plasmids are present in only a few copies and **relaxed replicated** plasmids are present in many copies, sometimes up to 200. In addition, some relaxed plasmids continue to be produced even after the antibiotic chloramphenicol is used to inhibit chromosomal DNA synthesis in the host cell. Under these conditions, many copies of the plasmid DNA may be produced (up to 2000 or 3000) and may accumulate to 30 to 40% of the total cellular DNA.

The ideal plasmid cloning vehicle has the following properties:

1. The plasmid should replicate in a relaxed fashion so that many copies are produced.

2. The plasmid should be small; then it is easier to separate from the larger chromosomal DNA, easier to handle without physical damage, and probably contains very few sites for attack by restriction nucleases.

Figure E14.2

Insertion of a DNA fragment into a linearized plasmid using cohesive ends and DNA ligase.

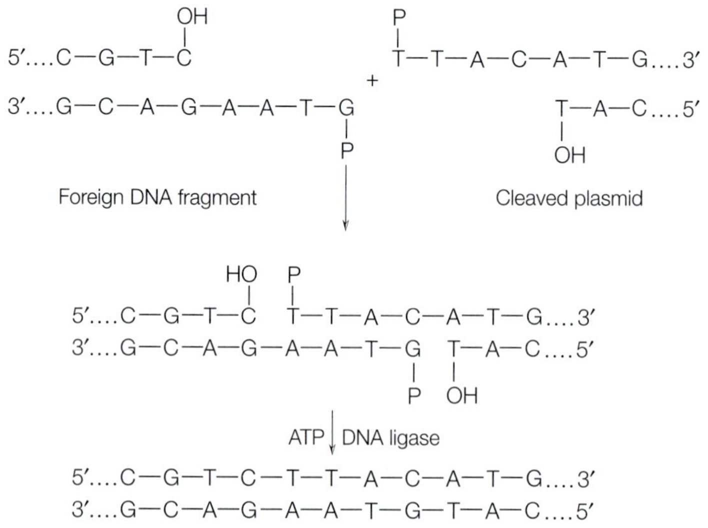

3.   The plasmid should contain identifiable markers so that it is possible to screen progeny for the presence of the plasmid. At least two selective markers are desirable, a primary one to confirm the presence of the plasmid and a secondary marker to confirm the insertion of foreign DNA. Resistance to antibiotics is a convenient type of marker.

4.   The plasmid should have only one cleavage site for a specific restriction endonuclease. This provides only two "ends" to which the foreign DNA can be attached. Ideally, the single restriction site should be within a gene, so that insertion of the foreign DNA will inactivate the gene (called insertional marker inactivation).

Among the most widely used *E. coli* plasmids are derivatives of the replicon plasmid ColE1. This plasmid carries a resistance gene against the antibiotic colicin E. The plasmid is under relaxed control, and up to 3000 copies may be produced when the proper *E. coli* strain is grown in the presence of chloramphenicol. One especially useful derivative plasmid of ColE1 is pBR322. It has all the properties previously outlined; in addition, its nucleotide sequence of 4363 base pairs is known, and it contains several different restriction endonuclease cleavage sites where foreign DNA can be inserted. For example, pBR322 has a single restriction site for the restriction enzyme *Eco*RI. *In vitro* insertion of a foreign DNA fragment into the *Eco*RI cleavage site and incorporation into a host cell (transformation) lead to immunity of the host cell to colicin E1, but the cell is unable to produce colicin.

Strains of *E. coli* are relatively easy to grow and maintain. The rate of growth is exponential and can be monitored by measuring the absorbance of a culture sample at 600 nm. One absorbance unit corresponds to a cell density of approximately 8×10^8 cells/mL. Some *E. coli* strains that harbor the ColE1 plasmids are RR1, HB101, GM48, 294, SK1592, JC411Thy⁻/ColE1, and CR34/ColE1. The typical procedure for growth and amplification of plasmids is, first, to establish the cells in normal medium for several hours. An aliquot of this culture is then used to inoculate medium containing the appropriate antibiotic. After overnight growth, a new portion of medium containing the antibiotic is inoculated with an aliquot of overnight culture. After the culture has been firmly established, a solution of chloramphenicol is added to inhibit chromosomal DNA synthesis. The ColE1 plasmids continue to replicate. The culture is then incubated for 12 to 18 hours and harvested by centrifugation.

Isolation of Plasmid DNA

The presence of ColE1 and other plasmids in a bacterial cell may be confirmed by genetic screening of antibiotic resistance. However, it is sometimes necessary to isolate plasmid DNA for further characterization and manipulation. Isolation and purification of plasmids are usually carried out for one of the following reasons:

1.   Construction of new recombinant plasmids

2.   Analysis of molecular size by agarose gel electrophoresis

3. Electrophoretic analyses of restriction enzyme digests and construction of a restriction enzyme map (see Experiment 15)

4. Sequence analysis of nucleotides by the Sanger or Maxam-Gilbert method

Several methods for isolating plasmid DNA have been developed; some lead to a more highly purified product than others. All isolation methods have the same objective, separation of plasmid DNA from chromosomal DNA. Plasmid DNA has two major structural differences from chromosomal DNA: (1) plasmid DNA is almost always extracted in a covalently closed circular configuration, whereas isolated chromosomal DNA usually consists of sheared linear fragments, and (2) ColE1 plasmids and other potential vehicles are much smaller than chromosomal DNA.

The structural differences cause physicochemical differences that can be exploited to separate the two types of molecules. Methods for isolating plasmid DNA fall into three major categories:

1. Methods that rely on specific interaction between plasmid DNA and a solid support. Examples are adsorption to nitrocellulose microfilters and hydroxyapatite columns.

2. Methods that cause selective precipitation of chromosomal DNA by various agents. These methods exploit the relative resistance of covalently closed circular DNA to extremes of pH, temperature, or other denaturing agents.

3. Methods based on differences in sedimentation behavior between the two types of DNA. This is the approach of choice if highly purified plasmid DNA is required.

Two isolation procedures based on methods in category 2 are described in this experiment, a large-scale and a microscale method. Each procedure yields plasmid DNA that is sufficiently pure for size analysis by agarose electrophoresis and for digestion by restriction enzymes as described in Experiment 15.

Separation of Plasmid DNA by Boiling (Holmes and Quigley, 1981)

The total cellular DNA must first be released by lysis of the bacterial cells. This is brought about by incubation with the enzyme lysozyme in the presence of reagents that inhibit nucleases (see Experiment 13). Chromosomal DNA is then separated from the plasmid DNA by boiling the lysis mixture for a brief period, followed by centrifugation. In contrast to closed circular plasmid DNA, linear chromosomal DNA becomes irreversibly denatured by heating and forms an insoluble gel, which sediments during centrifugation. Even though plasmid DNA may become partially denatured during boiling, the closed circular double helix reforms on cooling. The boiling serves a second purpose, that of denaturing deoxyribonucleases and other proteins.

Microscale Isolation of Plasmid DNA by Alkaline Lysis (Birnboim, 1983)

Often it is necessary to detect and analyze plasmid DNA in a large number of small bacterial samples. Several procedures are available for the rapid, small-scale isolation of plasmid DNA. Many of these provide plasmids of sufficient quantity and quality for characterization by agarose gel electrophoresis and restriction enzyme digestion. Besides providing rapid screening of plasmids from several bacterial samples, small-scale procedures require only minimal handling of potentially hazardous plasmids. One widely used microscale method is the alkaline lysis extraction process. For this procedure, host bacterial cells harboring the plasmid are grown in small culture volumes (1–5 mL) or in single colonies on agar plates. The cells are lysed and their contents denatured by alkaline sodium dodecyl sulfate (SDS). Proteins and high-molecular-weight chromosomal DNA denatured under these conditions precipitate as a gel that can be centrifuged from the supernatant, which contains plasmid DNA and bacterial RNA.

Characterization of DNA by Agarose Gel Electrophoresis

Several techniques for the characterization of nucleic acids have been introduced in this manual, but the standard method for separation and analysis of plasmids and other smaller DNA molecules is agarose gel electrophoresis. The method has several advantages, including ease of operation, sensitive staining procedures, high resolution, and a wide range of molecular weights ($0.6–100 \times 10^6$) that can be analyzed. As discussed in Chapter 4, the mobility of nucleic acids in agarose gels is influenced by the agarose concentration, the molecular size of the DNA, and the molecular shape of the DNA. In general, the lower the agarose concentration in the gels, the larger the DNA that can be analyzed. A practical lower limit of agarose concentration is reached at 0.3% agarose, below which gels become too fragile for ordinary use. This lower limit of 0.3% agarose in the gel allows analysis of linear double-stranded DNA within the range of 5 and 60 kilobase pairs (up to 150×10^6 in molecular weight). Gels with an agarose concentration of 0.8% can separate DNA in the range of 0.5–10 kilobase pairs and 2% agarose gels are used to separate smaller DNA fragments (0.1–3 kilobase pairs).

Nucleic acids migrate in an agarose medium at a rate that is inversely proportional to their size (kilobase pairs or molecular weights). In fact, a linear relationship exists between mobility and the logarithm of kilobase pairs (or molecular weight) of a DNA fragment. A standard curve may be prepared by including on the gel a sample containing DNA fragments of known molecular weights (see Figure E14.3).

Agarose gel electrophoresis is an ideal technique for analysis of DNA fragments. In addition to the positive characteristics discussed, the technique is simple, rapid, and relatively inexpensive. Fragments that differ in molecular weight by as little as 1% can be resolved on agarose gels, and as little as 1 ng of DNA can be detected on a gel. Nucleic acids are visualized after electrophoresis by soaking the gel in ethidium bromide or by performing the electrophoresis with ethidium bromide incorporated in the gel and buffer.

Overview of the Experiment

This experiment is divided into three parts: (A) growth of bacteria harboring a plasmid, (B) isolation of plasmid DNA, and (C) analysis of the plasmid DNA by agarose electrophoresis. Part A introduces students to some of the microbiological procedures required in molecular cloning. All media and supplies, except antibiotics, must be autoclaved before use. Sterile transfers are required at all stages. It is important to organize carefully the time commitments for this experiment. The total amount of actual working time is not large, but several hours of incubation time are required for a suitable culture.

The isolation of plasmid DNA by the boiling method (part B-1) yields a product of sufficient purity for analysis in Experiments 14 and 15. The procedure works well with chloramphenicol-amplified or untreated cell cultures. If cell cultures are available, the isolation method

Figure E14.3

Standard curve obtained by electrophoresis of λ phage DNA fragments from *Eco*RI cleavage.

- -

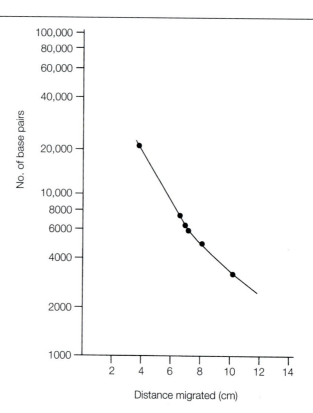

can easily be completed in about 2 hours. The yield is 2–3 mg of DNA/liter of culture.

As an alternative to the boiling method, students may perform the microscale isolation method. This version is more convenient to set up and complete, especially if facilities are limited. The procedure works well with unamplified or chloroamphenicol-amplified cultures. Each milliliter of an unamplified, overnight culture will yield approximately 1–2 μg of plasmid DNA. This isolation procedure requires $1^1/_2$ to 2 hours of student lab time if a cell culture is prepared in advance.

Students may be instructed to prepare their own agarose gels or precast gels may be purchased. If precast gels are provided, part C requires approximately 1 hour. An additional hour is required if gels are made by students.

The experiment may be carried out in one of the following ways:

1. Part A completed as a class project, preparing enough cell culture for all members to use in parts B and C. This option will require a total time of about 4 hours not counting cell growth time.

2. Completed in groups of four to six students. At least one student in each group should be experienced in standard microbiological procedures so that part A can be completed correctly. Approximately 4 hours are required if this option is completed.

3. The instructor or teaching assistant can complete part A and provide cell cultures to students for plasmid isolation and electrophoresis. Approximately 3–4 hours are required for parts B and C.

II.   MATERIALS AND SUPPLIES

A. Growth of Bacteria

–   *Escherichia coli* slant, slabs, plate or liquid culture. Recommended strains are RR1, HB101, GM48, 294, SK1591, JC411Thy⁻/ColE1, or CR34/ColE1. The strain must contain a plasmid such as ColE1, pBR322, pUR290, or the pUC series of plasmids.

–   Luria-Bertani (LB) medium. Each liter contains 10 g Bacto-tryptone, 5 g Bactoyeast extract, and 10 g NaCl. Adjust to pH 7.5 with NaOH.

–   Ampicillin. A solution of the sodium salt is prepared in water (25mg/mL) and sterilized by passage through a microfilter (e.g., a 0.22 μ Millipore filter). Store the solution in a freezer.

–   Chloramphenicol, 25 mg/mL in 100% ethanol

–   Autoclave

–   Flask shaker in environmental room at 37°C

–   Spectrophotometer and cuvettes for absorbance measurements at 600 nm

– Standard microbiological equipment: sterile flasks, pipets, test tubes, etc.
– Centrifuge, capable of $4000 \times g$

B. Isolation of Plasmid DNA

B-1. Large-Scale Boiling Method

– *E. coli* cells, 100 mL of culture grown as described in part A
– Standard Tris buffer, 0.01 M Tris-HCl, containing 0.1 M NaCl and 0.001 M EDTA, 5% Triton X-100, pH 8.0. Keep cold.
– Lysozyme solution, 10 mg/mL in 0.01 M Tris-HCl, pH 8.0. Must be freshly prepared.
– Bunsen burner
– Boiling-water bath in 1-liter beaker
– Centrifuge, capable of $12,000 \times g$

B-2. Microscale Alkaline Lysis Method

– Bacterial culture (2–3 mL) from part A
– STET buffer, containing 0.05 M glucose, 0.01 M EDTA, 0.025 M Tris-HCl, pH 8.0, and containing lysozyme (1 mg/mL)
– Alkaline-SDS solution, 0.2 M NaOH containing 1% (w/v) SDS
– Microcentrifuge, capable of $14,000 \times g$, at 4°C
– Potassium acetate, 5 M
– Ethanol
– Dry ice–acetone bath, approximately -18°C
– Tris-EDTA buffer, 0.01 M Tris-HCl, 0.001 M EDTA, pH 8.0.

C. Agarose Gel Electrophoresis

– Precast agarose plates, 1% agarose, or prepare gels:
  Agarose powder, electrophoresis grade
  Agarose slurry buffer, 0.04 M Tris-acetate, 0.002 M EDTA, pH 7.8
– Ethidium bromide solution, 10 mg/mL in H_2O
– Electrophoresis buffer, Tris-acetate buffer listed above containing ethidium bromide, 0.5 μg/mL
– Gel-loading buffer, 0.04 M Tris-acetate containing 50% glycerol and 0.25% bromophenol blue tracking dye, pH 8.0.
– Electrophoresis chamber. A setup for horizontal slab gel is recommended.
– Power supply

– UV lamp
– DNA molecular weight marker (DNA ladder); a solution of DNA fragments in the 50–3000 base pair range.

III. EXPERIMENTAL PROCEDURE

> **CAUTION**
>
> Experimentation with bacterial cells presents a potential biohazard. Do not pipet any solutions by mouth. Wash your hands well with hot water and soap before eating or drinking.

A. Growth of Bacteria

1. Using sterile techniques, transfer the appropriate bacterial cells from slant to a test tube or flask containing 10 mL of sterilized LB medium and 40 μL of the ampicillin solution. Incubate the culture with vigorous shaking at 37°C for 12 to 15 hours or overnight.

2. Prepare a 125-mL Erlenmeyer flask for transfer by adding 25 mL of sterile LB medium and 0.1 mL of the ampicillin solution. Transfer 0.1 mL of the overnight culture (from step 1) to the flask. Incubate at 37°C with vigorous shaking. At various time periods, remove 1-mL aliquots of the culture, transfer to a cuvette, and determine the cell density by measuring the absorbance at 600 nm.

3. When A_{600} reaches 0.6, transfer the entire culture from step 2 to a flask with 500 mL of sterile LB medium containing 2 mL of the ampicillin solution. Incubate at 37°C for 2.5 hours with vigorous shaking.

4. When A_{600} reaches about 0.4, add 4.0 mL of the chloramphenicol-ethanol solution to the growing culture.

5. Finally, incubate, with shaking, at 37°C for 12 to 16 hours or overnight and proceed to the isolation procedure in part B.

B. Isolation of Plasmid DNA

B-1. Large-Scale Boiling Method

The procedure outlined here can be used for bacterial culture ranging from 10 mL to 1 liter. Instructions will be given for 100 mL of cell culture. If only very small cultures (less than 10 mL) are available, follow the microscale procedure, B-2.

1. Obtain 100 mL of cell culture and centrifuge at 4000 $\times$ g for 10 minutes at 4°C. Discard the supernatant.

2. Wash the cells by adding 100 mL of standard Tris buffer containing NaCl and EDTA. Centrifuge as in step 1 and discard the supernatant.

3. Suspend the pellet in 15 mL of standard Tris buffer and transfer to a 50-mL Erlenmeyer flask.

4. Add 2.0 mL of lysozyme solution.

5. With a clamp, hold the flask over a Bunsen burner flame until the solution begins to boil. Shake constantly.

6. Incubate the flask in a boiling-water bath for 40 seconds. The bath should be set up with a 1-L beaker of water and a hot plate or flame.

7. Cool the flask in ice water.

8. Transfer the viscous lysate to a centrifuge tube and centrifuge at 12,000 $\times g$ for 20 minutes at 4°C.

9. Remove the cleared lysate (supernatant) and transfer it to a plastic test tube for storage. The present lysate containing plasmids may be analyzed directly by gel electrophoresis (part C) without further purification and used in Experiment 15.

B-2. Microscale Alkaline Lysis Method

1. Transfer 1.5 mL of an overnight culture to a microfuge tube. Centrifuge the tube at 4000 $\times g$ for 5 minutes.

2. Decant the supernatant and resuspend the pellet in 100 μL of sterile STET buffer containing lysozyme. Incubate at room temperature for 10 minutes.

3. Add 200 μL of 0.2 M NaOH containing 1% (w/v) SDS to the cell resuspension and mix thoroughly by inverting and tapping the tube gently. The suspension should become clearer as the cells are lysed. Incubate the suspension on ice for 5 minutes.

4. Add 150 μL of 5 M potassium acetate and mix thoroughly. Incubate on ice for 5 minutes.

5. Centrifuge the mixture at 14,000 $\times g$ for 10 minutes and transfer the supernatant to a fresh microtube.

6. Add 800 μL of chilled 95% ethanol and incubate on dry ice or in a freezer for 10 minutes.

7. Centrifuge for 15 minutes at 14,000 $\times g$ and remove the supernatant.

8. Remove excess ethanol from the pellet by air-drying or placing on a 37° hot block.

9. Resuspend the pellet of DNA in 25 μL Tris-EDTA, pH 8.0 buffer. Save the plasmid prep for agarose electrophoresis (part C) and use in Experiment 15.

C. Agarose Gel Electrophoresis

Many varieties of electrophoresis chambers are available. Specific details for the use of the electrophoresis unit should be obtained from the accompanying instructions or from your instructor.

Do not touch the electrophoresis chamber or wires while the electrophoretic operation is in progress. High voltages are required and shocks may be fatal.

Ethidium bromide is a mutagen. Always wear gloves when handling solutions or agarose gels containing the compound. Dispose of the final electrophoresis buffer containing ethidium bromide according to directions from your instructor. Avoid looking into UV light during the detection of DNA fragments on the agarose gel. Wear goggles!

1. Prepare a 1% (w/v) slurry of agarose in the agarose slurry buffer. The actual amount of solution to use depends on the size of the slab gel to be prepared. Follow the instructions given by your instructor.

2. Heat the slurry in a boiling-water bath or in a microwave oven until the agarose dissolves.

3. After the agarose solution has cooled to 50°C, add ethidium bromide solution to give a final dye concentration of 0.5 μg/mL.

4. Quickly pour the agarose solution over the glass plate to a depth of 2 to 4 mm. Insert the comb to make the sample wells.

5. Allow the gel to set for approximately 30 minutes and carefully remove the comb.

6. Clamp the agarose slab into the electrophoresis chamber and allow it to set for another 30 minutes.

7. Load one sample into each of the sample wells in the following manner. Mix each 20 μL of DNA plasmid and molecular weight standard marker (DNA ladder) with 10 μL of gel-loading buffer and apply one 10-μL sample to each well.

8. Add electrophoresis buffer to the reservoirs and connect the power supply.

9. Carry out the electrophoresis at approximately 3 V/cm or about 50 to 70 volts total.

10. Turn off the power supply when the tracking dye has migrated to the opposite edge of the gel.

11. Remove the gel from the chamber and glass plate, and examine it under a UV light. DNA fragments will appear as red-orange fluorescent bands. Take a photograph or draw a picture of the slab gel, showing the position of each DNA band.

IV.   ANALYSIS OF RESULTS

A. Growth of Bacteria

Prepare a flowchart showing each step carried out in the growth of the bacteria. Construct a growth curve of A_{600} vs. time, and identify any lag, log, and stationary phases of growth. Identify on the curve where chloramphenicol was added.

B. Isolation of Plasmid DNA

Prepare a flowchart outlining each isolation step. Write a brief explanation of the purpose of each reagent and procedural step.

C. Agarose Gel Electrophoresis

Examine the picture of the slab gel after electrophoresis. Note the number of fragments as represented by red-orange fluorescent bands. (Describe the action of ethidium bromide in formation of the color.) Measure the distance from the origin migrated by each band. Use the data from the DNA molecular weight standard (DNA ladder) to prepare a curve for size determination. Use a computer software program such as EXCEL to prepare the curve. Alternatively, semilog graph paper may be used to plot the size of each fragment on the log axis vs. the distance migrated by each fragment (see Figure 14.3). Use the standard curve to estimate the molecular size of each plasmid.

Study Problems

➤ 1. Explain the action of ampicillin as an inducer of plasmid replication.

➤ 2. How does chloramphenicol inhibit protein synthesis?

3. Why is a wavelength of 600 nm used to measure growth of bacteria? Could other wavelengths be used?

➤ 4. What is the purpose of the ethanol precipitation step in the preparation of plasmids?

5. Compare the isolation procedures used in this experiment with that used for chromosomal DNA (Experiment 13). Explain the differences.

6. Compare the two plasmid DNA isolation methods described in this experiment. Emphasize the differences.

➤ 7. Why are glycerol and bromophenol blue dye added to the gel-loading buffer?

➤ 8. Why is polyacrylamide electrophoresis not suitable for analysis of most plasmid DNA?

9. What assumption is made about the relative electrophoretic mobility of bromophenol blue dye and plasmid DNA?

10. What is the purpose of ethidium bromide in the gel electrophoresis?

Further Reading

H. Ahern, *The Scientist,* May 15, pp. 17–19 (1995). "Technology Makes DNA Isolation, Purification Simple and Swift."

H. Birnboim, *Methods Enzymol.* **100,** 243–255 (1983). "A Rapid Alkaline Extraction Method for the Isolation of Plasmid DNA."

R. Boyer, *Concepts in Biochemistry* (1999), Brooks/Cole (Pacific Grove, CA), pp. 384–407. An introduction to recombinant DNA.

M. Brockman et al., *J. Chem. Educ.* **73,** 542–543 (1996). A biotech lab project for recombinant protein expression in bacteria.

R. Garrett and C. Grisham, *Biochemistry,* 2nd ed. (1999), W. B. Saunders (Orlando, FL), pp. 395–422. An introduction to recombinant DNA.

J. Hammond and G. Spanswick, *Biochem. Educ.* **25,** 109–111 (1997). "A Demonstration of Genomic DNA Profiling by RAPD Analysis."

D. Holmes and M. Quigley, *Anal. Biochem.* **114,** 193–197 (1981). "A Rapid Boiling Method for the Preparation of Bacterial Plasmids."

A. Lehninger, D. Nelson, and M. Cox, *Principles of Biochemistry,* 3rd ed. (1999), Worth Publishers (New York), pp. 1119–1152. Recombinant DNA.

C. Mathews, K. van Holde, and K. Ahern, *Biochemistry,* 3rd ed. (2000), Benjamin/Cummings (San Francisco), pp. 970–973. Plasmids and recombinant DNA.

H. Miller, *Methods Enzymol.* **152,** 145–172 (1987). "Practical Aspects of Preparing Phage and Plasmid DNA: Growth, Maintenance and Storage of Bacteria and Bacteriophage."

L. Perez-Pons and E. Querol, *Biochem. Educ.* **24,** 54–56 (1996). A laboratory experiment illustrating basic principles of DNA cloning and molecular biology techniques.

J. Sambrook, E. Fritsch, and T. Maniatis, *Molecular Cloning, A Laboratory Manual,* 2nd ed. (1989), Cold Spring Harbor Laboratory (Cold Spring Harbor, NY), Vols. I, II, and III, pp. 1.21–1.52. Techniques in recombinant DNA.

L. Stryer, *Biochemistry,* 4th ed. (1995), W. H. Freeman (New York), pp. 119–142. Recombinant DNA.

D. Voet and J. Voet, *Biochemistry,* 2nd ed. (1995), John Wiley & Sons (New York), pp. 897–909. Recombinant DNA.

D. Voet, J. Voet, and C. Pratt, *Fundamentals of Biochemistry* (1999), John Wiley & Sons (New York), pp. 64–74. Recombinant DNA.

DNA on the Web

http://www.tgbiotech.com/methods.html
    Methods for DNA preparation.

http://research.nwfsc.noaa.gov/protocols.html
    Methods for DNA isolation and characterization.

http://res.agr.ca/winn/ethidium.htm
    Decontamination of ethidium bromide solutions.

15

THE ACTION OF RESTRICTION ENDONUCLEASES ON PLASMID OR VIRAL DNA

- **Recommended Reading**

Chapter 4; Experiments 13 and 14.

- **Synopsis**

Restriction endonucleases catalyze the hydrolysis of specific phosphodiester bonds in double-stranded DNA. These enzymes are often used to linearize a circular plasmid for hybrid DNA construction. They have also found use in the analysis of DNA and the construction of restriction maps. In this experiment, students will incubate various restriction enzymes with plasmid or viral DNA and analyze the product DNA fragments by agarose gel electrophoresis.

I.   INTRODUCTION AND THEORY

Experiment 14 introduced students to some principles and techniques involved in recombinant DNA research. Specifically, the experiment outlined the replication, isolation, and analysis of bacterial plasmid vehicles for molecular cloning experiments. This experiment describes another tool that is essential in hybrid plasmid construction and analysis—restriction enzyme action. The procedures introduced here have also found widespread use in the analysis and characterization of all DNA molecules.

The Action of Restriction Endonucleases

Bacterial cells produce many enzymes that act to degrade various forms of DNA. Of special interest are the **restriction endonucleases,** enzymes that recognize specific base sequences in double-stranded DNA and catalyze hydrolytic cleavage of the two strands in or near that specific region. The biological function of these enzymes is to degrade or restrict foreign DNA molecules. Host DNA is protected from hydrolysis because some bases near the cleavage sites are methylated. The action of the restriction enzyme *Eco*RI is shown in Equation E15.1.

>>

Equation E15.1

$$5'...G{-}\overset{\downarrow}{G}{-}A{-}T{-}C{-}C...3' \quad \xrightarrow[\text{H}_2\text{O}]{EcoRI} \quad 5'...G \qquad\qquad G{-}A{-}T{-}C{-}C...3'$$
$$3'...C{-}C{-}T{-}A{-}\underset{\uparrow}{G}{-}G...5' \qquad\qquad 3'...C{-}C{-}T{-}A{-}G \qquad + \qquad G...5'$$

The site of action of *Eco*RI is a specific hexanucleotide sequence. Two phosphodiester bonds are hydrolyzed (see arrows), resulting in fragmentation of both strands. Note the twofold rotational symmetry feature at the recognition site and the formation of cohesive ends. The weak base pairing between the cohesive ends is not sufficient to hold the two fragments together.

Several hundred restriction enzymes have been isolated and characterized. Nomenclature for the enzymes consists of a three-letter abbreviation representing the source (*Eco* = *E. coli*), a letter representing the strain (R), and a roman numeral designating the order of discovery. *Eco*RI is the first to be isolated from *E. coli* (strain R) and characterized. Table E15.1 lists several other restriction enzymes, their recognition sequence for cleavage, and optimum reaction conditions.

Restriction enzymes are used extensively in nucleic acid chemistry. They may be used to cleave large DNA molecules into smaller fragments that are more amenable to analysis. For example, λ phage DNA, a linear, double-stranded molecule of 48,502 base pairs (molecular weight 31×10^6), is cleaved into six fragments by *Eco*RI or into more than 50 fragments by *Hinf* I (*Haemophilus influenzae*, serotype f). The base sequence recognized by a restriction enzyme is likely to occur only a very few times in any particular DNA molecule; therefore, the smaller the DNA molecule, the fewer the number of specific sites. The λ phage DNA is cleaved into 0 to 50 or more fragments, depending on the restriction enzyme used, whereas larger bacterial or animal DNA will most likely have many recognition sites and be cleaved into hundreds of fragments. Smaller DNA molecules, therefore, have a much greater chance of producing a unique set of fragments with a particular restriction enzyme. It is unlikely that this set of fragments will be the same for any two different DNA molecules, so the fragmentation pattern is unique and can be considered a "fingerprint" of the DNA substrate. The fragments are readily separated and sized by agarose gel electrophoresis as described later in this experiment.

| Table E15.1 |

Specificity and Optimal Conditions for Several Restriction Endonucleases.

Name	Recognition Sequence 5'........3'	T (°C)	pH	Tris (mM)	NaCl (mM)	MgCl$_2$ (mM)	DTT[2] (mM)
AatII	G—A—C—G—T↓C	37	7.9	20	50	10	10
AluI	A—G↓C—T	37	7.5	10	50	10	10
BalI	T—G—G↓C—C—A	37	7.9	6	—	6	6
BamHI	G↓G—A—T—C—C	37	8.0	20	100	0.7	1
BclI	T↓G—A—T—C—A	60	7.5	10	50	10	1
EcoRI	G↓A—A—T—T—C	37	7.5	10	100	10	1
HaeII	Pu[1]—G—C—G—C↓Py[1]	37	7.5	10	50	10	10
HindIII	A↓A—G—C—T—T	37–55	7.5	10	60	10	1
HpaI	G—T—T↓A—A—C	37	7.5	10	50	10	1
MseI	T↓T—A—A	37	7.9	10	50	10	1
NotI	G—C↓G—G—C—C—G—C	37	7.9	10	150	10	—
SalI	G↓T—C—G—A—C	37	8.0	10	150	10	1
ScaI	A—G—T↓A—C—T	37	7.4	10	100	10	1
TaqI	T↓C—G—A	65	8.4	10	100	10	10

[1] Pu = a purine base; Py = a pyrimidine base.
[2] DTT = dithiothreitol.

Restriction endonucleases are also valuable tools in the construction of hybrid DNA molecules. Several restriction enzymes act at a single site on bacterial plasmid vehicles. This linearizes the circular plasmid and allows the insertion of a foreign DNA fragment. For example, the popular plasmid vehicle pBR322 has a single restriction site for *BamHI* (*Bacillus amyloliquefaciens*, H) that is within the tetracycline resistance gene. The enzyme not only opens the plasmid for insertion of a DNA fragment but also destroys a phenotype; this fact aids in the selection of transformed bacteria.

Restriction enzymes can be used to obtain physical maps of DNA molecules. Important information can be obtained from maps of DNA, whether the molecules are small and contain only a few genes, such as viral or plasmid DNA, or large and complex, such as bacterial or eukaryotic chromosomal DNA molecules. An understanding of the genetics, metabolism, and regulation of an organism requires knowledge of the precise arrangement of its genetic material. The construction of a **restriction enzyme map** for a DNA molecule provides some of this information. The map displays the sites of cleavage by restriction endonucleases and the number of fragments obtained after digestion with each enzyme.

A map is constructed by first digesting plasmid DNA with restriction endonucleases that yield only a few fragments. Each digest of DNA obtained with a single nuclease is analyzed by agarose gel electrophoresis, using standards for molecular weight determination. These are referred to as the primary digests. Second, each of the primary digests is treated with a series of additional restriction enzymes, and the digests are analyzed by agarose gel electrophoresis. The restriction map is then constructed by combining the various fragments by trial and error and logic, much as in sequencing a protein by proteolytic digestion by several enzymes and searching for overlap regions in the fragments. In the case of the restriction endonuclease digests, two characteristics of the DNA fragments are known: the approximate molecular weights (from electrophoresis) and the nature of the fragment ends (from the known selectivity of the individual restriction enzymes). The fragments may also be sequenced by the Sanger or Maxam-Gilbert method. A restriction enzyme map for the plasmid pBR322 is shown in Figure E15.1.

Practical Aspects of Restriction Enzyme Use

Restriction enzymes are heat-labile and expensive biochemical reagents. Their use requires considerable planning and care. Each restriction nuclease has been examined for optimal reaction conditions in regard to specific pH range, buffer composition, and incubation temperature. This information for each enzyme is readily available from the commercial supplier of the enzyme or from the literature. Table E15.1 gives important reaction information for several enzymes. The temperature range and pH optima for most restriction nucleases are similar (37°C, 7.5–8.0); however, optimal buffer composition is variable. Typical buffer components are Tris, NaCl, $MgCl_2$, and a sulfhydryl reagent (β-mercaptoethanol or dithiothreitol). Proper reaction conditions are crucial for optimal reaction rate; but, more important, changing reaction conditions have been shown to alter the specificity of some restriction enzymes.

Although restriction enzymes are very unstable reagents, they can be stored at $-20°C$ in buffer containing 50% glycerol. They are usually prepared in an appropriate buffer and shipped in packages containing dry ice.

Disposable gloves should be worn when you are handling the enzyme container. Remove the enzyme from the freezer just before you need it. Store the enzyme in an ice bucket when it is outside the freezer. The enzyme should never be stored at room temperature. Because of high cost, digestion by restriction enzymes is carried out on a microscale level. A typical reaction mixture will contain about 1 μg or less of DNA and 1 unit of enzyme in the appropriate incubation buffer. One unit is the amount of enzyme that will degrade 1 μg of λ phage DNA in 1 hour at the optimal temperature and pH. The total reaction volume is usually between 20 and 50 μL. Incubation is most often carried out at the recommended temperature for about 1 hour. The reaction is stopped by adding EDTA solution, which complexes divalent metal ions essential for nuclease activity.

Figure E15.1

A restriction enzyme map for the plasmid pBR322. From *Molecular Cloning, A Laboratory Manual,* by T. Maniatis, E. Fritsch, and J. Sambrook. Cold Spring Harbor Laboratory, 1982.

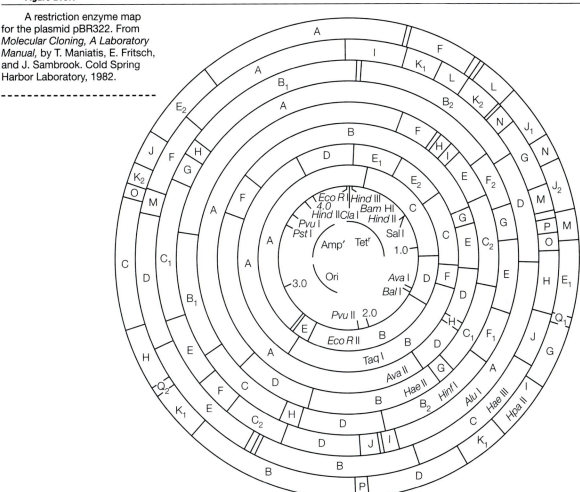

Reaction mixtures from restriction enzyme digestion may be analyzed directly by agarose gel electrophoresis. This technique combines high resolving power and sensitive detection to allow the analysis of minute amounts of DNA fragments.

Characterization of DNA by Agarose Gel Electrophoresis

Agarose gels are the standard media for separating and characterizing DNA molecules. As discussed in Chapter 4, the mobility of nucleic acids in agarose gels is influenced by the agarose concentration, the molecular size of the DNA, and the molecular shape of the DNA. In general, the lower the agarose concentration in the gels, the larger the DNA that can be analyzed. A practical lower limit of agarose concentration is reached at 0.3% agarose, below which gels become too fragile for ordinary use. This lower limit of 0.3% agarose in the gel allows analysis of linear double-stranded DNA within the range of 5 and 60 kilobase pairs (up to 150×10^6 in molecular weight). Gels with an agarose concentration of 0.8% can separate DNA in the range of 0.5–10 kilobase pairs and 2% agarose gels are used to separate smaller DNA fragments (0.1–3 kilobase pairs). For most routine analysis of restriction enzyme fragments, agarose concentrations in the range of 0.7 to 1% are most appropriate.

Nucleic acids migrate in an agarose medium at a rate that is inversely proportional to their size (kilobase pairs or molecular weights). In fact, a linear relationship exists between mobility and the logarithm of kilobase pairs (or molecular weight) of a DNA fragment. After electrophoresis of a restriction enzyme digest containing DNA fragments of known size, a standard curve may be prepared (see Figure E14.3). The standard reaction mixture shown here is the *Eco*RI digestion of λ phage DNA. The mixture contains six DNA fragments with the following sizes: 21226, 7421, 5804, 5643, 4878, and 3530 base pairs.

The influence of molecular shape on electrophoretic mobility is not a critical factor to consider in the experiment. The action of restriction enzymes on both covalent closed circular DNA and linear DNA results in the formation of linear fragments.

Agarose gel electrophoresis is an ideal technique for analysis of DNA fragments. In addition to the positive characteristics discussed, the technique is simple, rapid, and relatively inexpensive. Fragments that differ in molecular weight by as little as 1% can be resolved on agarose gels, and as little as 1 ng of DNA can be detected on a gel. Nucleic acids are visualized after electrophoresis by soaking the gel in ethidium bromide or by performing the electrophoresis with ethidium bromide incorporated in the gel and buffer.

Overview of the Experiment

The objective of the experiment is to evaluate the action of restriction enzymes on bacterial plasmids, λ phage DNA, or viral DNA. The DNA will be incubated under the appropriate conditions with selected restriction enzymes. The reaction mixtures will be subjected to agarose gel electrophoresis in order to determine the number and molecular size of the restriction fragments.

The following options are available for this experiment:

1. Students may use isolated and purified plasmids from Experiment 14. Alternatively, several purified plasmids, including ColE1, pBR322, and pUB110, are available from commercial sources. The plasmids may be cleaved with various restriction enzymes and the products analyzed by agarose gel electrophoresis.

2.   Students may characterize λ phage or adenovirus 2 DNA. Both are standard substrates for assays of restriction endonucleases.

Approximately 2 hours are required to prepare and incubate the restriction enzyme reaction mixtures. The agarose slab gel may be prepared during the incubation period. Up to 1 to 2 hours are required to prepare the gel slab. Alternatively, precast agarose gels may be provided. The electrophoresis may be completed as a group project. Most electrophoresis chambers accommodate slab gels with 15 to 20 sample wells.

II.   MATERIALS AND SUPPLIES

A. Digestion of DNA with Restriction Enzymes

–   DNA: Plasmid, commercial or from Experiment 14; λ phage; adenovirus (Ad 2); approximately 0.2–0.4 mg/mL in Tris-HCl, NaCl, EDTA, pH 7.0.

–   Restriction enzymes, 1 unit/μL. Keep stored in freezer until ready to use. Recommended enzymes are *Bam*HI and *Eco*RI for λ phage DNA, *Eco*RI and *Taq*I for plasmids (pBR322 or ColE1), and *Hpa*I for Ad 2 DNA.

–   Sterilized H_2O

–   Incubation buffers

   *Bam*HI: 0.2 M Tris-HCl, pH 8.0, 0.07 M MgCl$_2$, 1 M NaCl, 0.01 M dithiothreitol

   *Eco*RI: 0.1 M Tris-HCl, pH 7.5, 0.1 M MgCl$_2$, 1 M NaCl, 0.01 M dithiothreitol

   *Hpa*I: 0.1 M Tris-HCl, pH 7.5, 0.1 M MgCl$_2$, 0.5 M NaCl, 0.01 M dithiothreitol

   *Taq*I: 0.1 M Tris-HCl, pH 7.4, 0.1 M MgCl$_2$, 1 M NaCl, and 0.01 M dithiothreitol

   Quench buffer, 0.1 M EDTA, pH 7

B. Agarose Gel Electrophoresis

–   Precast agarose gels

–   Agarose powder, electrophoresis grade

–   Agarose slurry buffer, 0.04 M Tris-acetate, 0.002 M EDTA, pH 7.8

–   Ethidium bromide solution, 10 mg/mL in H_2O

–   Electrophoresis buffer, Tris-acetate buffer listed above containing ethidium bromide, 0.5 μg/mL

–   Gel-loading buffer, 0.04 M Tris-acetate containing 50% glycerol and 0.25% bromophenol blue tracking dye, pH 8.0.

- Electrophoresis chamber. A setup for horizontal slab gel is recommended.
- Power supply
- UV lamp
- Standard restriction enzyme digest, λ phage DNA digested by *Eco*RI
- Constant-temperature water bath at 37°C

III.   EXPERIMENTAL PROCEDURE

A. Digestion of DNA with Restriction Enzymes

The reaction conditions are specific for each restriction enzyme. The commercial supplier of the enzyme will provide an information sheet giving details of the reaction conditions. For a detailed listing of optimal conditions for many restriction enzymes, see Brooks (1987) or Table E15.1.

*Bam*HI

1. Obtain a small microcentrifuge tube for each reaction mixture and add the following components:

    15 μL of sterile H_2O

    2 μL of DNA solution

    2 μL of *Bam*HI incubation buffer; mix well by tapping.

    1 unit of *Bam*HI restriction enzyme. The actual volume depends on the concentration of enzyme. In most cases this will be 1 μL of solution. Mix the tube by gentle tapping.

2. Incubate the reaction mixture at 37°C for 1 hour.

3. Stop the reaction by adding 2 μL of EDTA quench buffer.

4. Store the reaction mixture on ice until agarose gel electrophoresis.

*Eco*RI

1. Obtain a small test tube for each reaction mixture and add the following components:

    15 μL of sterile H_2O

    2 μL of DNA solution

    2 μL of *Eco*RI incubation buffer; mix well by tapping.

    Add 1 unit of *Eco*RI restriction enzyme. The volume depends on the concentration of enzyme. For most enzymes this will be 1 μL of solution. Mix the tube with gentle tapping.

2. Incubate the reaction mixture at 37°C for 1 hour.

3. Add 2 μL of EDTA quench buffer.

4. Store the reaction mixture on ice until agarose gel electrophoresis.

*Hpa*I

1. Obtain a small test tube for each reaction mixture and add the following components:

    15 μL of sterile H_2O

    2 μL of the DNA solution

    2 μL of *Hpa*I incubation buffer; mix well by tapping.

    1 unit of *Hpa*I restriction enzyme. In most cases this will be 1 μL.

    Mix well by gentle tapping.

2. Incubate the reaction mixture at 37°C for 1 hour.

3. Add 2 μL of EDTA quench buffer.

4. Store the reaction mixture on ice until agarose gel electrophoresis.

*Taq*I

1. Obtain a small microcentrifuge tube for each reaction mixture and add each of the following components:

    15 μL of sterile H_2O

    2 μL of DNA solution

    2 μL of *Taq*I incubation buffer; mix well by tapping

    1 unit of *Taq*I restriction enzyme

2. Incubate the reaction mixture at 37°C for 1 hour.

3. Stop the reaction by adding 2 μL of EDTA quench buffer.

4. Store the reaction mixture on ice until gel electrophoresis.

B. Agarose Gel Electrophoresis

Many varieties of electrophoresis chambers are available. Specific details for the use of the electrophoresis unit should be obtained from the accompanying instructions or from your instructor.

CAUTION

Do not touch the electrophoresis chamber or wires while the electrophoretic operation is in progress. High voltages are required and shocks may be fatal.

Ethidium bromide is a mutagen. Always wear gloves when handling solutions or agarose gels containing the compound. Dispose of the final electrophoresis buffer containing ethidium bromide according to directions from your instructor. Avoid looking into UV light during the detection of DNA fragments on the agarose gel. Wear goggles!

1. Prepare a 1% (w/v) slurry of agarose in the agarose slurry buffer. The actual amount of solution to use depends on the size of the slab gel to be prepared. Follow the instructions given by your instructor.

2. Heat the slurry in a boiling-water bath until the agarose dissolves.

3. After the agarose solution has cooled to 50°C, add ethidium bromide solution to give a final dye concentration of 0.5 μg/mL.

4. Quickly pour the agarose solution over the glass plate to a depth of 2 to 4 mm. Insert the comb to make the sample wells.

5. Allow the gel to set for approximately 30 minutes and carefully remove the comb.

6. Clamp the agarose slab into the electrophoresis chamber and allow it to set for another 30 minutes.

7. Load one sample into each of the sample wells in the following manner. Mix each 20 μL of reaction mixture from the restriction enzyme with 10 μL of gel-loading buffer and apply one 10 μL sample to each well. Put 10 μL of *Eco*RI standard digest of λ phage DNA into one sample well.

8. Add electrophoresis buffer to the reservoirs and connect the power supply.

9. Carry out the electrophoresis at approximately 3 V/cm or about 50 to 70 volts total.

10. Turn off the power supply when the tracking dye has migrated to the bottom edge of the gel.

11. Remove the gel from the chamber and glass plate, and examine it under a UV light. DNA fragments will appear as red-orange fluorescent bands.

12. Take a photograph or draw a picture of the slab gel, showing the position of each DNA band.

IV.   ANALYSIS OF RESULTS

Examine the picture of the slab gel after electrophoresis. Note the number of fragments as represented by red-orange fluorescent bands. (Describe the action of ethidium bromide in formation of the color.) Measure the distance from the origin migrated by each band. Use the data from the *Eco*RI standard digest of λ phage DNA to prepare a standard curve for size determination. Using semilog graph paper, plot the size of each fragment on the log axis vs. the distance migrated by each fragment. The sizes of the fragments are 21226, 7421, 5804, 5643, 4878, and 3530 base pairs. Use the standard curve to estimate the molecular weight of each restriction enzyme fragment.

Study Problems

▸ 1. How does the EDTA quench buffer stop the restriction enzyme reaction?

▶ 2. Why are glycerol and bromophenol blue dye added to the gel-loading buffer?

3. Describe the pattern obtained from electrophoresis of a standard *Eco*RI digest of λ phage DNA on 2% agarose gel.

▶ 4. Why is polyacrylamide electrophoresis not suitable for analysis of restriction enzyme fragments of most DNA?

▶ 5. How many DNA fragments result from the action of the restriction enzyme *Hae*II on the plasmid pBR322?

6. Assume the reaction digest from Problem 5 is analyzed by agarose electrophoresis. Draw an electrophoresis gel and show the approximate location of each DNA fragment.

7. Explain the action of ethidium bromide in locating the DNA fragments after electrophoresis.

8. Ethidium bromide is a mutagen. Explain all precautions you took during this experiment in order to avoid contact with the chemical.

▶ 9. What assumptions must be made about the relative mobility of bromophenol blue dye and DNA fragments during electrophoresis?

▶ 10. Which of the following base sequences are probably not recognition sites for cleavage by restriction endonucleases? Why not?

(a) 5′GAATTC3′

(b) 5′CATTAG3′

(c) 5′CATATG3′

(d) 5′CAATTG3′

Further Reading

R. Boyer, *Concepts in Biochemistry* (1999), Brooks/Cole (Pacific Grove, CA), pp. 300–302. DNA restriction enzymes.

J. Brooks, in *Methods in Enzymology,* Vol. 152, S. Berger and A. Kimmel, Editors (1987), Academic Press (Orlando, FL), pp. 113–129. "Properties and Uses of Restriction Endonucleases."

S. Farrell, *J. Chem. Educ.* **71,** 1095–1096 (1994). "A Fast Restriction Enzyme Experiment for the Undergraduate Biochemistry Lab."

R. Garrett and C. Grisham, *Biochemistry,* 2nd ed. (1999), W. B. Saunders (Orlando, FL), pp. 350–355; 396–417. Introduction to restriction enzymes and DNA libraries.

A. Lehninger, D. Nelson, and M. Cox, *Principles of Biochemistry,* 3rd ed. (1999), Worth Publishers (New York), pp. 1120–1124. Restriction enzymes.

C. Mathews, K. van Holde, and K. Ahern, *Biochemistry,* 3rd ed. (2000), Benjamin/Cummings (San Francisco), pp. 929–934. Introduction to restriction enzymes.

D. Micklos and G. Freyer, *DNA Science: A First Course in Recombinant DNA Technology* (1990), Cold Spring Harbor Laboratory (Cold Spring Harbor, NY). An excellent reference book for beginning students.

J. Perez-Ortin, M. Del Olmo, E. Matallana, and V. Tordera, *Biochem. Educ.* **25,** 237–242 (1997). "Making Your Own Gene Library."

J. Sambrook, E. Fritsch, and T. Maniatis, *Molecular Cloning, A Laboratory Manual,* 2nd ed. (1989), Cold Spring Harbor Laboratory (Cold Spring Harbor, NY), Vols. I, II, and III. An excellent lab manual for recombinant DNA.

L. Stryer, *Biochemistry,* 4th ed. (1995), W. H. Freeman (New York), pp. 119–144. Introduction to restriction enzymes.

D. Voet and J. Voet, *Biochemistry,* 2nd ed. (1995), John Wiley & Sons (New York), pp. 882–888. Restriction enzymes.

D. Voet, J. Voet, and C. Pratt, *Fundamentals of Biochemistry* (1999), John Wiley & Sons (New York), pp. 56–58; 752–753. Restriction enzymes.

Restriction Enzymes on the Web

http://www.cgcsci.com/cminfo.shtml
    Restriction map of pBR322 in circular and linear format. Designed by CGC Scientific, a provider of software tools for biologists.

http://www.fermentas.com
    Click on Product Profiles; then Restriction endonucleases for a listing of enzymes. Click on individual enzymes for specific data including cleavage site and reaction conditions.

http://www.fhsu.edu/chemistry/twiese/360/molbio/index.htm
    Click on "Recombinant DNA Technology" for a PowerPoint presentation.

PROPERTIES OF COMMON
ACIDS AND BASES

Compound	Formula	Molecular Weight	Specific Gravity	% by Weight	Molarity (*M*)
Acetic acid, glacial	CH_3COOH	60.1	1.05	99.5	17.4
Ammonium hydroxide	NH_4OH	35.0	0.89	28	14.8
Formic acid	$HCOOH$	46.0	1.20	90	23.4
Hydrochloric acid	HCl	36.5	1.18	36	11.6
Nitric acid	HNO_3	63.0	1.42	71	16.0
Perchloric acid	$HClO_4$	100.5	1.67	70	11.6
Phosphoric acid	H_3PO_4	98.0	1.70	85	18.1
Sulfuric acid	H_2SO_4	98.1	1.84	96	18.0

II

PROPERTIES OF COMMON BUFFER COMPOUNDS

Compound	Abbreviation	Molecular Weight	pK_1	pK_2 (20°C)	pK_3	pK_4
N-(2-Acetamido)-2-aminoethanesulfonic acid	ACES	182.2	6.9	—	—	—
N-(2-Acetamido)-2-iminodiacetic acid	ADA	212.2	6.60	—	—	—
Acetic acid		60.1	4.76	—	—	—
Arginine	Arg	174.2	2.17	9.04	12.48	—
Barbituric acid		128.1	3.79	—	—	—
N,N-bis (2-Hydroxyethyl)-2-aminoethane-sulfonic acid	BES	213.1	7.15	—	—	—
N,N-bis (2-Hydroxyethyl)glycine	Bicine	163.2	8.35	—	—	—
Boric acid		61.8	9.23	12.74	13.80	—
Citric acid		210.1	3.10	4.75	6.40	—
Ethylenediaminetetraacetic acid	EDTA	292.3	2.00	2.67	6.24	10.88
Formic acid		46.03	3.75	—	—	—
Fumaric acid		116.1	3.02	4.39	—	—
Glycine	Gly	75.1	2.45	9.60	—	—
Glycylglycine		132.1	3.15	8.13	—	—
N-2-Hydroxyethylpiperazine-N'-2-ethanesulfonic acid	HEPES	238.3	7.55	—	—	—
N-2-Hydroxyethylpiperazine-N'-3-propanesulfonic acid	HEPPS	252.3	8.0	—	—	—
Histidine	His	209.7	1.82	6.00	9.17	—
Imidazole		68.1	6.95	—	—	—
2-(N-Morpholino) ethanesulfonic acid	MES	195	6.15	—	—	—
3-(N-Morpholino) propanesulfonic acid	MOPS	209.3	7.20	—	—	—
Phosphoric acid		98.0	2.12	7.21	12.32	—
Succinic acid		118.1	4.18	5.60	—	—
3-Tris (hydroxymethyl)aminopropanesulfonic acid	TAPS	243.2	8.40	—	—	—
N-Tris (hydroxymethyl)methyl-2-aminoethanesulfonic acid	TES	229.2	7.50	—	—	—
N-Tris (hydroxymethyl)methylglycine	Tricine	179	8.15	—	—	—
Tris (hydroxymethyl)aminomethane	Tris	121.1	8.30	—	—	—

III

pK_a Values and pH$_I$ Values of Amino Acids

Name	Abbreviations		pK$_1$ (α-carboxyl)	pK$_2$ (α-amino)	pK$_R$ (side chain)	pH$_I$
Alanine	Ala	A	2.3	9.7	—	6.0
Arginine	Arg	R	2.2	9.0	12.5	10.8
Asparagine	Asn	N	2.0	8.8	—	5.4
Aspartate	Asp	D	2.1	9.8	3.9	3.0
Cysteine	Cys	C	1.7	10.8	8.3	5.0
Glutamate	Glu	E	2.2	9.7	4.3	3.2
Glutamine	Gln	Q	2.2	9.1	—	5.7
Glycine	Gly	G	2.3	9.6	—	6.0
Histidine	His	H	1.8	9.2	6.0	7.6
Isoleucine	Ile	I	2.4	9.7	—	6.1
Leucine	Leu	L	2.4	9.6	—	6.0
Lysine	Lys	K	2.2	9.0	10.5	9.8
Methionine	Met	M	2.3	9.2	—	5.8
Phenylalanine	Phe	F	1.8	9.1	—	5.5
Proline	Pro	P	2.0	10.6	—	6.3
Serine	Ser	S	2.2	9.2	—	5.7
Threonine	Thr	T	2.6	10.4	—	6.5
Tryptophan	Trp	W	2.4	9.4	—	5.9
Tyrosine	Tyr	Y	2.2	9.1	10.1	5.7
Valine	Val	V	2.3	9.6	—	6.0

MOLECULAR WEIGHTS OF SOME COMMON PROTEINS

Name (Source)	Molecular Weight (daltons)
Albumin (bovine serum)	65,400
Albumin (egg white)	45,000
Carboxypeptidase A (bovine pancreas)	35,268
Carboxypeptidase B (porcine)	34,300
Catalase (bovine liver)	250,000
Chymotrypsinogen (bovine pancreas)	23,200
Cytochrome *c*	13,000
Hemoglobin (bovine)	64,500
Insulin (bovine)	5,700
α-Lactalbumin (bovine milk)	14,200
Lysozyme (egg white)	14,600
Malate dehydrogenase (pig heart mitochondria)	70,000
Myoglobin (horse heart)	16,900
Pepsin (porcine)	35,000
Peroxidase (horseradish)	40,000
Ribonuclease I (bovine pancreas)	12,600
Trypsinogen (bovine pancreas)	24,000
Trypsin (bovine pancreas)	23,800
Tyrosinase (mushroom)	128,000
Uricase (pig liver)	125,000

Common Abbreviations Used in This Text

A	adenine or absorbance
AMP, ADP, ATP	adenosine mono-, di-, or triphosphate
AP	alkaline phosphatase
BCA	bicinchoninic acid
bp	base pairs
BPG	bisphosphoglycerate
Bq	Becquerel
BSA	bovine serum albumin
C	cytosine
CE	capillary electrophoresis
Ci	Curie
CM cellulose	carboxymethyl cellulose
COSY	correlation spectroscopy
CPM	counts per minute
DABITC	4-*N*,*N*-dimethylaminoazobenzene-4′-isothiocyanate
Da	dalton
DANSYL	1,1-dimethylaminonaphthalene-5-sulfonyl chloride
DCIP	dichlorophenolindophenol
DEAE cellulose	diethylaminoethyl cellulose
DNA	deoxyribonucleic acid
DNP	dinitrophenyl
DOPA	dihydroxyphenylalanine
E	absorption coefficient
EDTA	ethylenediaminetetraacetic acid
ELISA	enzyme linked immunosorbent assay
FAB	fast atom bombardment
FAD(H$_2$)	flavin adenine dinucleotide (reduced form)
FAME	fatty acid methyl ester
FMN(H$_2$)	flavin mononucleotide (reduced form)
FMOC	9-fluorenylmethyl chloroformate
FPLC	fast protein liquid chromatography
FT	Fourier transform
G	guanine
GC	gas chromatography
GMP, GDP, GTP	guanosine mono-, di-, or triphosphate
Hb	hemoglobin
HDL	high-density lipoprotein
HMIS	Hazardous Materials Identification System
HPLC	high-performance liquid chromatography
HRP	horseradish peroxidase
IE	immunoelectrophoresis
IEF	isoelectric focusing
IHP	inositol hexaphosphate
IMAC	immobilized metal-ion affinity chromatography

IR	infrared
IVS	intervening sequence
kb	kilobase pairs
LDL	low-density lipoprotein
MAO	monoamine oxidase
MS	mass spectrometry
MSDS	material safety data sheet
NAD(H)	nicotinamide adenine dinucleotide (reduced form)
NADP(H)	nicotinamide adenine dinucleotide phosphate (reduced form)
NBT	nitroblue tetrazolium
NMR	nuclear magnetic resonance
NOESY	nuclear Overhauser effect spectroscopy
OSHA	Occupational Safety and Health Administration
PAGE	polyacrylamide gel electrophoresis
PCR	polymerase chain reaction
PEG	polyethylene glycol
PFGE	pulsed field gel electrophoresis
PMS	phenazine methosulfate
PMT	photomultiplier tube
POPOP	1,4-bis[5-phenyl-2-oxazolyl]benzene
PPO	2,5-diphenyloxazole
PTH	phenylthiohydantoin
RCF	relative centrifugal force
RFLP	restriction fragment length polymorphism
RIA	radioimmunoassay
RNA	ribonucleic acid
s	sedimentation coefficient
S	Svedberg (10^{-13}s)
SDS	sodium dodecyl sulfate
SMP	submitochondrial particles
T	thymine
TAG	triacylglycerol
TEMED	N,N,N',N'-tetramethylethylenediamine
TLC	thin-layer chromatography
Tris	tris(hydroxymethyl)aminomethane
TROSY	transverse relaxation-optimized spectroscopy
U	uracil
UV	ultraviolet
VIS	visible
VLDL	very low-density lipoprotein

Units of Measurement

The International System of Units (SI)

Quantity	Unit	Abbreviation
length	meter	m
mass	kilogram	kg
time	second	s
temperature	kelvin	K
electric current	ampere	A
amount of substance	mole	mol
radioactivity	becquerel	Bq
volume	liter	L

Metric Prefixes

Name	Abbreviation	Multiplication Factor (relative to "1")
atto	a	10^{-18}
femto	f	10^{-15}
pico	p ($\mu\mu$)	10^{-12}
nano	n (mμ)	10^{-9}
micro	μ	10^{-6}
milli	m	10^{-3}
centi	c	10^{-2}
deci	d	10^{-1}
deca	da	10
kilo	k	10^{3}
mega	M	10^{6}
giga	G	10^{9}

Units of Length

Name	Abbreviation	Multiplication Factor (relative to meter)
kilometer	km	10^3
meter	m	1
centimeter	cm	10^{-2}
millimeter	mm	10^{-3}
micrometer	μm	10^{-6}
nanometer	nm	10^{-9}
Angstrom	Å	10^{-10}

Units of Mass

Name	Abbreviation	Multiplication Factor (relative to gram)
kilogram	kg	10^3
gram	g	1
milligram	mg	10^{-3}
microgram	μg	10^{-6}
nanogram	ng	10^{-9}

Units of Volume

Name	Abbreviation	Multiplication Factor (relative to liter)
liter	L	1
deciliter	dL	10^{-1}
milliliter	mL	10^{-3}
microliter	μL	10^{-6}

TABLE OF THE ELEMENTS*

	Symbol	Atomic No.	Atomic Mass		Symbol	Atomic No.	Atomic Mass
Actinium	Ac	89	227.0278	Mercury	Hg	80	200.59
Aluminum	Al	13	26.98154	Molybdenum	Mo	42	95.94
Americium	Am	95	[243]†	Neodymium	Nd	60	144.24
Antimony	Sb	51	121.75	Neon	Ne	10	20.179
Argon	Ar	18	39.948	Neptunium	Np	93	237.0482
Arsenic	As	33	74.9216	Nickel	Ni	28	58.70
Astatine	At	85	[210]	Niobium	Nb	41	92.9064
Barium	Ba	56	137.33	Nitrogen	N	7	14.0067
Berkelium	Bk	97	[247]	Nobelium	No	102	[259]
Beryllium	Be	4	9.01218	Osmium	Os	76	190.2
Bismuth	Bi	83	208.9804	Oxygen	O	8	15.9994
Boron	B	5	10.81	Palladium	Pd	46	106.4
Bromine	Br	35	79.904	Phosphorus	P	15	30.97376
Cadmium	Cd	48	112.41	Platinum	Pt	78	195.09
Calcium	Ca	20	40.08	Plutonium	Pu	94	[244]
Californium	Cf	98	[251]	Polonium	Po	84	[209]
Carbon	C	6	12.011	Potassium	K	19	39.0983
Cerium	Ce	58	140.12	Praseodymium	Pr	59	140.9077
Cesium	Cs	55	132.9054	Promethium	Pm	61	[145]
Chlorine	Cl	17	35.453	Protactinium	Pa	91	231.0359
Chromium	Cr	24	51.996	Radium	Ra	88	226.0254
Cobalt	Co	27	58.9332	Radon	Rn	86	[222]
Copper	Cu	29	63.546	Rhenium	Re	75	186.207
Curium	Cm	96	[247]	Rhodium	Rh	45	102.9055
Dysprosium	Dy	66	162.50	Rubidium	Rb	37	85.4678
Einsteinium	Es	99	[252]	Ruthenium	Ru	44	101.07
Erbium	Er	68	167.26	Samarium	Sm	62	150.4
Europium	Eu	63	151.96	Scandium	Sc	21	44.9559
Fermium	Fm	100	[257]	Selenium	Se	34	78.96
Fluorine	F	9	18.998403	Silicon	Si	14	28.0855
Francium	Fr	87	[223]	Silver	Ag	47	107.868
Gadolinium	Gd	64	157.25	Sodium	Na	11	22.98977
Gallium	Ga	31	69.72	Strontium	Sr	38	87.62
Germanium	Ge	32	72.59	Sulfur	S	16	32.06

*Atomic masses are based on carbon-12.
†A value given in brackets denotes the mass number of the longest-lived or best-known isotope.

	Symbol	Atomic No	Atomic Mass		Symbol	Atomic No.	Atomic Mass
Gold	Au	79	196.9665	Tantalum	Ta	73	180.9479
Hafnium	Hf	72	178.49	Technetium	Tc	43	[98]
Helium	He	2	4.00260	Tellurium	Te	52	127.60
Holmium	Ho	67	164.9304	Terbium	Tb	65	158.9254
Hydrogen	H	1	1.0079	Thallium	Tl	81	204.37
Indium	In	49	114.82	Thorium	Th	90	232.0381
Iodine	I	53	126.9045	Thulium	Tm	69	168.9342
Iridium	Ir	77	192.22	Tin	Sn	50	118.69
Iron	Fe	26	55.847	Titanium	Ti	22	47.90
Krypton	Kr	36	83.80	Tungsten	W	74	183.85
Lanthanum	La	57	138.9055	Uranium	U	92	238.029
Lawrencium	Lr	103	[260]	Vanadium	V	23	50.9415
Lead	Pb	82	207.2	Xenon	Xe	54	131.30
Lithium	Li	3	6.941	Ytterbium	Yb	70	173.04
Lutetium	Lu	71	174.967	Yttrium	Y	39	88.9059
Magnesium	Mg	12	24.305	Zinc	Zn	30	65.38
Manganese	Mn	25	54.9380	Zirconium	Zr	40	91.22
Mendelevium	Md	101	[258]				

VIII

Values of *T* for Analysis of Statistical Confidence Limits

			Probability of Larger Value of *t*, Sign Ignored				
d.f.	0.05	0.02	0.01	d.f.	0.05	0.02	0.01
1	12.706	31.821	63.657	14	2.145	2.624	2.977
2	4.303	6.0965	9.925	15	2.131	2.602	2.947
3	3.182	4.541	5.841	16	2.120	2.583	2.921
4	2.776	3.747	4.604	17	2.110	2.567	2.898
5	2.571	3.365	4.032	18	2.101	2.552	2.878
6	2.447	3.143	3.707	19	2.093	2.539	2.861
7	2.365	2.998	3.499	20	2.086	2.528	2.845
8	2.306	2.896	3.355	21	2.080	2.518	2.831
9	2.262	2.821	3.250	22	2.074	2.508	2.819
10	2.228	2.764	3.169	23	2.069	2.500	2.807
11	2.201	2.718	3.106	24	2.064	2.492	2.797
12	2.179	2.681	3.055	25	2.060	2.485	2.787
13	2.160	2.650	3.012				

From *Experimental Techniques in Biochemistry* by J. Brewer, A. Pesce, and R. Ashworth, p. 332. Copyright © 1974, Reprinted by permission of Prentice-Hall, Inc., Englewood Cliffs, NJ.

$$IX$$

Answers to Selected Study Problems

Chapter 1

2. Personal Protection Index H: splash goggles, gloves, synthetic apron, vapor respirator.

4. (a) Add 75.1 g of zwitterionic glycine to a 1-liter volumetric flask. Add purified water to mark.

   (b) Add 90 g of glucose to a 1-liter volumetric flask and add water to mark.

   (c) Add 0.46 g of ethanol to a 1-liter volumetric flask and add water to mark.

   (d) Add 6.5 mg of hemoglobin to a 1-liter volumetric flask and add water to mark.

5. (a) Use 0.75 g of glycine.

   (b) Use 0.9 g of glucose.

   (c) Use 4.6×10^{-3} g of ethanol.

   (d) Use 0.065 mg of hemoglobin.

6. 100 mM

7. (a) 0.56 mM; 560 μM

   (b) 220 mM; 220,000 μM

8. 58.5 mM; 20 mg/mL; 2%

9. 5.17 mM; 4.17 mM; 3.33 mM

10. (a) Sample mean = $+3.21°$

    (b) Standard deviation = $\pm 0.043°$

    (c) 95% confidence limits = $+3.21° \pm 0.03°$ at a probability of 0.05.

Chapter 2

1. 0.31 mole NaH_2PO_4

   0.19 mole Na_2HPO_4

37.2 g NaH_2PO_4

26.9 g Na_2HPO_4

2. Weigh out 0.5 mole of NaH_2PO_4 and dissolve in about 900–950 mL of purified water. Monitor the pH of the solution and add, dropwise, a concentrated solution of NaOH until the pH is 7.0. Add water to 1 liter mark. Check final pH.

3. Use 0.03 mole of glycine and 0.07 mole of sodium glycinate.

4. Dissolve 24.2 g of Tris base in about 900–950 mL of purified water. Add concentrated hydrochloric acid dropwise until the pH is 8.0. Add water to 1 liter.

5.

Tube No.	1	2	3	4	5	6
μg BSA	0	10	20	40	80	100
A_{595}	0.0	0.08	0.16	0.32	0.64	0.80

6. 65 μg/mL

7. 17.5 μg/mL

9. Tyr and Trp; no.

10. Most proteins that contain phenylalanine and other aromatic amino acids.

11. (a) Citrate

    (b) Imidazole

    (c) Glycine

Chapter 3

1. (a) Asp, Gly, His

    (b) Glu, Ala, Arg

    (c) Glu, Phe, His

2. Ser, Lys, Ala, Val, Leu

3. Cyt c, myoglobin = hemoglobin, serum albumin, egg albumin, pepsin

4. Myosin, catalase, serum albumin, chymotrypsinogen, myoglobin, cytochrome c

5. They must be converted to volatile derivatives such as esters.

6. Differential refractometer–essentially all molecules are detected, but this method is not especially sensitive. Photometric detector–molecules that absorb in the ultraviolet or visible light region. Fluorescence detector–molecules that fluoresce.

7. Malate dehydrogenase, alcohol dehydrogenase, glucokinase

9. A dilute solution of NAD^+ should elute the enzyme from the affinity gel.

Chapter 4

1. Charge, size

2. From top to bottom: serum albumin, egg white albumin, chymotrypsin, lysozyme

4. (a) Monomer for polymeric gel matrix

   (b) Monomer for adding cross-linking to gel matrix

   (c) Catalyst for polymerization process.

   (d) Detergent that denatures proteins for electrophoresis

   (e) Dye used to stain proteins after gel electrophoresis

   (f) Molecule used as a "tracking dye" during electrophoresis

5. The gel matrix in slab gels is more uniform than column gels, which are made individually.

6. Polyacrylamide gels may be used for nucleic acids up to 2000 base pairs.

Chapter 5

1. (a) X-ray

   (b) Ultraviolet

   (c) Visible

2. $10.7 \times 10^{-4} \ M$

3. (1) a

   (2) a

   (3) c

   (4) d

   (5) b

4. a, b, and d.

6. Glass does not allow the transmission of UV light.

7. All the molecules contain alternating double bonds (are highly conjugated).

8. Fluorescing light is measured at right angles to the light irradiating the sample.

Chapter 6

1. 100 minutes

2. 0.049 day^{-1}

3. Stable nuclei have an equal number of protons and neutrons.

4. 77 days

5. ^{32}P

7. See Table 6.3.

8. Must be β emitters: ^3H, ^{14}C, ^{32}P, ^{35}S, ^{36}Cl

10. One curie $= 3.70 \times 10^{10}$ becquerels

Chapter 7

1. a, b, d

2. 48,000

3. Hemoglobin

4. When substrate is bound, the enzyme molecule folds into a more compact or spherical shape.

5. There are four 10,000 subunits and two 30,000 subunits.

6. Cell nuclei would be present in the sediment after centrifugation at $600 \times g$.

7. Magnesium ions bind to DNA in place of smaller protons, causing it to spread out and thus become less dense.

8. Yes, the two forms have different densities. Supercoiled DNA is usually more compact.

9. $125,000 \times g$

10. Most mitochondria sediment at $20,000 \times g$.

PART II EXPERIMENTS

Experiment 2

1. Edman method vs. dansyl chloride. Because of the cyclization-cleavage reaction that occurs, the Edman method can be used for sequential degradation of a peptide or protein. NH_2-terminal residues can be removed, one at a time, and identified chromatographically. Dansyl chloride can be used only for single NH_2-terminal analysis. Dansyl chloride, however, has some advantages. Dansyl amino acids can be more readily separated by chromatographic techniques and detection by a UV lamp is more sensitive than for phenylthiohydantoin amino acids (PTH-amino acids). Also, work-up of the dansyl chloride reaction is less cumbersome and, hence, more rapid. Excess phenylisothiocyanate reagent and coupling products must be extracted from the Edman reaction mixture. This requires more work-up steps. In the dansyl chloride reaction, excess reagent is hydrolyzed to the corresponding sulfonic acid, which does not interfere with chromatographic analysis.

2. See Figures E2.1 and E2.4.

4. From first to last: Gly, Ala, Val, Phe, Leu

5. (a) FMOC-Ser, FMOC-Ala

(b) FMOC-Asp, FMOC-Pro

(c) FMOC-Gly, FMOC-Pro, FMOC-Leu

6. If the solvent level in a chromatography jar is above the application spots, the compounds spotted will be dissolved in the solvent and diffuse throughout the chromatography solvent. The compounds will no longer be highly concentrated at a single area on the plate. Even if there is sufficient compound remaining on the plate for detection, the final developed spot will most likely be too large for an accurate R_f measurement.

7. Dansic acid is very polar; in fact, it may be ionized during TLC analysis.

8. Gly + Ala

Experiment 3

2. From a plot of L added (*x*-axis) and L bound (*y*-axis), the amount of ligand that causes saturation is about 400 μM. The number of binding sites, *n*, is about 10 binding sites per protein molecule. K_f, the concentration of ligand that leads to one-half saturation, is 85 μM.

3. The tryptophan residue must be close to a site on the human albumin where sugars bind. Bound sugars interact with the tryptophan.

4. Plot of [X] vs. $\bar{v}$: $n = 2$; $K_f = 1.1$ mM

Plot of $\bar{v}$ vs. $\bar{v}$ /[X]: $n = 2.5$; $K_f = 0.6$ mM

Experiment 4

2. Gel electrophoresis

3. The mobility of bromophenol blue must be greater than that of any proteins in the sample.

4. (a) Slower mobility

(b) Probably no effect

(c) No effect

(d) No effect

(e) Slower mobility

(f) No effect on mobility; however, actual distance moved would be less.

5. Read Chapter 2, Section D, and Experiment 5.

6. Read Chapter 5, Section A.

7. Read Chapter 2, Section D.

8. Read Chapter 2, Section D.

9. Read Chapter 2, Section D.

10. His, Trp, Cys, Ser, and others with electron-donating groups.

Experiment 5

1. Derivation of the Eadie-Hofstee equation:

$$\frac{1}{v_0} = \frac{K_M}{V_{max}} \frac{1}{[S]} + \frac{1}{V_{max}}$$

Multiply by V_{max}:

$$\frac{V_{max}}{v_0} = \frac{K_M}{[S]} + 1$$

Multiply by v_0:

$$V_{max} = \frac{K_M}{[S]} v_0 + v_0$$

Rearrange:

$$v_0 = -\frac{K_M}{[S]} v_0 + V_{max}$$

A plot of v_0 vs. $v_0/[S]$ yields a straight line. The intercept on the v_0 axis is V_{max}. The intercept on the $v_0/[S]$ axis is V_{max}/K_M.

2. From Michaelis-Menten plot:

$$V_{max} = 140 \ \mu mole/min$$

$$K_M = 4.0\text{--}4.5 \times 10^{-5} \ M$$

3. The turnover number, k_3, is defined as follows:

$$k_3 = \frac{V_{max}}{[E_t]}$$

From Problem 2, $V_{max} = 140 \ \mu mole/min$

$$k_3 = \frac{140 \ \mu mole/min}{5 \times 10^{-7} M} = 280 \ min^{-1}$$

5. (a) Sodium azide–noncompetitive

 Phenylalanine–competitive

 Tryptophan–competitive

 Cysteine–noncompetitive

 4-Chlororesorcinol–competitive

Sodium cyanide–noncompetitive

8-Hydroxyquinoline–noncompetitive

Diethyldithiocarbamate–noncompetitive

Thiourea–noncompetitive

Phenylacetate–competitive

(b) The reverse of part a.

6.   Type of inhibition: competitive

$$K_I = 5.9 \times 10^{-4}\ M$$

7. (a) Increase

(b) Increase

(c) Decrease

(d) Decrease

(e) Decrease

(f) Increase

8. $v_0 = \dfrac{V_{max}\ [S]}{K_M + [S]}$

take reciprocal:

$\dfrac{1}{v_0} = \dfrac{K_M + [S]}{V_{max}\ [S]}$; rearrange:

$\dfrac{1}{v_0} = \dfrac{K_M}{V_{max}} \dfrac{1}{[S]} + \dfrac{1}{V_{max}}$

9. About 2 minutes.

10. 0.1 mg/mL; 100 μg/mL

Experiment 6

1. After completing the experiment, you should recognize that the lipid is a triacylglycerol. Hydrolyze the fatty acid esters of glycerol by reaction with NaOH. Convert the fatty acids to methyl esters and analyze by gas chromatography.

2. Lipids that have both polar and nonpolar portions (fatty acids, etc.) tend to "streak" or "tail" during thin-layer chromatography development. This leads to long, narrow spots after iodine treatment. Addition of a small amount of a very polar solvent (acetic acid) greatly reduces this tailing and results in a more circular, better resolved spot.

3. The order of migration of the standard lipids during chromatography depends primarily on polarity. The less polar the lipid, the greater its affinity for the nonpolar solvent (compared to the polar silica gel) and hence the greater the migration.

4. Iodine interacts with many organic compounds to form pi complexes that are colored. This method of detection is especially useful for lipids containing double bonds. A lipid with several double bonds will give a darker spot with iodine. The darker spots may also be due to a higher concentration of lipid.

5. $R_f = \dfrac{63\ mm}{85\ mm} = 0.74$

7. For chemical substances to be analyzed by gas chromatography, they must be relatively nonpolar and volatile. Fatty acids are not readily vaporized at temperatures attainable in a gas chromatograph (up to 200°C). FAMEs have much lower boiling points than fatty acids. Polar fatty acids also have very long retention times unless extremely nonpolar column packings are used.

8. Trimyristin

9. BF_3 is a Lewis acid that catalyzes ester formation.

10. (a) 12:0, 14:0, 16:0, 18:0, 20:0

Experiment 7

1. Western blotting—used to identify a specific protein or group of proteins by immunoblotting (detection by antibodies).

   Southern blotting—used to identify a specific base sequence in DNA.

   Northern blotting—used to identify specific base sequences in RNA.

2. No. Two highly specific interactions between antibody and protein occur in the traditional Western blot, which uses a primary and secondary antibody. With Con A–HRP, only one specific interaction takes place. Only a specific class of proteins is identified in this experiment, not a specific protein.

3. Milk has a relatively high content of the protein casein, which acts as an inexpensive blocking agent.

4. Nylon membranes because they are cationic and strongly bind acidic proteins.

5. Infection by the AIDS virus will cause the production of antibodies in the patient's serum. Serum proteins are separated by PAGE and a detection system must be developed to recognize those antibodies.

6. If SDS-PAGE is used for separation, the proteins to be blotted are denatured. The antibodies used for the detection process must be able to recognize denatured proteins.

7. Ovalbumin, a glycoprotein that is often present in standard molecular weight mixtures, gives a positive result with the Con A–HRP detection system.

8. The assumption must be made that the dye moves faster than any of the sample proteins.

10. The spinach protein is hydrophobic and will probably bind strongly to a nitrocellulose membrane.

Experiment 8

1. Methanol, ethanol, 2-propanol.
2. (a) Chlorophyll *a*, chlorophyll *b*

   (b) β-Carotene, lutein, violaxanthin, neoxanthin.
3. (a) Chlorophyll *a*

   (b) β-Carotene

   (c) Chlorophyll *b*
4. Aqueous-based buffers would be best for extracting intact protein-pigment complexes. Molecular weights could be determined by gel electrophoresis using appropriate standards. See Chapter 4, Section B.
5. They are very highly conjugated.
6. The aqueous mixture of acetone is more polar than pure acetone, therefore less of the low-polarity compounds, chlorophyll *a* and β-carotene, are extracted.
7. Larger quantities of the leaves could be used with a blender.
8. Summer, when leaves produce more chlorophyll.
9. The starting point of chlorophyll is the reaction between two molecules of δ-aminolevulinic acid. δ-Aminolevulinic acid in higher plants is produced from glutamic acid.
10. PUFAs are not conjugated.

Experiment 9

2. Chlorophyll is a nonpolar molecule so it is soluble in most organic solvents. Acetone disrupts protein-pigment complexes (see Experiment 8). A more efficient extraction could be achieved if several extractions were carried out and the extracts pooled.

3. ATP formation may also be measured by monitoring the uptake of inorganic phosphorus. This may be done with ^{32}P or by colorimetric measurement.

5. Ion-selective electrodes may be used to measure chloroplast exchange of K^+, Mg^{2+}, or Ca^{2+}.

8. 0.004 mg/mL.

9. Proton uptake is most efficient when light is absorbed by chlorophyll in the ranges 400–500 nm and 650–700 nm.

10. (a) $NADP^+$

    (b) H_2O

    (c) Light ($h\nu$)

Experiment 10

1. Mitochondria are most stable in an environment that is relatively polar but not ionic. Although sucrose is polar, it is not ionic; therefore it will not disrupt the structure of mitochondria and cause excessive loss of protein material.

2. Sodium deoxycholate is a detergent that releases proteins from membranes.

4. Electron transport and ADP phosphorylation may be studied by incubating the mitochondrial fraction with various substrates including malate, succinate, and ascorbate. Oxidative phosphorylation is monitored by measuring the uptake of oxygen, using an oxygen electrode. Alternatively, the uptake of inorganic phosphate is measured with a colorimetric assay.

6. See Chapter 7, Section C.

8. Mitochondrial matrix.

9. (a) 6 mg/mL

    (b) 0.17 mL of mitochondrial fraction plus 0.83 mL of buffer.

10. $\Delta A_{340} = 1.86$

Experiment 11

1. 5.4%

2. For the single standard method to be valid, there must be a linear relationship between absorbance and cholesterol concentration. The cholesterol assay is linear up to about 500 mg/100 mL.

3. Analyze a smaller serum sample, perhaps 50–75% of the standard amount.

5. 69 mg of vitamin C in 100 mL of orange juice.

6. Compounds containing the sulfhydryl group, shown as R–SH below, react with DCIP:

   2 R–SH + DCIP (oxidized) → R–S–S–R + DCIPH$_2$ (reduced)

7. 19 mg

8. Carbohydrate

9. The phosphotungstate reagent causes the precipitation of VLDL and LDL but HDL-cholesterol remains in solution.

10. Hydrolase, oxidoreductase, oxidoreductase.

Experiment 12

1. To remove horseradish peroxidase that is not trapped in gel. Analyze rinse water for the presence of the enzyme.

2. See Experiment 5.

3. See Problem 1. Increase the percent of cross-linking agent, methylene bisacrylamide, in the gel.

4. Should be a linear graph.

5. Ammonium persulfate initiates radical formation to begin gel polymerization. Riboflavin in the presence of light also produces radicals.

6. 0.00005 units/mg.

8. (a) AH$_2$

   (b) Reduced

   (c) Heme

Experiment 13

1. DNA, which forms viscous solutions, is released from the cells to the medium.

2. Add a portion of the enzyme ribonuclease.

3. SDS is a detergent that dissociates protein-lipid complexes in cell membranes.

4. Highly purified DNA may be obtained by repeating the chloroform–isoamyl alcohol extraction several times. The alcohol precipitation step may also be carried out many times. Various chromatographic methods including ion exchange have been applied to the purification of nucleic acids.

5. 28 μg/mL

6. $A_{270} = 0.70$

7. (a) Lower T_m

(b) Lower T_m

(c) Lower T_m

(d) Probably little or no effect

8. Sample I, because of quick cooling, did not form a complete double helix. Slow cooling of sample II allowed more complete double-helix formation.

9. Yes

Experiment 14

1. Many plasmids contain genes that carry messages for the synthesis of proteins that protect microorganisms against antibiotics. The presence of certain antibiotics is a signal to the microorganism that more of these proteins are necessary. The microorganism responds by increasing the rate of production of plasmids.

2. See the section on protein synthesis in your biochemistry textbook.

4. Ethanol is added to precipitate plasmid DNA from solution. Most RNA remains in solution; hence, ethanol precipitation provides an effective technique for removing contaminating RNA from DNA extracts.

7. Glycerol is a dense, viscous chemical that aids in the application of samples to the gel. Bromophenol blue dye acts as a marker. It migrates very rapidly in electrophoresis.

8. Natural DNA molecules and restriction fragments are too large to penetrate polyacrylamide gels. Even gels with low percentage cross-linking are not useful.

Experiment 15

1. Restriction endonucleases require the presence of Mg^{2+} for activity. The quench buffer contains EDTA, which complexes transition metal ions in the solution. The metal ions are no longer available for binding by the nuclease molecules and enzyme activity is inhibited.

2. Glycerol is a dense, viscous chemical that aids in the application of samples to the gel. Bromophenol blue dye acts as a marker. It migrates very rapidly in electrophoresis.

4. Natural DNA molecules and restriction fragments are usually too large to penetrate polyacrylamide gels. Even gels with low percentage cross-linking are not useful.

5. 10 fragments

9. Bromophenol blue dye must move faster during electrophoresis than any of the DNA molecules analyzed.

10. b

Index